Polymer Nanocomposites

Polymer Nanocomposites

Special Issue Editor

Giuliana Gorrasi

MDPI • Basel • Beijing • Wuhan • Barcelona • Belgrade

MDPI

Special Issue Editor
Giuliana Gorrasi
University of Salerno
Italy

Editorial Office
MDPI
St. Alban-Anlage 66
Basel, Switzerland

This is a reprint of articles from the Special Issue published online in the open access journal *Nanomaterials* (ISSN 2079-4991) from 2017 to 2018 (available at: https://www.mdpi.com/journal/nanomaterials/special_issues/polymer_nanocomposites)

For citation purposes, cite each article independently as indicated on the article page online and as indicated below:

LastName, A.A.; LastName, B.B.; LastName, C.C. Article Title. *Journal Name* **Year**, *Article Number*, Page Range.

ISBN 978-3-03897-326-3 (Pbk)
ISBN 978-3-03897-327-0 (PDF)

Cover image courtesy of Tifeng Jiao.

Contents

About the Special Issue Editor vii

Preface to "Polymer Nanocomposites" ix

Lorenzo Massimo Polgar, Francesco Criscitiello, Machiel van Essen, Rodrigo Araya-Hermosilla, Nicola Migliore, Mattia Lenti, Patrizio Raffa, Francesco Picchioni and Andrea Pucci
Thermoreversibly Cross-Linked EPM Rubber Nanocomposites with Carbon Nanotubes
Reprinted from: *Nanomaterials* **2018**, *8*, 58, doi: 10.3390/nano8020058 1

Sithiprumnea Dul, Luca Fambri and Alessandro Pegoretti
Filaments Production and Fused Deposition Modelling of ABS/Carbon Nanotubes Composites
Reprinted from: *Nanomaterials* **2018**, *8*, 49, doi: 10.3390/nano8010049 19

Shujahadeen B. Aziz
Morphological and Optical Characteristics of Chitosan$_{(1-x)}$:Cu$^o{}_x$ ($4 \leq x \leq 12$) Based Polymer Nano-Composites: Optical Dielectric Loss as an Alternative Method for Tauc's Model
Reprinted from: *Nanomaterials* **2017**, *7*, 444, doi: 10.3390/nano7120444 44

Lik-ho Tam and Chao Wu
Molecular Mechanics of the Moisture Effect on Epoxy/Carbon Nanotube Nanocomposites
Reprinted from: *Nanomaterials* **2017**, *7*, 324, doi: 10.3390/nano7100324 59

Rong Guo, Tifeng Jiao, Ruirui Xing, Yan Chen, Wanchun Guo, Jingxin Zhou, Lexin Zhang and Qiuming Peng
Hierarchical AuNPs-Loaded Fe$_3$O$_4$/Polymers Nanocomposites Constructed by Electrospinning with Enhanced and Magnetically Recyclable Catalytic Capacities
Reprinted from: *Nanomaterials* **2017**, *7*, 317, doi: 10.3390/nano7100317 79

Sébastien Livi, Luanda Chaves Lins, Jakub Peter, Hynek Benes, Jana Kredatusova, Ricardo K. Donato and Sébastien Pruvost
Ionic Liquids as Surfactants for Layered Double Hydroxide Fillers: Effect on the Final Properties of Poly(Butylene Adipate-Co-Terephthalate)
Reprinted from: *Nanomaterials* **2017**, *7*, 297, doi: 10.3390/nano7100297 95

Ilke Uysal Unalan, Derya Boyacı, Silvia Trabattoni, Silvia Tavazzi and Stefano Farris
Transparent Pullulan/Mica Nanocomposite Coatings with Outstanding Oxygen Barrier Properties
Reprinted from: *Nanomaterials* **2017**, *7*, 281, doi: 10.3390/nano7090281 111

Mario Abbate and Loredana D'Orazio
Water Diffusion through a Titanium Dioxide/Poly(Carbonate Urethane) Nanocomposite for Protecting Cultural Heritage: Interactions and Viscoelastic Behavior
Reprinted from: *Nanomaterials* **2017**, *7*, 271, doi: 10.3390/nano7090271 125

Valeria Bugatti, Gianluca Viscusi, Carlo Naddeo and Giuliana Gorrasi
Nanocomposites Based on PCL and Halloysite Nanotubes Filled with Lysozyme: Effect of Draw Ratio on the Physical Properties and Release Analysis
Reprinted from: *Nanomaterials* **2017**, *7*, 213, doi: 10.3390/nano7080213 143

Albanelly Soto-Quintero, Ángel Romo-Uribe, Víctor H. Bermúdez-Morales,
Isabel Quijada-Garrido and Nekane Guarrotxena
3D-Hydrogel Based Polymeric Nanoreactors for Silver Nano-Antimicrobial
Composites Generation
Reprinted from: *Nanomaterials* **2017**, 7, 209, doi: 10.3390/nano7080209 . . . . **155**

Giuseppe Cavallaro, Anna A. Danilushkina, Vladimir G. Evtugyn, Giuseppe Lazzara,
Stefana Milioto, Filippo Parisi, Elvira V. Rozhina and Rawil F. Fakhrullin
Halloysite Nanotubes: Controlled Access and Release by Smart Gates
Reprinted from: *Nanomaterials* **2017**, 7, 199, doi: 10.3390/nano7080199 . . . . **173**

Tolesa Fita Chala, Chang-Mou Wu, Min-Hui Chou, Molla Bahiru Gebeyehu and
Kuo-Bing Cheng
Highly Efficient Near Infrared Photothermal Conversion Properties of Reduced Tungsten
Oxide/Polyurethane Nanocomposites
Reprinted from: *Nanomaterials* **2017**, 7, 191, doi: 10.3390/nano7070191 . . . . **185**

Jordina Fornell, Jorge Soriano, Miguel Guerrero, Juan de Dios Sirvent,
Marta Ferran-Marqués, Elena Ibáñez, Leonardo Barrios, Maria Dolors Baró,
Santiago Suriñach, Carme Nogués, Jordi Sort and Eva Pellicer
Biodegradable FeMnSi Sputter-Coated Macroporous Polypropylene Membranes for the
Sustained Release of Drugs
Reprinted from: *Nanomaterials* **2017**, 7, 155, doi: 10.3390/nano7070155 . . . . **198**

Caitlin Brocker, Hannah Kim, Daniel Smith and Sutapa Barua
Heteromer Nanostars by Spontaneous Self-Assembly
Reprinted from: *Nanomaterials* **2017**, 7, 127, doi: 10.3390/nano7060127 . . . . **210**

About the Special Issue Editor

Giuliana Gorrasi is Associate Professor of Chemistry at the Department of Industrial Engineering of University of Salerno (Italy). The teaching activitiy regards the teachings of General and Inorganic Chemistry at all Engineering courses, with particular reference to the chemical bases of technology. Hes research activity is focused on the study of the correlation between structural organization and physical properties of polymeric materials, composites and nanocomposites. She is the author of several publications in international peer review journals with high impact factor (www.scopus.com), several book chapters on the invitation of the Editor, 4 patents, and presented original contributions in many national and international conferences. Her research activity is mainly devoted to the preparation and characterization of structural and functional polymeric composites and nanocomposites. Schematically, the research activity can be grouped into two macro-themes:

1. Relationships between microstructure and structural organization and physical and transport properties of polymers and blends, thermoplastic copolymers, subjected to thermal, mechanical and solvent treatments.

2. Preparation, structural characterization and study of the physical and transport properties of structural and functional polymeric nanocomposites.

The innovative and original contribution of scientific production within this theme is represented by the use of mechanical milling technology (MM) as an ecological and economic alternative to obtain a homogeneous dispersion of nano-fillers inside polymeric matrices (biodegradable and not). The advantage of working at low temperatures, without the use of solvents, and with a wide variety of polymeric matrices, opens up new and interesting scenarios for the preparation of innovative structural and functional materials. The use of MM involves several advantages:

- strong reduction in the disposal of substances harmful to the environment, such as solvents

- control of degradation processes deriving from the use of high temperatures

- possibility of compatibilizing mixtures of incompatible materials

The simultaneous production and dispersion of nano-particles, the promotion of mixing processes that can occur mechanically-chemically and the possibility of manipulating thermosensitive organic molecules, such as antimicrobials, oxygen-scavengers, and molecules with pharmacological activity, has allowed get new materials for targeted applications. Moreover, this technology has been proved to be particularly useful and efficient for the preparation of new nanocomposites based on natural polymers and from renewable sources, for which both the in situ polymerization method and the mixing in the melt are impracticable.

Preface to "Polymer Nanocomposites"

Polymer nanocomposites are hybrid inorganic-organic materials that represent a fast expanding area of research, either basic or applied with unique and promising physical properties. They are materials mixed at the nanometer scale, that combine the best properties of each of the components, often unknown in the constituent materials, with great expectations in terms of advanced applications. Significant effort has been focused on the possibility to deeply control the nanoscale structures via innovative manufacturing approaches. The properties of polymeric nanocomposite depend not only on the properties of their individual components but also on their morphology and interfacial characteristics. Experimental work demonstrated that genrally all types of nanocomposite materials lead to new and improved properties, when compared to their macrocomposite counterparts. It was shown to be significantly improved the electrical conductivity and thermal conductivity of the pristine polymers, as well as the mechanical properties (i.e., strength, modulus, and dimensional stability). Other properties that might be improved are the permeability to gases, water and hydrocarbons, the thermal stability and chemical resistance, in some cases also the surface appearance and optical clarity. Therefore, polymeric nanocomposites promise new applications in many fields such as mechanically reinforced lightweight components, nonlinear optics, battery cathodes and solid state ionics, nanowires, sensors, and many others. Much effort is going on to develop more efficient combinations of new polymers and fillers and to impart multifunctionalities to the novel materials obtained. In this chapter are collected the most recent research in the field of polymeric nanocomposites, with particular emphasis to the role of the functional fillers with respect to the final properties of the materias.

Giuliana Gorrasi
Special Issue Editor

nanomaterials

MDPI

Article

Thermoreversibly Cross-Linked EPM Rubber Nanocomposites with Carbon Nanotubes

Lorenzo Massimo Polgar [1,2], Francesco Criscitiello [3], Machiel van Essen [1], Rodrigo Araya-Hermosilla [1], Nicola Migliore [1], Mattia Lenti [1,3], Patrizio Raffa [1], Francesco Picchioni [1,2,*] and Andrea Pucci [3]

[1] Department of Chemical Engineering, University of Groningen, Nijenborgh 4, 9747 AG Groningen, The Netherlands; L.m.polgar@rug.nl (L.M.P.); machielvanessen@live.nl (M.v.E.); r.a.araya.hermosilla@rug.nl (R.A.-H.); nicola_migliore@hotmail.it (N.M.); mattia.lenti@ymail.com (M.L.); p.raffa@rug.nl (P.R.)
[2] Dutch Polymer Institute (DPI), P.O. Box 902, 5600 AX Eindhoven, The Netherlands
[3] Department of Chemistry and Industrial Chemistry, University of Pisa, Via Moruzzi 13, I-56124 Pisa, Italy; francrisci86@gmail.com (F.C.); andrea.pucci@unipi.it (A.P.)
* Correspondence: f.picchioni@rug.nl; Tel.: +31-50-36-34333

Received: 29 November 2017; Accepted: 12 January 2018; Published: 23 January 2018

Abstract: Conductive rubber nanocomposites were prepared by dispersing conductive nanotubes (CNT) in thermoreversibly cross-linked ethylene propylene rubbers grafted with furan groups (EPM-g-furan) rubbers. Their features were studied with a strong focus on conductive and mechanical properties relevant for strain-sensor applications. The Diels-Alder chemistry used for thermoreversible cross-linking allows for the preparation of fully recyclable, homogeneous, and conductive nanocomposites. CNT modified with compatible furan groups provided nanocomposites with a relatively large tensile strength and small elongation at break. High and low sensitivity deformation experiments of nanocomposites with 5 wt % CNT (at the percolation threshold) displayed an initially linear sensitivity to deformation. Notably, only fresh samples displayed a linear response of their electrical resistivity to deformations as the resistance variation collapsed already after one cycle of elongation. Notwithstanding this mediocre performance as a strain sensor, the advantages of using thermoreversible chemistry in a conductive rubber nanocomposite were highlighted by demonstrating crack-healing by welding due to the joule effect on the surface and the bulk of the material. This will open up new technological opportunities for the design of novel strain-sensors based on recyclable rubbers.

Keywords: strain sensor; rubber nanocomposite; thermoreversible cross-linking; Joule effect; crack-healing

---

1. Introduction

"Smart rubbers" are defined as elastomeric materials that respond to external stimuli through a macroscopic output in which the energy of the stimulus is transduced appropriately as a function of external interference [1]. Polymer or rubber nanocomposites have gained scientific and technological interest because they often exhibit enhanced or novel properties compared with the neat polymer or conventional composites at the same filler loading. The incorporation of carbon nanotubes (CNT) into such polymer matrices yields nanocomposites with high strength and electrical conductivity [2]. These nanocomposites have found their way into a variety of applications, especially in the field of electrically-conductive plastic networks [3–5].

Thermoplastic elastomers (TPE) are attractive supporting materials for CNT because they are easily processed and fabricated into solid-state forms, such as thin films that are often required for applications. Such TPE/CNT nanocomposites can be prepared by either melt blending or in situ polymerization [6–8], but solution mixing is the most effective process to produce them at a small sample level. In this case a solvent is used to disperse CNT, generally attained by ultrasonication—recognizing that significant damage of their structures as well as shortening occur thus limiting the full potential of CNTs as additives in polymer nanocomposites—and/or opportune amounts of surfactants to produce a metastable suspension of nanotubes. The polymer, dissolved separately in the same solvent, is then added to the mixture. The final nanocomposite is obtained after solvent evaporation at reduced time by spin-coating the suspension, thereby reducing the typical CNT re-aggregation.

TPE nanocomposites containing CNT have received considerable attention in the literature due to the development of stretchable resistivity-strain sensors for detecting dangerous deformations and vibrations of mechanical parts in many fields of science and engineering [9–12]. In these nanocomposites, the applied strain induces carbon nanotube displacement/sliding on the microscale, as well as tensile deformation applied locally to individual CNT. These responses give rise to piezo resistive behavior as applied tensile strains yield measurable changes in electrical resistivity across the composite length. Nanocomposites in which 0.01–5 wt % of CNT (the corresponding percolation threshold) were dispersed in a polymeric matrix of styrene-butadiene-styrene rubber (SBS) [13], polymethyl methacrylate (PMMA) [14], polystyrene (PS) [15], thermoplastic polyurethane (TPU) [12,16–18] or combinations thereof [19] all display a similar behavior as their surface resistivity was correlated with the applied strains and observed to increase with increasing tensile strain. This behavior was addressed to the reduction in conductive network density and increase in inter-tube distances induced by deformation.

Ethylene propylene diene rubbers (EPDM) are one of the most frequently used materials in such TPE and can be found in window profiles, automotive, and roofing applications. Cross-linked EPDM rubbers are ideal candidates for low-cost elastomeric-based stress-strain sensors containing CNT as they display relatively high moduli, strengths, and elasticities, and are renowned for their good weather, temperature, chemical, ozone, and stress cracking resistance [20]. Cross-linking the rubber matrix also helps to overcome the general complications associated with the utilization of CNT and the strong van der Waals interactions between individual nanotubes that make achieving a uniformly dispersed composite at the nanoscale difficult. Unfortunately, the excellent properties of these typically sulfur vulcanized and peroxide cured EPDM rubber compounds are associated with the practical impossibility of reprocessing them after their product life. A recently developed alternative to these conventional, irreversible cross-linking techniques is found in thermoreversible cross-linking via Diels-Alder (DA) chemistry [21–23]. A good example is found with bismaleimide (BM) cross-linking of furan-functionalized EPM rubbers [21] as the resulting covalent cross-links yield a material with properties similar to those of conventionally cross-linked EPDM gum rubbers that are retained upon reprocessing. This material would therefore be an excellent candidate for the preparation of nanocomposites containing well-dispersed carbon nanotubes. While the system itself allows for the preparation of a fully cradle-to-cradle recyclable, conductive nanocomposite, the functional groups on the polymer backbone allow for various interactions with (defects in the) CNT [24] that may yield durability for subsequent cycles of measurements (Figure 1).

The goal of this work was to study the material properties (with a strong focus on conductive and mechanical properties that are relevant for strain-sensor applications) and reprocessability of nanocomposites based on thermoreversibly cross-linked EPM rubbers and CNT. This is done in the context of developing materials for strain-sensor applications. First, the effect of adding various amounts of CNT to thermoreversibly cross-linked EPM rubbers on their dispersion throughout the rubber matrix and the material properties of the resulting nanocomposites is studied and their ultimate use as strain-sensors evaluated. Only multi-walled CNT are used for this purpose as it was found that these yield nanocomposites with a higher electrical conductivity and piezoresistive sensitivity than single-walled CNT as a result of their metallic character [18,19]. Secondly, both the rubber matrix and the CNT are chemically functionalized to stimulate the formation of primary or secondary interactions between them. The effects of such interactions on the dispersion of the CNT is studied as this may improve the compatibility of both components and thereby enhance the material properties of the resulting nanocomposite. This may also affect the material properties of the rubber nanocomposites with respect to their application as strain sensor. Finally, the advantages of using thermoreversible chemistry in a conductive rubber nanocomposite are highlighted by demonstrating crack-healing by welding due to the joule effect on the surface and the bulk of the material.

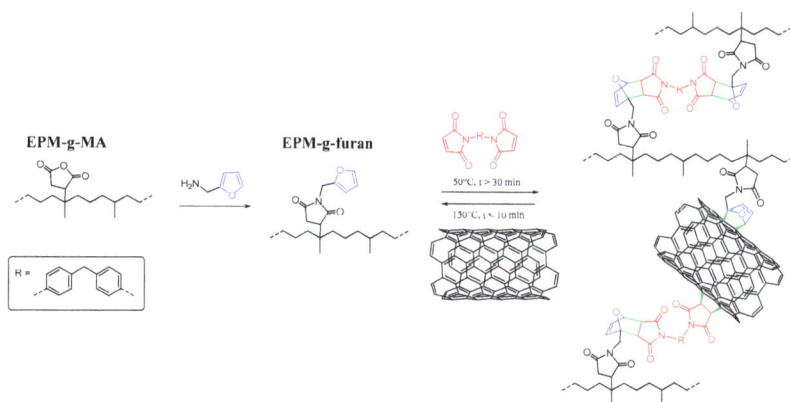

Figure 1. Furan functionalization and bismaleimide (BM) cross-linking of EPM-g-furan and the integration of CNT fillers via covalent interactions in the thermoreversibly cross-linked network of the nanocomposite.

2. Materials and Methods

2.1. Materials

A maleated EPM (EPM-g-MA, Keltan DE5005, 49 wt % ethylene, 2.1 wt % MA, M_n = 50 kg/mol, polydispersity index = 2.0) were kindly provided by ARLANXEO Performance Elastomers. Furfurylamine (FFA, Sigma-Aldrich, St. Louis, MO, USA, ≥99%) was freshly distillated. Multi-walled carbon nanotubes (CNT, Sigma-Aldrich, diameter × L 6–9 nm × 5 μm, >95% (carbon)) were used as additive and cross-linking agent. 1,1'-methylenedi-4,1-phenylene)bismaleimide (BM, Sigma-Aldrich, ≥97%) and dicumyl peroxide (DCP, Sigma-Aldrich, 98%) were used as reversible and irreversible cross-linking agents, respectively. 3-azido-1-propanamine (90%), octadecyl-1-(3,5-di-tert-butyl-4- hydroxyphenyl) propionate (anti-oxidant, 99%), 1-methyl-2-pyrrolidinone (NMP, 99.5%), tetrahydrofuran (THF, >99.9%), decahydro naphthalene (decalin, mixture of *cis* + *trans*, >98%) and acetone (>99.5%) were all bought from Sigma-Aldrich and used as received.

2.2. Methods

2.2.1. Furan-Functionalization of EPM-g-MA

Prior to the reaction, EPM-g-MA was dried in a vacuum oven for 1 h at 175 °C to convert the present diacids into anhydrides [21]. The EPM-g-MA precursor was then converted into EPM-g-furan using FFA according to a reported procedure [21].

2.2.2. Amine Modification of CNT

An amount of 3.00 g CNT was dispersed in 240 mL NMP by sonication for 30 min. Then 11.33 g 3-azido-1-propanamine was added and refluxed at 160 °C for 24 h under a N_2 atmosphere. The resulting solution was diluted with 250 mL acetone and centrifuged for 15 min. The solvents were removed to recover the modified CNT. Then 480 mL acetone was added to the CNTs and the suspension was sonicated for 30 min. Again, the mixture was centrifuged at 4500 rpm for 15 min and the solvent was removed. This washing cycle was repeated 5 times. Finally, the product was dried in an oven at 70 °C for 2 days to yield 2.57 g of amine modified CNT. The amine-modification of CNT was analyzed by elemental analysis (EA: 2.64 wt % N, 92.9 wt % C, and 0.65 wt % H). The modified CNT display a functionalization degree of 0.94 mmol/g, which is comparable to values found in literature in an acceptable range (1 added in 10 to 100 carbon atoms) [2].

2.2.3. Solution Mixing and Cross-Linking of Nanocomposites

Typically, 5.0 g of EPM-g-furan rubber was dissolved in 50 mL THF. Meanwhile, 0.5 to 10 wt % of CNT (with respect to EPM-g-furan) was exfoliated by suspending in 50 mL THF and sonicating for 30 min. Both solutions were then mixed and homogenized by stirring for 15 min and sonicating for 30 min. Then 0.5 molar equivalent (based on the furan content of EPM-g-furan) of cross-linking agent (BM or DCP) 1000 ppm phenolic anti-oxidant were dissolved in approximately 2 mL of THF and added to the mixture before refluxing it for 24 h. After mixing all components, the solvent was removed and the remaining product was dried in an oven at 50 °C for 24 h. Finally, the resulting nanocomposite was compression molded at 150 °C and 100 bar for 30 min and thermally annealed in a 50 °C oven for 3 days. Samples were reprocessed by grinding them into a ball mill at −195 °C and compression molding the resulting powder into new sample bars at 150 °C and 100 bar for 30 min and thermally annealing them in a 50 °C oven for 3 days.

2.3. Characterization

The conversion of EPM-g-MA to EPM-g-furan was followed by Fourier Transform Infrared spectroscopy (FT-IR) and EA. FT-IR spectra were recorded on a Perkin-Elmer Spectrum 2000 (Perkin Elmer, Waltham, MA, USA). Rubber films with a thickness of 0.1 mm were compression molded at 150 °C and 100 bar for 30 min, thermally annealed to ensure maximum DA cross-linking and measured in a KBr tablet holder. Measurements were performed over a spectral range from 4000 to 600 cm^{-1} at a resolution of 4 cm^{-1}, co-averaging 32 scans. Deconvolution was used to quantify the areas under the individual FT-IR peaks ($R^2 > 0.95$). The differences in relative peak areas were used to calculate the reaction conversion. The methyl rocking vibration peak at 723 cm^{-1} was used as an internal reference, as it originates from the EPM backbone and is not affected by chemical modification. The decrease of the absorbance of the C=O symmetrical stretch vibration of the anhydride groups at 1856 cm^{-1} was used to calculate the conversion of the reaction from EPM-g-MA to EPM-g-furan, according to a reported procedure [21]. The decrease of the characteristic C–O–C symmetrical stretch vibration of the furan groups at 1013 cm^{-1} was used to determine the conversion of the cross-linking reaction in the same way. EA for the elements N, C and H was performed on a Euro EA elemental analyzer. The nitrogen content was related to the furan-functionalization according to a reported procedure [21] and to the amine functionalization of CNT as no nitrogen is present in the non-modified CNT.

Gel Permeation Chromatography (GPC) was performed using triple detection with refractive index, viscosity, and light scattering detectors, i.e., a Viscotek Ralls detector (Malvern Instruments Ltd., Malvern, UK), a Viscotek Viscometer Model H502 (Malvern Instruments Ltd., Malvern, UK) and a Shodex RI-71 Refractive Index detector (Showa Denko Europe GmbH, Munich, Germany), respectively. The separation was carried out using a guard column (PL-gel 5 μm Guard, 50 mm) and two columns (PL-gel 5 μm MIXED-C, 300 mm) from Agilent Technologies (Amstelveen, The Netherlands) at 30 °C. THF 99+%, stabilized with butylated hydroxytoluene, was used as the eluent at a flow rate of 1.0 mL/min. The samples (~2 mg/mL) were filtered over a 0.45 μm PTFE filter prior to injection. Four GPC measurements were performed on each sample. Data acquisition and calculations were performed using Viscotek OmniSec software version 4.6.1 (Malvern Instruments Ltd., Malvern, UK), using a refractive index increment (d_n/d_c) of 0.052. Molecular weights were determined using a universal calibration curve, generated from narrow polydispersity polystyrene standards (Agilent and Polymer Laboratories, Santa Clara, CA 95051, USA).

Equilibrium swelling experiments were performed in decalin at room temperature. The rubber sample (approximately 500 mg) was weighed in 20 mL vials (W_0) and immersed in 15 mL solvent until equilibrium swelling was reached (3 days). The sample was then weighed after removing the solvent on the surface with a tissue (W_1) and was dried in a vacuum oven at 110 °C until a constant weight was reached (W_2). The gel content of the gum rubber samples is defined as (W_2/W_0) 100%. The apparent cross-link density [XLD] was calculated from W_1 and W_2 using the Flory-Rehner Equation (1) [25–27]. It is noted that the Flory-Rehner equation is only applicable for homogeneous rubber samples with difunctional cross-links, whereas the samples in this study are rubber nanocomposites containing up to 10 wt % of CNT. The calculated values therefore only represent apparent cross-link densities.

$$[\text{XLD}] = \frac{\ln(1 - V_R) + V_R + \chi V_R^2}{2V_S \left(0.5V_R - V_R^{\frac{1}{3}}\right)} \text{ with } V_R = \frac{W_2}{W_2 + (W_1 - W_2) \cdot \frac{\rho_{\text{EPM-g-furan}}}{\rho_{\text{decalin}}}} \tag{1}$$

V_R Volume fraction of rubber in swollen sample.
V_S Molar volume of solvent (decalin: 154 mL/mol at room temperature).
χ Flory-Huggins interaction parameter (decalin-EPDM: 0.121 + 0.278V_R) [28].
ρ Density (0.860 g/mL for EPM-g-furan and 0.896 g/mL for decalin).

Thermographic analysis (TGA) was performed using a Mettler Toledo TGA/SDTA851e (Mettler Toledo, Columbus, OH, USA), connected to an auto robot TS0801RO with a Mettler Toledo TS0800GC1 Gas Control unit. The samples were heated from 20 °C to 600 °C at 10 °C per min under nitrogen to pyrolyze the rubber part of the residue, while leaving the CNT unaffected.

The surfaces of the nanocomposites were characterized by scanning electron microscopy (SEM) imaging using a Philips XL30 Environmental SEM FEG instrument (Philips, Amsterdam, The Netherlands). Samples were prepared by cryogenic fracture in order to create a surface with exfoliated CNTs.

X-ray photoelectron spectroscopy (XPS) was performed on a SSX-100 spectrometer (Surface Science Instrument, Fisons plc, Ipswich, Suffolk, UK) equipped with a monochromatic Al Kα X-ray source (h_v = 1486.6 eV) that operates at a base pressure of 3×10^{-10} mbar. The CNTs samples were prepared by re-suspending in toluene and drop-casting on golden substrates. After evaporation of the solvent, the samples were transferred into an ultra-high vacuum system.

Tensile strength (T_b) and elongation at break (E_b) were measured on an Instron 5565 (Instron, High Wycombe, UK) with a clamp length of 15 mm, according to the ASTM D412 standard. A displacement rate of 500 ± 50 mm/min was applied. For each measurement 10 samples were tested and the two outliers were excluded to calculate the averages. The median stress-strain curves are shown in the figures. Cyclic hysteresis tests were performed on the same instrument with a clamp length of ± 3 cm. Samples were subjected to 5 cycles of 5%, 10%, 15%, and 20% strain with a strain rate of 10% of the sample length per minute.

The percolation threshold was determined by measuring the conductive behavior as function of the CNT loading. The electrical resistance of each sample was measured 3 times at various places of the sample at a length of ± 1 cm with a Keithley multimeter (model 2010, Keithley Instruments, Cleveland, OH, USA).

Low sensitivity deformation of the nanocomposites was tested by measuring their cyclic conductive behavior under strain on an Instron 4464 (Instron, High Wycombe, UK) with a clamp length of 3 cm. Samples were subjected to 5 cycles of 5%, 10%, 15% and 20% strain with a strain rate of 10% of the sample length per minute. The resistance ($R_m = (R_{\text{ext}} \times V_{\text{out}})/(V_{\text{in}} - V_{\text{out}})$) was digitally monitored according to a specific circuit (Figure 2) where R_m is the sample resistance and R_{ext} is an external resistance of equal magnitude.

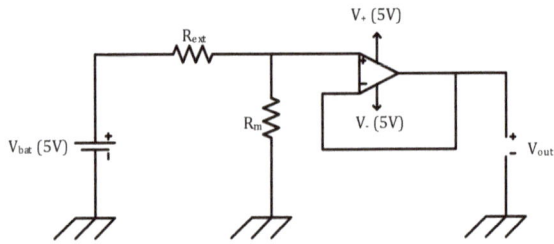

Figure 2. Schematic illustration of the circuit used for determining the sample resistance under strain.

High sensitivity deformation testing of the CNT filled rubber nanocomposites was performed at around the percolation threshold. The conductive behavior of the nanocomposites under strain was measured on a Tinius Olsen H25KT tensile tester (Tinius Olsen TMC, Horsham, PA, USA) with a clamp length of 1 cm. Nanocomposite sample bars were clamped in between copper sheets stretched manually in steps of ±2 mm and holding them in position for 30 s to measure the resistance with a Gossen Metrawatt Metrahit 18S multimeter (GMC-I Messtechnik GmbH, Nürnberg, Germany). Three deformation cycles were performed for each sample after at least 24 h from deformation, thus allowing the complete elastic recovery of the specimen.

The Joule effect was visualized by collecting thermographic images with a Fluke Ti10 IR Fusion Technology camera (Fluke Corporation, Everett, WA, USA) at steady state heat generation. The thereby enabled crack-healing by welding of the nanocomposites was demonstrated via a scratch test and by re-annealing broken tensile test samples. Scratch tests were performed by polymer solution casting of a nanocomposite film on a glass microscope slide, making a microscopic scratch on the surface with a scalpel. The film was then exposed to a potential source (7 V and 0.05 A) for 30 min by clamping the metal wires on the edges of the film in between the glass microscope slide substrate and another one covering it. The slow disappearance of the scratch was observed using an Zeiss Axioskop with HCS MX5 framegrabber (Zeiss, Oberkochen, Germany). Current induced welding of the bulk of the rubber nanocomposites was performed by cutting sample bars in half and pushing the freshly cut surfaces of the two halves together in a home-made device (Figure 3). The sample in the device was exposed to a potential source (7 V and 0.05 A) for 90 min. The welded samples were left at room temperature for 30 min before re-examining them.

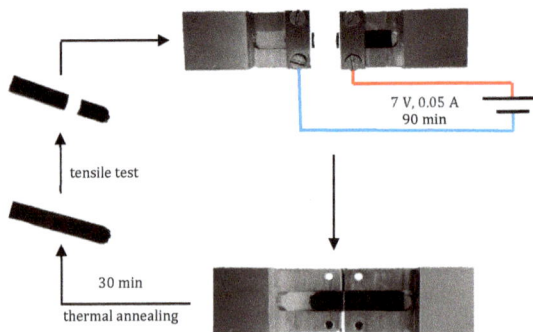

Figure 3. Illustration of the practical procedure to perform welding tests using a home-made device and exposing the freshly cut sample bar to a source potential (I).

3. Results and Discussion

3.1. The Reinforcing Effect of CNT in Rubber Nanocomposites

3.1.1. Chemical Characterization

The conversion of EPM-g-MA into EPM-g-furan was successful with high yields (>95%) according to FT-IR and EA [21]. Differential scanning calorimetry of the (non-cross-linked and cross-linked) rubbers showed a T_g of approximately −61 °C with the addition of CNT resulting in an expected increase in T_g of up to merely 3 °C for a 10 wt % CNT loading (data not shown for brevity) as a result of the decrease in segmental mobility of the polymer chains [20]. TGA thermograms all display a strong weight loss around 450 °C attributed to the degradation of the polymer matrix. This transition temperature is ~10 °C higher for samples loaded with CNT, which is attributed to the scavenging properties of graphitic fillers [27]. The amount of residue remaining at the end of analysis (600 °C) corresponds to the CNT content expressed in percentage by weight (Table 1). These values correspond to the amount of loaded CNT, indicating that the developed experimental methodology allows a complete transfer of the entire graphitic mass into the polymer matrix. The gel content of the EPM-g-furan nanocomposites is systematically larger than that of EPM-g-MA at the same CNT loading. This suggests some special interaction between the CNT surface and the furan groups that are grafted on the polymeric backbone [2,24,29]. The relatively high gel contents of the BM cross-link samples indicate that all chains are part of the rubber network. The systematic increase in the apparent cross-link density with the CNT loading of all BM cross-linked rubbers indicates that the CNT fillers participate in the formed rubber network.

Table 1. Composition and properties of a homologous series of rubber/carbon nanotube composites.

EPM * Rubber	CNT (wt %)	BM † (g)	TGA ‡ Residue at 600 °C (%)	Gel Content (%)	Cross-Link Density (10^{-4} mol/mL)	CNT (wt %)
EPM-g-MA	-	-	0.0	0	-	-
3.023 g EPM-g-MA	2.4	-	1.7	25	0.23	2.4
3.047 g EPM-g-MA	3.5	-	9.4	55	0.64	3.5
EPM-g-furan	-	-	0.0	0	-	-
3.019 g EPM-g-furan	2.4	-	1.6	68	0.75	2.4
3.007 g EPM-g-furan	3.4	-	9.5	82	0.95	3.4
3.010 g EPM-g-furan	-	-	0.0	93	2.1	-
3.012 g EPM-g-furan	2.4	-	1.5	95	5.2	2.4
3.036 g EPM-g-furan	4.0	0.114	4.3	96	3.6	4.0
3.024 g EPM-g-furan	4.8	0.114	4.1	98	4.0	4.8
3.146 g EPM-g-furan	5.6	0.118	10.1	99	4.7	5.6
3.028 g EPM-g-furan	6.5	0.113	7.7	99	6.3	6.5

* Ethylene propylene rubber. † Bismaleimide. ‡ Thermographic analysis.

3.1.2. Morphological Characterization

SEM micrographs of the fractured surface of the CNT filled polymeric matrices display CNT as white filaments (Figure 4). From the micrographs, it is evident that the CNT are distributed in homogeneously dispersed bundles throughout the polymer matrix. The relatively large diameter of these bundles (30–50 nm with respect to 6–9 nm for single CNT) may also imply that the surface of the CNT is covered with a layer of polymer as has previously been observed for polycarbonate/CNT composites [3].

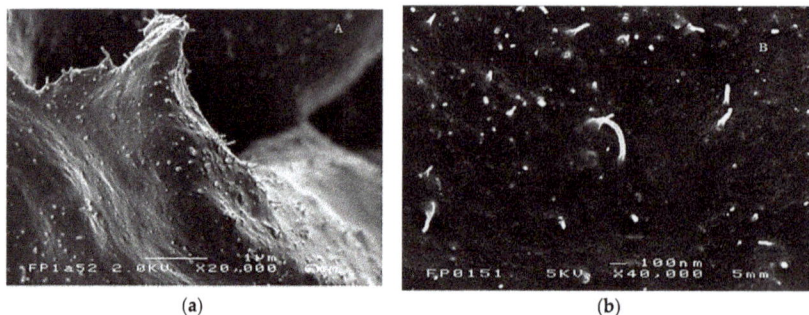

Figure 4. Scanning electron microscopy (SEM) micrograph of EPM-g-furan/CNT nanocomposites with (**a**) 7 wt % and (**b**) 10 wt % CNT loading.

3.1.3. Tensile Properties of BM Cross-Linked EPM-g-Furan Nanocomposites with CNT

The stress-strain curve of EPM-g-furan is typical for a non-cross-linked rubber with an extremely large E_b and a very low T_b (Figure 5). As expected, the T_b increases and the E_b break decreases upon BM cross-linking. The Young's modulus and T_b evidently increase and the E_b decreases upon the addition of CNT as reinforcing additives to both non-cross-linked EPM-g-furan (stress-strain curves not shown for brevity) and BM cross-linked EPM-g-furan. This means that the CNT are successfully incorporated into the rubber matrix, i.e., they display their characteristic toughening and reinforcing ability. Finally, reprocessing of the BM cross-linked EPM-g-furan sample bars yielded new coherent samples (impossible for the peroxide cured reference samples) with material properties that are similar (approximately 90% retention of properties) to those of the original samples. This is evidence that the addition of CNT does not significantly affect the reprocessability of the BM cross-linked EPM-g-furan rubbers.

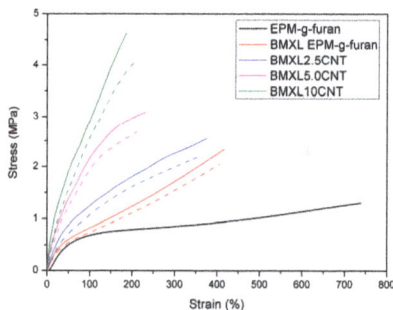

Figure 5. Median stress-strain curves of (BM cross-linked) EPM-g-furan/CNT nanocomposites with various CNT loading before (solid lines) and after reprocessing (dashed lines).

All samples used for cyclic tensile tests display elastic hysteresis (Figure 6). Deforming the composite by loading and unloading the material with force therefore results in an internal deformation and rearrangement of the Amatrix and dispersion and stabilization of the CNT [24]. At the first stage of extension a relatively large amount of force is required to overcome any physical interactions and to align the polymer chains and CNT. Unloading the material results in a reversed behavior as initially the applied force per decrease in elongation decreases, evidencing the retraction and energetically favored rearrangement of the polymer chains. These internal rearrangements and deformations cause dissipation of energy for every tensile cycle. Less force is therefore required to reach the same level of elongation when applying multiple cycles. The softening effect (decrease in toughness) is decreasingly visible for every cycle and increases with the exerted extension. This indicates that the CNT in the matrix gradually disconnect from each other, making the nanocomposites possibly suitable for sensor applications. Toughening of the samples is directly correlated to the CNT content and is especially evident for the BM cross-linked samples.

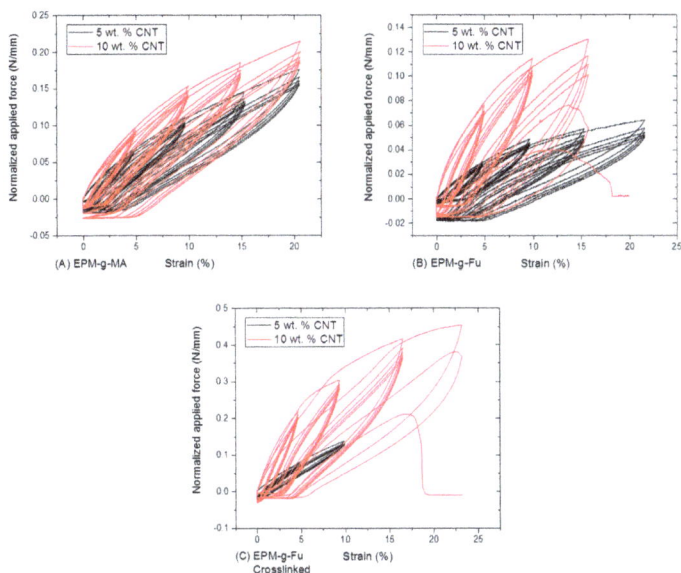

Figure 6. Cyclic tensile test results of CNT nanocomposites with (**A**) EPM-g-MA, (**B**) EPM-g-furan and (**C**) BM cross-linked EPM-g-furan (right) containing 5 and 10 wt % of CNT.

3.1.4. Tensile Properties of Rubber Nanocomposites with Various (Modified) CNT

The CNT loading appears to directly correlate to an increase in T_b and a decrease in EB for both the regular and the modified CNT (Table 2). The amine modification of CNT was also successful (2 wt % of FFA attached) according to FT-IR (Figure A1), TGA (Figure A2), XPS (Figure A3) and SEM (Figure A4). The amino functionalized CNTs appear to result in nanocomposites with a larger T_b and smaller E_b than with the original CNT, possibly due to the formation of effective interactions with the polymer matrix. Considering in a similar the modifications provided by the modified CNT to the mechanical properties of EPM-g-MA and EPM-g-furan, these interactions might be addressed to secondary interactions only. The amino functionalities could indeed react with EPM-g-MA thus generating amide or imide covalent linkages, which would be even more effective in mechanical properties modifications [30].

The small difference in material properties between the filled EPM-g-MA and EPM-g-furan samples may imply the presence of some interactions between the furan groups grafted onto the polymer backbone and the CNT. Similar interactions between polymer-linked furan groups and CNT have been described in the literature [24,31–33]. The effect of the addition of CNT on the material properties of the BM cross-linked EPM-g-furan is also more evident than for the non-cross-linked precursors. This well-known effect is also attributed to the interactions between the CNT and the polymer chains in the rubber matrix [9,34]. These interactions evidently increase the effective degree of cross-linking (Table 1), resulting in the expected effect on the tensile properties. To investigate such interactions, BM and (exfoliated) CNT were mixed in THF, refluxed for 1 day, filtered, washed, and dried in an oven. It appeared that approximately 0.05 mmol BM was absorbed per g CNT. These (reversible) interactions between furan or BM and CNT are even more prevalent in the nanocomposites with modified CNT fillers as is evident from the larger T_b and smaller E_b of all rubber samples with the same filler loading.

Table 2. Tensile strength and elongation at break of EPM-g-MA, EPM-g-furan, BM cross-linked EPM-g-furan and reprocessed BM cross-linked EPM-g-furan with different amounts of (modified) CNT. Typical standard deviations $\pm 10\%$ and ± 15–20% of the original value of the tensile strength (T_b) and elongation at break (E_b), respectively.

Filler Loading (wt %)		EPM-g-MA		EPM-g-Furan		BM Cross-Linked EPM-g-Furan		Reprocessed, Cross-Linked EPM-g-Furan	
		T_b (MPa)	E_b (%)	T_b (MPa)	E_b (%)	T_b (MPa)	E_b (%)	T_b (MPa)	E_b (%)
No CNT		1.2	950	1.3	750	2.2	210	1.7	190
CNT	2	2.5	250	2.5	250	3.1	200	2.9	190
	5	3.7	200	3.8	180	5.2	140	5	130
	10	5	130	5.1	130	7	80	6.5	90
modified CNT	2	2.5	240	2.6	220	3.0	210	2.9	200
	5	5.3	180	5.7	150	6.3	120	6.2	110
	10	7.2	110	7.5	100	7.8	100	7.5	90

3.2. Conductive Behaviour of BM Cross-Linked Rubber Nanocomposites

3.2.1. Percolation Threshold

Above a certain concentration, CNT create percolative paths in the insulating polymeric matrix in which the electrons can move with minimal resistance [35]. The percolation threshold must be studied to understand how the dispersion of the CNT affects the electrical properties of the rubber nanocomposite [36]. A substantial decrease in resistance is particularly evident for CNT concentrations close to the percolation threshold until a plateau is reached and the subsequent formation of percolation paths no longer influences the conductivity of the medium, as these do not give rise to more paths but only to alternative routes to those already existing. This large decrease in resistivity is evident in the range between 2.5% and 5% of CNT (Figure 7), which is comparable to the percolation threshold of that of other elastomeric CNT reinforced composites in the literature [2,17,34].

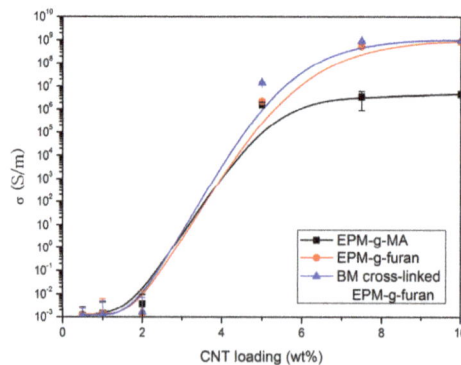

Figure 7. Conductivity of nanocomposites of EPM-g-MA (black), EPM-g-furan (red) or BM cross-linked EPM-g-furan (blue) with CNT as a function of the CNT loading. The error bars indicate ± 2 standard deviations.

3.2.2. Sensitivity Deformation Testing

Having determined the percolation threshold, the samples are used at its lower limit for low sensitivity deformation testing (high step strain with respect to high sensitivity). To illustrate the relevance of measuring samples at around their percolation threshold, high sensitivity deformation sensing measurements were performed on BM cross-linked EPM-g-furan samples containing 5 and 10 wt % of CNT with the vertical red lines indicating the end of one elongation and relaxation cycle (Figure 8). The sample containing 10 wt % CNT, i.e., well above the percolation threshold (Figure 7), displays a typical linear resistance-deformation dependence as the material reaches a maximum resistivity at full strain, that is the middle of each period. The excess of CNT in the nanocomposite allows for a large number of interconnected percolative pathways, which are gradually disrupted upon straining the sample as the CNT are pulled apart from each other. The sample containing 5 wt % of CNT, which is in the upper limit of the percolation threshold, displays non-linear resistance-deformation dependence. Unlike the previously observed "Λ"-like resistance pattern, this sample displays a "W"-like resistance pattern in each cycle of elongation and relaxation. This may be a result of the limited number of percolative pathways in this sample. The CNT that partake in these pathways are brought closer together by the tightening of the sample, orientate themselves and become more aligned by uniaxial stretching of the sample, resulting in an initial decrease in resistivity.

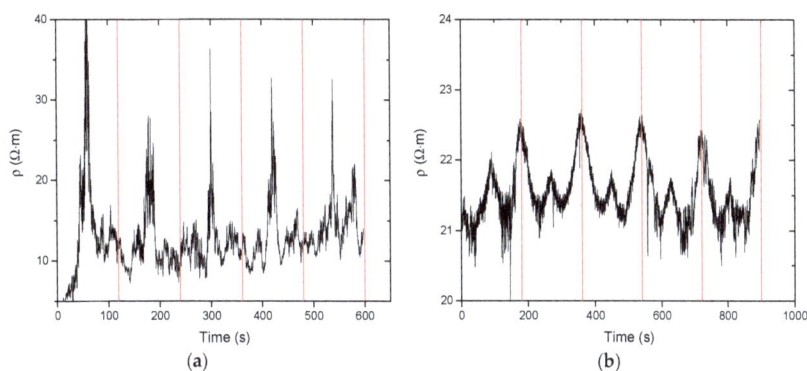

Figure 8. Electrical resistivity versus time over 5 cycles of 10% strain on BM cross-linked EPM-g-furan samples containing (**a**) 10 wt % and (**b**) 5 wt % of CNT. Each red line indicates the end of one cycle of strain.

Non-linear resistance-deformation dependence cannot be observed in the low sensitivity experiments on BM cross-linked EPM-g-furan rubber nanocomposites containing 5 wt % of CNT (Figure 9). The reason is that these are performed stepwise instead of continuously, with large steps in strain, by which relocation of the CNT in the polymer matrix is disfavored. As expected, the electrical resistance of both samples increases gradually upon stretching because of the induced progressive breakdown of interconnected filler networks [34,35,37,38]. The relationship between the variation in resistance and the elongation represents the sensitivity of the nanocomposite as a greater variation indicates that the sensor material can detect moderate deformations. For the second and third stretching cycles of both nanocomposites, the resistance variation was therefore calculated as $(R - R_0)/R_0$ where R is the measured resistance to elongation and R_0 the initial resistance of the sample.

Unfortunately, the resistance variation did not remain constant over the subsequent stretch cycles as the gauge factor decreased from 0.4 to below 0.05 after the first stretch cycle. This behavior can be attributed to the poor dispersion of the CNT (see Figure 4) and hysteresis phenomena that occur after progressive deformation cycles, scilicet once the stress is removed, the elastomeric chains gradually return to their initial state while the CNT retain the newly acquired anisotropic distribution throughout the matrix and a more effective percolation network that is maintained (larger R_0) even after removal of the mechanical stress.

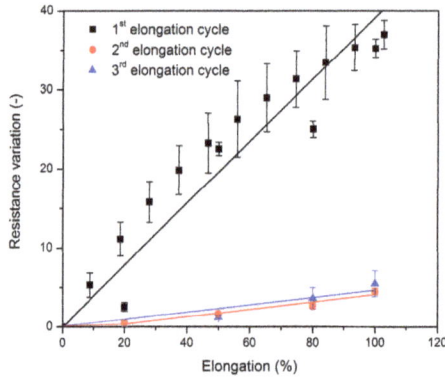

Figure 9. Resistance variation as a function of elongation for sample bars of BM cross-linked EPM-g-furan nanocomposites at around their percolation threshold (containing 5 wt % of CNT) for up to three cycles of elongation to 100%. The error bars indicate ± 2 standard deviations.

3.3. Crack-Healing by Welding

As the prepared rubber nanocomposites are semi-conductors, exposing them to a potential source current will result in heating up of the material via the Joule-effect [2]. This resistive heating of the sample was monitored by collecting themographic images over time (Figure 10). The effect of the increased temperature on the resistance of the material was neglected as the temperature of all tested nanocomposites reached a steady state. The dissipation of heat throughout the matrix becomes lower over time and more localized at the point of highest resistance, i.e., the interface between the CNT.

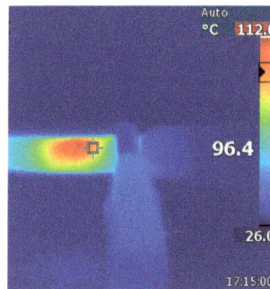

Figure 10. Thermal image of BM cross-linked EPM-g-furan containing 5wt % of CNT at 7 V and ~50 mA.

The temperatures observed as a consequence of autonomous heating of the materials via the Joule-effect are relatively high (>90 °C for all samples). The intrinsic temperatures at the conjugated DA cross-links may therefore be above the threshold temperature for DA re-cross-linking [39]. Crack-healing by welding due to the Joule effect [40] was therefore successfully applied to the nanocomposites at around their percolation threshold (5 wt % CNT). Microscopic scratches on the surface of the nanocomposites disappear within 30 min when exposed to a potential source of 7 V and 0.05 A (Figure 11) while broken tensile test bars would fully re-attach and regain their original tensile properties by exposing them to a current of 7 V for 90 min and compressing them in the home-made device. While the T_b and E_b of the welded sample are not different from that of the original sample bar, the shape of the stress strain curve is. This was not observed for the reprocessed samples and may therefore be due to the initial aligning of the CNT in the nanocomposite that took place during the tensile test of the original sample. This would result in an initially increased stress at relatively low strains for aligning the CNT in the original sample and a much lower stress at the same strain for the second tensile test as the CNT are already aligned.

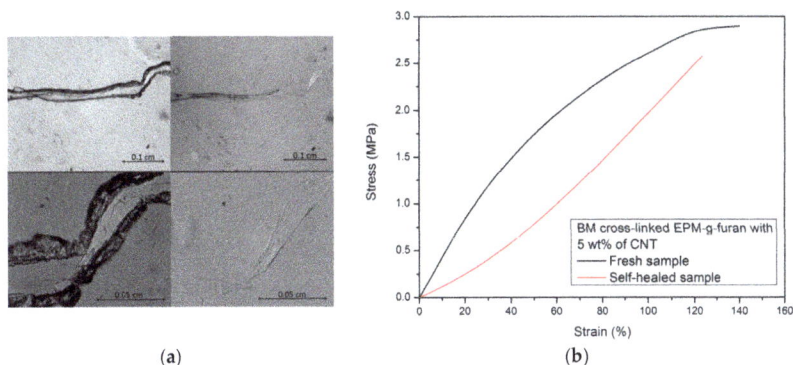

(a)   (b)

Figure 11. Crack-healing by welding of nancomposites of BM cross-linked EPM-g-furan with 5 wt % CNT performed by exposing the sample materials to a potential source of 7 V and 0.05. (**a**) A macroscopic scratch before and after 30 min of crack-healing by welding and (**b**) the tensile test of a nanocomposite sample bar before and after re-attaching the broken parts by crack-healing by welding for 90 min.

4. Conclusions

In this study, conductive rubber nanocomposites were prepared by dispersing CNT in thermoreversibly cross-linked EPM-g-furan rubbers for the preparation of recyclable and reusable devices able to detect a mechanical stress via a resistive output.

Diels-Alder chemistry was effectively used for thermoreversible cross-linking of the polymer matrix and although fully cradle-to-cradle materials were not provided, recyclable, well homogeneous, and conductive nanocomposites were obtained. Among the different strategies designed for enhancing the CNT dispersion, the rubber and the chemical modifications of the CNT appeared the most representative. An apparent increase of the cross-link density with CNT content weas obtained flanked by a high retention of tensile properties upon reprocessing. Functionalized CNT provided nanocomposites with a larger T_b and smaller E_b than with the original CNT, thus possibly suggesting their increased interaction with the polymer matrix. The percolation threshold of these nanocomposites appeared to be at a CNT loading ranging from 2.5 to 5 wt %. Low sensitivity deformation testing experiments demonstrated the necessity of using nanocomposite samples at their percolation threshold for strain-sensor applications. Only fresh samples displayed a linear response of their electrical resistivity to deformations as the resistance variation collapsed already after one cycle of elongation. This issue could be overcome by the advantages of using thermoreversible chemistry in a conductive rubber nanocomposite that were crack healed by welding due to the joule effect.

Acknowledgments: This research forms part of the research program of the Dutch Polymer Institute, project #749. Angela Grassi, Scuola Superiore Sant'Anna, Pontedera, is kindly thanked for her help with performing the strain sensor measurements and the visualization of the Joule effect. Machiel van Essen and Mattia Lenti would like to acknowledge the Erasmus exchange program for the financial support in their exchange program.

Author Contributions: Lorenzo Massimo Polgar coordinated the research and wrote the paper under supervision of Francesco Picchioni; Francesco Criscitiello designed and performed all experiments at the University of Pisa under supervision of Andrea Pucci; Machiel van Essen performed experiments at and initated the collaboration for this project between the University of Pisa and the University of Groningen; Nicola Migliore and Mattia Lenti performed all experiments at the University of Groningen; Rodrigo Araya-Hermosilla and Patrizio Raffa helped in analyzing the data and supervising students performing the experiments.

Conflicts of Interest: The authors declare no conflict of interest.

Appendix A

FT-IR was used to qualitatively determine the success of the modification of the CNT (Figure A1). Notably, the following bands can be detected:

- 3400 cm^{-1}, characteristic of the N-H bond stretching;
- 1060 cm^{-1}, characteristic of a stretching of the C–O bond;
- 1404 cm^{-1} band relating to the bending of C–O bond;
- 1399 cm^{-1} relative to the stretching of the C–N bond;
- 801 cm^{-1} band relative to the bending of the N–H bond;
- 1561 and 1560 cm^{-1} can be observed the band of the C=C stretching of the graphitic structure.

The functionalized CNT were also analysed by TGA (Figure A2). Pristine CNT are thermally stable in the presence of nitrogen up to 400–500 °C. A slight weight loss of about 2 wt % was detected above this temperature and was attributed to the more reactive portions (imperfections and structural defects) of the nanotube. The curve of functionalized CNT on the other hand, displays a clear weight loss for temperatures above 200 °C. The first derivatives of these experimental curves can be used to learn something about the nature of these weight losses.

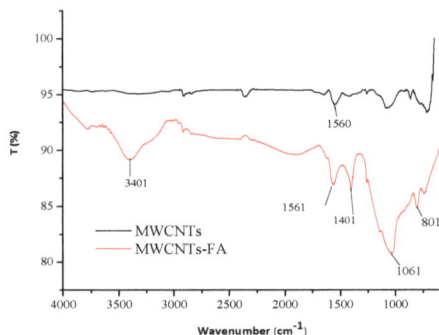

Figure A1. Comparison between the FT-IR spectra of pristine carbon nanotubes (black) and functionalized carbon nanotubes (red).

The curve corresponding to the functionalized CNT displays a first weight loss at a temperature of 240 °C. This was attributed to a desorption of the physisorbed FFA from the nanotube walls. Conversely, weight losses observed at higher temperatures were attributed to FFA bound to the surface of the nanotube via primary interactions, such as covalent bonds. It was speculated that this loss could be preliminarily attributed to FFA intercalated in the internal cavity of the CNT. The resulting very effective interactions between the FFA and the CNT may yield a weight loss at temperatures as high as the observed 527 °C. This phenomenon is likely to occur because the internal diameters of the CNT are approximately 4 nm while the FFA molecule has an average molecular size of about 6.9 Å.

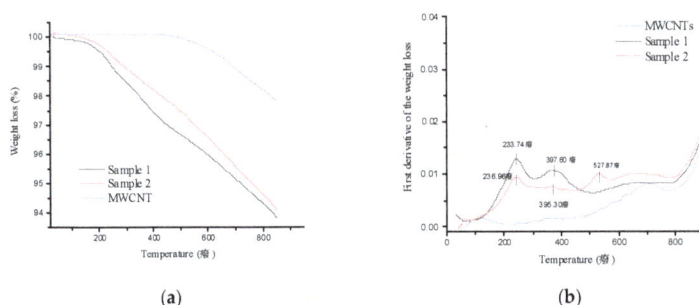

Figure A2. (**a**) TGA analysis under nitrogen atmosphere of pristine MWCNTs (blue) and two different batches of MWCNT -FAs (red and black) FA obtained under the same conditions of functionalization and (**b**) the first derivative of these curves.

XPS evidenced the success of the functionalization process due to the presence in the photoemission spectrum of peaks attributable to the nitrogen atom at about 400 eV (Figure A3). A precise measurement of the energy of each peak provides information on the chemical state of the corresponding chemical element. The amounts of each element present in the samples were obtained from the integral of the area for each signal (Table A1). It was found that the photoemission peak of the 1s orbital of N is composed of two contributions:

- Nitrogen of the amine with an energy of 401 eV;
- Nitrogen in the form of an oxidized species (or with a different chemical environment like an adsorbed amine) with energy equal to 399 eV.

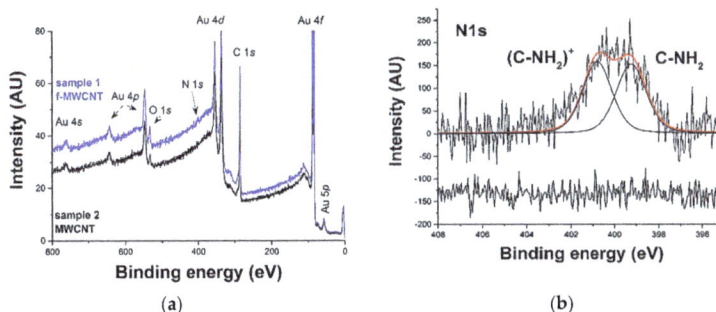

Figure A3. (**a**) XPS spectra of non-functionalized (black/down) and FFA functionalized (blue/up) CNT and (**b**) the first derivative of those spectra in the region relative to the peak photoemission on the N 1s orbital at around 400 eV. The signal was analyzed by deconvolution operated by the instrument software.

Table A1. The percentage of each element present in the non-functionalized and FFA functionalized CNT. Each analysis was repeated on three different portions of the sample analyzed.

	C	O	N
non-functionalized CNT	91	9	0
	90	10	0
	89	11	0
FFA functionalized CNT	93	4.9	2.0
	93	5.1	1.6
	93	5.3	1.6

SEM was performed to evaluate the morphology of the carbon nanotubes after functionalization (Figure A4). All recorded images evidenced the undamaged structure of CNT with retention of the natural flexibility of the graphitic structure and their aspect ratio that still corresponds to the size parameters given by the producer. The unaffected integrity of the CNT confirms the soft nature of the functionalization process.

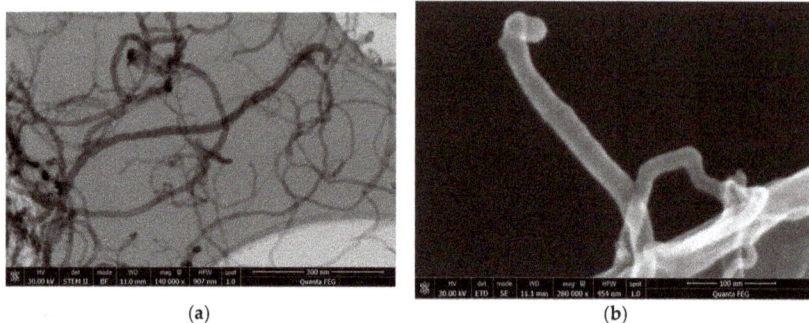

(a)

(b)

Figure A4. SEM images at of a FFA functionalized CNT, with an increase in magnification in going from (**a**) to (**b**), display that both the length (500 nm) and the diameter (20 nm) are still in agreement with the dimensions assessed by the producer.

References

1. Polgar, L.M.; van Essen, M.; Pucci, A.; Picchioni, F. *Smart Rubbers: Synthesis and Applications*; Smithers Rapra: Shawbury, UK, 2017; ISBN 978-1-91108-823-3.
2. Ponnamma, D.; Sadasivuni, K.K.; Grohens, Y.; Guo, Q.; Thomas, S. Carbon nanotube based elastomer composites—An approach towards multifunctional materials. *J. Mater Chem. C* **2014**, *2*, 8446–8485. [CrossRef]
3. Ajayan, P.; Stephan, O.; Colliex, C.; Trauth, D. Aligned Carbon Nanotube Arrays Formed by Butting A Polymer Resin-Nanotube Composite. *Science* **1994**, *265*, 1212–1214. [CrossRef] [PubMed]
4. De Heer, W. Nanotubes and the Pursuit of Applications. *MRS Bull.* **2004**, *29*, 281–285. [CrossRef]
5. Ciardelli, F.; Coiai, S.; Passaglia, E.; Pucci, A.; Ruggeri, G. Nanocomposites based on polyolefins and functional thermoplastic materials. *Polym. Int.* **2008**, *57*, 805–836. [CrossRef]
6. Salavagione, H.J.; Diez-Pascual, A.M.; Lazaro, E.; Vera, S.; Gomez-Fatou, M.A. Chemical sensors based on polymer composites with carbon nanotubes and graphene: The role of the polymer. *J. Mater. Chem. A* **2014**, *2*, 14289–14328. [CrossRef]
7. Byrne, M.T.; Gun'ko, Y.K. Recent advances in research on carbon nanotube—Polymer composites. *Adv. Mater.* **2010**, *22*, 1672–1688. [CrossRef] [PubMed]
8. Grady, B.P. Recent developments concerning the dispersion of carbon nanotubes in polymers. *Macromol. Rapid Commun.* **2010**, *31*, 247–257. [CrossRef] [PubMed]
9. Toprakci, H.A.K.; Kalanadhabhatla, S.K.; Spontak, R.J.; Ghosh, T.K. Polymer Nanocomposites Containing Carbon Nanofibers as Soft Printable Sensors Exhibiting Strain-Reversible Piezoresistivity. *Adv. Funct. Mater.* **2013**, *23*, 5536–5542. [CrossRef]
10. Lin, L.; Liu, S.; Zhang, Q.; Li, X.; Ji, M.; Deng, H.; Fu, Q. Towards Tunable Sensitivity of Electrical Property to Strain for Conductive Polymer Composites Based on Thermoplastic Elastomer. *ACS Appl. Mater. Interfaces* **2013**, *5*, 5815–5824. [CrossRef] [PubMed]
11. Li, T.; Ma, L.; Bao, R.; Qi, G.; Yang, W.; Xie, B.; Yang, M. A new approach to construct segregated structures in thermoplastic polyolefin elastomers towards improved conductive and mechanical properties. *J. Mater. Chem. A* **2015**, *3*, 5482–5490. [CrossRef]
12. Bilotti, E.; Zhang, R.; Deng, H.; Baxendale, M.; Peijs, T. Fabrication and property prediction of conductive and strain sensing TPU/CNT nanocomposite fibres. *J. Mater. Chem.* **2010**, *20*, 9449–9455. [CrossRef]

13. Costa, P.; Ferreira, A.; Sencadas, V.; Viana, J.C.; Lanceros-Mendez, S. Electro-Mechanical Properties of Triblock Copolymer Styrene-Butadiene-Styrene/Carbon Nanotube Composites for Large Deformation Sensor Applications. *Sens. Actuators A Phys.* **2013**, *201*, 458–467. [CrossRef]

14. Pham, G.T.; Park, Y.; Liang, Z.; Zhang, C.; Wang, B. Processing and modeling of conductive thermoplastic/carbon nanotube films for strain sensing. *Compos. Part B Eng.* **2008**, *39*, 209–216. [CrossRef]

15. Robert, C.; Feller, J.F.; Castro, M. Sensing skin for strain monitoring made of PC-CNT conductive polymer nanocomposite sprayed layer by layer. *ACS Appl. Mater. Interfaces* **2012**, *4*, 3508–3516. [CrossRef] [PubMed]

16. Bilotti, E.; Zhang, H.; Deng, H.; Zhang, R.; Fu, Q.; Peijs, T. Controlling the dynamic percolation of carbon nanotube based conductive polymer composites by addition of secondary nanofillers: The effect on electrical conductivity and tuneable sensing behaviour. *Compos. Sci. Technol.* **2013**, *74*, 85–90. [CrossRef]

17. Zhang, R.; Deng, H.; Valenca, R.; Jin, J.; Fu, Q.; Bilotti, E.; Peijs, T. Carbon nanotube polymer coatings for textile yarns with good strain sensing capability. *Sens. Actuators A Phys.* **2012**, *179*, 83–91. [CrossRef]

18. Bautista-Quijano, J.R.; Aviles, F.; Cauich-Rodriguez, J.V.; Schoenfelder, R.; Bachmatiuk, A.; Gemming, T.; Ruemmeli, M.H. Tensile piezoresistivity and disruption of percolation in singlewall and multiwall carbon nanotube/polyurethane composites. *Synth. Met.* **2013**, *185*, 96–102. [CrossRef]

19. Deng, H.; Ji, M.; Yan, D.; Fu, S.; Duan, L.; Zhang, M.; Fu, Q. Towards tunable resistivity–strain behavior through construction of oriented and selectively distributed conductive networks in conductive polymer composites. *J. Mater. Chem. A* **2014**, *2*, 10048–10058. [CrossRef]

20. Sperling, L.H.; Mishra, V. Study on superabsorbent of maleic anhydride/acrylamide semi-interpenetrated with poly(vinyl alcohol). *Polym. Adv. Technol.* **1996**, *7*, 197–208. [CrossRef]

21. Polgar, L.M.; van Duin, M.; Broekhuis, A.A.; Picchioni, F. Use of Diels—Alder Chemistry for Thermoreversible Cross-Linking of Rubbers: The Next Step toward Recycling of Rubber Products? *Macromolecules* **2015**, *48*, 7096–7105. [CrossRef]

22. Araya-Hermosilla, R.; Broekhuis, A.A.; Picchioni, F. Reversible polymer networks containing covalent and hydrogen bonding interactions. *Eur. Polym. J.* **2014**, *50*, 127–134. [CrossRef]

23. Zhang, Y.; Broekhuis, A.A.; Picchioni, F. Thermally Self-Healing Polymeric Materials: The Next Step to Recycling Thermoset Polymers? *Macromolecules* **2009**, *42*, 1906–1912. [CrossRef]

24. Araya-Hermosilla, R.; Pucci, A.; Araya-Hermosilla, E.; Pescarmona, P.; Raffa, P.; Polgar, L.; Moreno-Villoslada, I.; Flores, M.; Fortunato, G.; Broekhuis, A.; et al. An easy synthetic way to exfoliate and stabilize MWCNTs in a thermoplastic pyrrole-containing matrix assisted by hydrogen bonds. *RSC Adv.* **2016**, *6*, 85829–85837. [CrossRef]

25. Posadas, P.; Fernandez-Torres, A.; Chamorro, C.; Mora-Barrantes, I.; Rodriguez, A.; Gonzalez, L.; Valentin, J.L. Study on peroxide vulcanization thermodynamics of ethylenevinyl acetate copolymer rubber using 2,2,6,6,-tetramethylpiperidinyloxy nitroxide. *Polym. Int.* **2013**, *62*, 909–918. [CrossRef]

26. Hirschl, C.; Biebl-Rydlo, M.; DeBiasio, M.; Muehleisen, W.; Neumaier, L.; Scherf, W.; Oreski, G.; Eder, G.; Chernev, B.; Schwab, W.; et al. Determining the degree of crosslinking of ethylene vinyl acetate photovoltaic module encapsulants—A comparative study. *Sol. Energy Mater Sol. Cells* **2013**, *116*, 203–218. [CrossRef]

27. Flory, P.J.; Rehner, J., Jr. The entropy of the rotational conformations of (poly)isoprene molecules and its relationship to rubber elasticity and temperature increase for moderate tensile or compressive strains. *J. Chem. Phys.* **1943**, *11*. [CrossRef]

28. Dikland, H.G.; Van Duin, M. Miscibility of EPM-EPDM Blends. *Rubber Chem. Technol.* **2003**, *76*, 495–506. [CrossRef]

29. Giuliani, A.; Placidi, M.; Francesco, F.; Pucci, A. A new polystyrene-based ionomer/MWCNT nanocomposite for wearable skin temperature sensors. *React. Funct. Polym.* **2014**, *76*, 57–62. [CrossRef]

30. Sahoo, N.G.; Rana, S.; Cho, J.W.; Li, L.; Chan, S.H. Polymer nanocomposites based on functionalized carbon nanotubes. *Prog. Polym. Sci.* **2010**, *35*, 837–867. [CrossRef]

31. Zydziak, N.; Huebner, C.; Bruns, M.; Barner-Kowollik, C. One-Step Functionalization of Single-Walled Carbon Nanotubes (SWCNTs) with Cyclopentadienyl-Capped Macromolecules via Diels—Alder Chemistry. *Macromolecules* **2011**, *44*, 3374–3380. [CrossRef]

32. Syrgiannis, Z.; Melchionna, M.; Prato, M. Covalent Carbon Nanotube Functionalization. In *Encyclopedia of Polymeric Nanomaterials*; Springer: Heidelberg, Germany, 2015; pp. 480–487, ISBN 978-3-642-29649-9.

33. Xue, P.; Wang, J.; Bao, Y.; Li, Q.; Wu, C. Synergistic effect between carbon black nanoparticles and polyimide on refractive indices of polyimide/carbon black nanocomposites. *New J. Chem.* **2012**, *36*, 903–910. [CrossRef]

34. Bokobza, L. Multiwall carbon nanotube elastomeric composites: A review. *Polymer* **2007**, *48*, 4907–4920. [CrossRef]

35. Bokobza, L.; Belin, C. Effect of strain on the properties of a styrene—Butadiene rubber filled with multiwall carbon nanotubes. *J. Appl. Polym. Sci.* **2007**, *105*, 2054–2061. [CrossRef]

36. Hirsch, A.; Backes, C. Carbon Nanotube Science. Synthesis, Properties and Applications. Von Peter J. F. Harris. *Angew. Chem.* **2009**, *122*, 1766–1767. [CrossRef]

37. Bhattacharyya, S.; Sinturel, C.; Bahloul, O.; Saboungi, M.; Thomas, S.; Salvetat, J. Improving reinforcement of natural rubber by networking of activated carbon nanotubes. *Carbon* **2008**, *46*, 1037–1045. [CrossRef]

38. Thostenson, E.; Ren, Z.; Chou, T. Advances in the science and technology of carbon nanotubes and their composites: A review. *Compos. Sci. Technol.* **2001**, *61*, 1899–1912. [CrossRef]

39. Polgar, L.M.; Kingma, A.; Roelfs, M.; van Essen, M.; van Duin, M.; Picchioni, F. Kinetics of cross-linking and de-cross-linking of EPM rubber with thermoreversible Diels-Alder chemistry. *Eur. Polym. J.* **2017**, *90*, 150–161. [CrossRef]

40. Binder, W.H. *Self-Healing Polymers: From Principles to Applications*; Wiley: Hoboken, NJ, USA, 2013; ISBN 978-3-527-67020-8.

nanomaterials

MDPI

Article

Filaments Production and Fused Deposition Modelling of ABS/Carbon Nanotubes Composites

Sithiprumnea Dul, Luca Fambri and Alessandro Pegoretti *

Department of Industrial Engineering and INSTM Research Unit, University of Trento, Via Sommarive 9, 38123 Trento, Italy; sithiprumnea.dul@unitn.it (S.D.); luca.fambri@unitn.it (L.F.)
* Correspondence: alessandro.pegoretti@unitn.it; Tel.: +39-0461-282452

Received: 1 December 2017; Accepted: 15 January 2018; Published: 18 January 2018

Abstract: Composite acrylonitrile–butadiene–styrene (ABS)/carbon nanotubes (CNT) filaments at 1, 2, 4, 6 and 8 wt %, suitable for fused deposition modelling (FDM) were obtained by using a completely solvent-free process based on direct melt compounding and extrusion. The optimal CNT content in the filaments for FDM was found to be 6 wt %; for this composite, a detailed investigation of the thermal, mechanical and electrical properties was performed. Presence of CNT in ABS filaments and 3D-printed parts resulted in a significant enhancement of the tensile modulus and strength, accompanied by a reduction of the elongation at break. As documented by dynamic mechanical thermal analysis, the stiffening effect of CNTs in ABS is particularly pronounced at high temperatures. Besides, the presence of CNT in 3D-printed parts accounts for better creep and thermal dimensional stabilities of 3D-printed parts, accompanied by a reduction of the coefficient of thermal expansion). 3D-printed nanocomposite samples with 6 wt % of CNT exhibited a good electrical conductivity, even if lower than pristine composite filaments.

Keywords: conductive composites; carbon nanotubes; fused deposition modelling; mechanical properties

1. Introduction

The development of nanocomposite materials for specific types of additional manufacturing has recently attracted remarkable interest because incorporated nanoparticles offer the potential to enhance various properties of 3D-printed parts [1–3]. In particular, filaments for fused deposition modelling (FDM)—which is a widely used 3D-printing technology—could be improved by the addition of nanofillers. In fact, the dispersion of conductive nanoparticles in a polymer matrix makes it possible to produce 3D-printed components for various applications such as electronic sensors [4–6], cases with good electromagnetic interference (EMI) shielding performances [7], circuits [8] and microbatteries [9].

To date, various conductive nanoparticles have been used in 3D printing, such as carbon black (CB) [4,10], graphene oxide (GO) [11,12], reduced graphene oxide (r-GO) [8], graphene [13,14] and carbon nanotubes [5,13,15–17]. However, very few studies have been focused on the production of nanocomposite filament feedstock for FDM. For example Zhang et al. [8] reported the resistivity of composite filaments with a diameter of 1.75 mm of r-GO/Polylactic acid (PLA) of 0.21 $\Omega \cdot cm$ (6 wt % r-GO), along with the superior mechanical properties of FDM parts. Zhang et al. [10] reported the effect of 15 wt % of CB on the resistivity of composite ABS feedstock filaments (about 2900 $\Omega \cdot cm$) and characterized the resistivity of 3D-printed parts under various FDM parameters. Wu et al. dispersed up to 3 wt % multi-walled carbon nanotubes (MWCNTs) in polyhydroxyalkanoate to produce feedstock filaments but the resistivity of filaments have not been reported [17]. Wei et al. [11] were able to produce 3D printed parts with 5.6 wt % of GO in ABS matrix but they did not investigate electrical or mechanical properties. Gnanasekaran et al. [13] reported on the 3D printing with polymer nanocomposites consisting of CNT- and graphene-based polybutylene terephthalate, finding that

3D-printed objects filled with CNT have better conductive and mechanical properties and better performance than those filled with graphene. In our recent study [14] we have used for the first time acrylonitrile–butadiene–styrene (ABS) matrix filled with 4 wt % graphene nanoplatelets (xGnP); the composite filaments were obtained by a solvent-free process consisting of melt compounding and extrusion.

In a previous paper [18], the main focus was the possibility of dispersing CNT in ABS by using a commercial masterbatch of ABS/CNT for the production of filaments with a non-standard diameter of 1.4 mm. Six wt % of CNT was found to be an optimal fraction for the production of composite filaments.

In this study, we have investigated the possibility to directly disperse CNT in ABS matrix in order to produce the ABS/CNT filaments suitable for the FDM process with a standard diameter of about 1.7 mm. Nanocomposite filaments were manufactured by using common industrial processing techniques such as internal mixer and twin-screw extruder to compound polymer pellets (without additives) with CNT nanofiller. Relatively higher viscosity ABS matrix and lower processing temperatures with respect to the previous paper have been properly selected in order to increase the processing shear stresses and to improve/facilitate CNT dispersion. Extensive thermal, mechanical and electrical characterization of the obtained filaments was carried out. Afterwards, selected filaments were used to feed a high-temperature FDM 3D printer to specify the effects of CNT on the properties 3D-printed components along various build orientations.

2. Materials and Methods

2.1. Materials

The acrylonitrile-butadiene-styrene (ABS) polymer (tradename Sinkral®F322) used in this study was kindly provided by Versalis S.p.A. (Mantova, Italy). According to producer's technical data sheet, the polymer is characterized by a density of 1.04 g/cm^3 and a melt volume rate of 14 cm^3/10 min (@220 °C/10 kg) [19]. Before processing, ABS chips were dried under vacuum at 80 °C for at least 2 h.

Multi-walled carbon nanotubes (CNTs) (tradename NC7000^{TM}) were provided by Nanocyl S.A. Sambreville, Belgium). The technical data sheet reports an average length of 1.5 µm, a diameter of 9.5 nm and a surface area of 250–300 m^2/g [20].

2.2. Materials Processing and Sample Preparations

2.2.1. Filament Extrusion

In order to maintain a proper distribution of nanofiller in the matrix, various amounts (1, 2, 4, 6 and 8 wt %) of CNTs were first melt blending with ABS matrix through a Thermo-Haake Polylab Rheomix counter-rotating internal mixer (Thermo Haake, Karlsruhe, Germany) at a temperature of 190 °C and rotor speed of 90 rpm for 15 min. The neat ABS was also processed under the same conditions. The resulting material was granulated in a Piovan grinder Model RN 166 (Piovan, S. Maria di Sala VE, Italy). Then the batches were used to feed a Thermo Haake PTW16 intermeshing, co-rotating twin screw extruder produced by Thermo Haake, Karlsruhe, Germany (screw diameter D = 16 mm; L/D ratio = 25, where L is screw length; rod die diameter 1.80 mm). The temperature profile was set as T_1 = 180 °C, T_2 = 205 °C, T_3 = 210 °C, T_4 = 215 °C and T_5 = 220 °C. Filaments with a nominal diameter of about 1.70 mm were collected by using a take-up unit Thermo Electron Type 002-5341 (Thermo Haake, Karlsruhe, Germany) at constant collection rate. The main parameters adopted for the filament production are summarized in Table 1.

Table 1. Processing parameters of twin screw extruder for the production of ABS and ABS/CNT nanocomposite filaments.

Samples	Pressure (bar)	Torque (Nm)	Screw Speed (rpm)	Collection Rate (m/min)	Output (g/h)
ABS	16.9	40.4	5	1.00	137.6
CNT1	17.1	38.4	5	1.00	134.6
CNT2	21.7	45.9	5	1.00	137.7
CNT4	28.0	66.8	5	1.00	138.6
CNT6	44.2	100.1	5	1.15	139.6
CNT8	45.7	119.6	4.5	0.88	122.1

ABS denotes filament of pure matrix; whereas CNT6 indicates composite filament with 6 wt % of CNTs.

2.2.2. 3D-Printed Fibers

3D-printed fibers were prepared, starting from extruded filament by using a prototype of a 3D printer for high temperature processing, Sharebot HT Next Generation desktop (Sharebot NG, Nibionno, LC, Italy). through a nozzle diameter of 0.40 mm at temperature of 250 °C or 280 °C for ABS or composites respectively. Fibers of about 100 cm length with a diameter of 0.50–0.65 mm were freely extruded at an extrusion speed of 40 mm/s for mechanical and electrical testing.

2.2.3. 3D-Printed Samples Preparation

3D-printed specimens were manufactured by feeding the 3D printer described in Section 2.2.2 with the filaments obtained as described in Section 2.2.1. As schematically depicted in Figure 1, dumbbell and parallelepiped specimens were built-up along different build orientation, i.e., horizontal concentric (HC (a)), horizontal 45° angle (H45 (b)) and vertical concentric (VC(c)). All samples were printed according to the following printing parameters: object infill 100%; no raft; nozzle diameter 0.40 mm; nozzle temperature 250 °C and 280 °C for neat ABS and nanocomposite respectively; bed temperature 110 °C, layer height 0.20 mm; infill speed 40 mm/s; raster angle and printing speed are reported in Table 2.

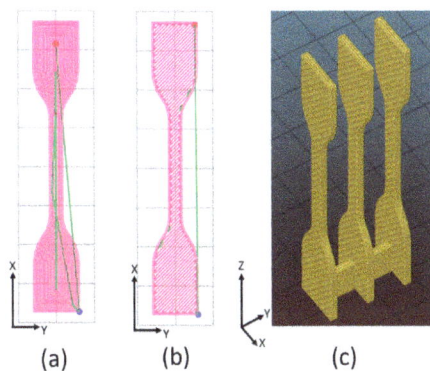

(a)   (b)   (c)

Figure 1. Schematic of 3D-printed dumbbell: (**a**) horizontal concentric (HC); (**b**) horizontal 45°angle (H45) and (**c**) vertical concentric (VC).

Table 2. 3D-printing parameters for each build orientation.

Samples	Infill Type	Raster Angle (°)	Printing Speed (%)
HC	Concentric	[0, 0]	100
H45	Rectangular	[+45, −45]	100
VC	Concentric	[0, 0]	40

Each specimen was individually printed, with the exception of VC samples for tensile, dynamic mechanical thermal analysis (DMTA) and creep tests (three specimens per printing session were simultaneously grown). 3D-printed samples were denoted indicating the material composition followed by build orientation. For example, CNT6-HC indicates the 3D-printed composite specimen with 6 wt % of CNT at HC build orientation.

2.3. Testing Techniques

2.3.1. Density Measurement

The density of the carbon nanotube was determined by using a Micromeritics®Accupyc 1330 helium pycnometry (Micromeritics, Norcross, GA, USA) at 23 °C within 300 measurements in a 10 cm^3 chamber; the average values and standard deviation of the last 200 measurements are reported (see details in Supplementary Materials).

Density measurement of bulk composite filaments was performed according to the standard ASTM D792-13. Moreover, the theoretical density and the voids content in nanocomposites were evaluated through the rule of mixture, as detailed in Supplementary Materials Section 2.

Linear density of the filament and the fiber was expressed in tex, according to ASTM D681-07, as the weight in grams of 1000 m of product and it was determined by weighting specimens of at least 90 mm in length. The results are the average of five measurements.

2.3.2. Melt Flow Index

The melt flow index (MFI) analysis of extruded filaments was carried out according to the ASTM D 1238 standard (procedure A), through a Kayeness Co. model 4003DE capillary rheometer (Morgantown, PA, USA) (barrel length of 162 mm and barrel diameter of 9.55 mm; die length of 8.000 mm and die diameter of 2.096 mm). About 5 g of chopped filament were tested at a temperature of 250 °C with an applied load of 10 kg, after the pre-heat and compact time of about 5 min. Pure ABS pellets were also tested at 220 °C and 10 kg. The results represent the average of at least five measurements (standard deviation is reported).

2.3.3. Scanning Electron Microscopy

Morphology of nanocomposites was studied by using a Carl Zeiss AG Supra 40 field emission scanning electron microscope (FESEM) (Carl Zeiss AG, Oberkochen, Germany). Specimens were fractured in liquid nitrogen and the fracture surfaces were observed at an acceleration voltage of 4 kV. Representative micrographs of filaments and 3D-printed dumbbell at 6 and 8 wt % of CNT were selected.

2.3.4. Thermogravimetric Analysis (TGA)

Thermal degradation was investigated through a Q5000 IR thermogravimetric analyzer (TA Instruments-Waters LLC, New Castle, DE, USA). The samples had a mass of about 10 mg and tests were performed in an air flow of 15 mL/min from 30 °C to 700 °C at a rate of 10 °C/min. The onset temperature of degradation (T_{onset}) was defined by the intersection point of the two tangent lines and the maximum temperatures (T_{max}) correspond to the maximum of the first derivative of weight loss. The residue at 475 °C, 575 °C and 700 °C were also reported in order to evaluate the content of CNT (C_{CNT}) according to the equation:

$$C_{CNT} = R_{comp} - R_{ABS} \tag{1}$$

where R_{comp} and R_{ABS} are the residue of composite and ABS respectively at the same temperature.

2.3.5. Differential Scanning Calorimetry (DSC)

DSC analyses were performed by a Mettler DSC 30 calorimeter (Mettler Toledo, Columbus, OH, USA) on samples with a mass of about 10 mg under a nitrogen flow of 100 mL/min. The samples were tested under heating-cooling-heating cycle from 30 °C to 260 °C at a rate of ±10 °C /min. Glass transition temperature (T_g) of styrene–acrylonitrile copolymer (SAN) phase was measured as the inflection point of the thermograms.

2.3.6. Quasi-Static Tensile Test

Uniaxial tensile test on filaments and 3D-printed samples were carried out at room temperature through an Instron® 5969 electromechanical tester (Norwood, MA, USA) equipped with a load cell of 50 kN. Yield and fracture properties were evaluated at a crosshead speed of 10 mm/min as an average value of at least three replicates. Filaments specimens had a length of 150 mm, a gauge length of 100 mm and a diameter of 1.70 mm. 3D-printed specimens (HC, H45 and VC) had a dumbbell geometry according to ISO 527 type 5A with a gauge length of 25 mm, a width of 4 mm and a thickness of 2 mm.

Tensile properties of 3D-printed fibers were determined by using an Instron® 5969 electromechanical tester equipped with a load cell of 100 N. Fiber specimens with a diameter between 500–650 micron and a gauge length of 20 mm were tested at a cross-head speed of 2 mm/min.

Elastic modulus of 3D-printed samples was determined at a cross-head speed of 1 mm/min by an electrical extensometer Instron® model 2620-601 (Norwood, MA, USA) with a gage length of 12.5 mm. Elastic modulus of filaments with a gage length of 100 mm and fiber with a gage length of 20 mm was tested at 10 mm/min and 2 mm/min, respectively, taking the system compliance into account. According to ISO 527 standard, the elastic modulus was determined as a secant value between strain levels of 0.05% and 0.25%.

2.3.7. Creep Test

Creep test was performed using a TA Instruments DMA Q800 (TA Instruments-Waters LLC, New Castle, DE, USA) at 30 °C up to 3600 s under a constant stress of 3.9 MPa on cylindrical extruded filament specimens and of 3 MPa on a 3D-printed rectangular sample with a width of 4 mm and a thickness of 1 mm. For all specimens, an overall the length of 25 mm was used and the adopted gauge length of all samples was 11.5 mm.

2.3.8. Dynamic Mechanical Thermal Analysis

Dynamic mechanical thermal analysis (DMTA) tests were performed under tensile mode by a TA Instruments DMA Q800 device (TA Instruments-Waters LLC, New Castle, DE, USA). 3D-printed specimens with a width of 4 mm and a thickness 2 mm were used. All specimens had an overall length of 5 mm and a gauge length of 11.5 mm. Tests were performed from 100 °C to 150 °C at a heating rate of 3 °C/min applying a dynamic maximum strain of 0.05% at a frequency of 1 Hz. Storage modulus (E'), loss modulus (E'') and loss tangent (tanδ) as a function of the temperature were reported. According to the manufacture data sheet, the precision on storage modulus is ±1% [21].

In order to evaluate the stiffness effect of the filler in nanocomposites above T_g, the reduction of the main transition R has been defined in Equation (2) as the ratio of storage modulus above T_g (i.e., at 130 °C) and storage modulus below T_g (i.e., at 90 °C), after modification of the equation S "intensity of transition" previously defined [22].

$$R = \frac{E'_{130°C}}{E'_{90°C}} \tag{2}$$

Moreover, following Equation (2), a *F*-factor has been defined as in Equation (3) as the relative *R* ratio between composite and matrix and it is formally derived from the inverse of *C*-factor reported in the literature [23,24]:

$$F = \frac{R_{composite}}{R_{matrix}} = \frac{(E'_{130°C}/E'_{90°C})_{composite}}{(E'_{130°C}/E'_{90°C})_{matrix}} \tag{3}$$

The coefficient of linear thermal expansion (CLTE) in four intervals below T_g (i.e., $-50~-20$ °C; $20~50$ °C; $70~90$ °C, $108~113$ °C) and a coefficient of linear thermal deformation (CLTD) above T_g in the range $130~150$ °C were determined according to Equation (4). CLTE or CLTD were obtained by linear-fitting the experimental data of thermal strain as a function of temperature.

$$CLTE \ or \ CLTD = \frac{\Delta L/L_0}{\Delta T_0} \tag{4}$$

where L_0 and ΔL are the initial specimen gauge length and the length variation and ΔT_0 is the selected temperature interval.

2.3.9. Electrical Resistivity Test

The test was carried out following ASTM D4496-04 standard for moderately conductive materials under a four-point contact configuration. Each specimen was subjected to a voltage in the range 2–24 V by using a direct current (DC) power supply IPS303DD produced by ISO-TECH (Milan, Italy) while the current flow across it between external electrodes was measured by using an ISO-TECH IDM 67 Pocket Multimeter electrometer (ISO-TECH, Milan, Italy). Composite filaments, fibers and 3D-printed samples (cross-section 6 mm × 2 mm) with a length of 25 mm were tested at 23 ± 1 °C at different voltage; resistivity values represent the average of at least three specimens. Due to the rough surface of 3D-printed samples, a conductive silver paint was applied between the specimen surfaces at the contact electrodes in order to reduce contact resistance. The electrical volume resistivity of the samples was evaluated as follows:

$$\varrho = R \cdot \frac{A}{L} \tag{5}$$

where *R* is the electrical resistance, *A* is the is the cross-section of the specimen and *L* is the distance between the internal electrodes (i.e., 3.69 mm).

The heating of a specimen generated by a current flow is known as resistive heating and it is described by the Joule's law. Surface temperature evolution induced by Joule's effect upon different applied voltages was measured by using a Flir E6 thermographic camera (FLIR System, Wilsonville, OR, USA). The voltages were applied by a DC power supply (IPS 303DD produced by ISO-TECH), while the samples were fixed with two metal clips with an external distance of 30 mm. In these tests, specimen length was 50 mm with different cross-sections for filaments (about 2.3 mm^2) and 3D-printed (12.0 mm^2) specimens. The surface temperature values were recorded for 120 s under the application of voltages of 12 V and 24 V.

3. Results and Discussion

The first step of composite preparation was performed following compounding procedure with the direct mixing of filler and polymeric matrix as previously reported in [14,25]. For the purpose to increase the processing shear stresses during CNT dispersion, an ABS matrix with MFI of 14.8 ± 1.0 g/cm^3 (220 °C and 10 kg) was properly selected, with viscosity higher than ABS with MF of 23 g/cm^3 (220 °C and 10 kg) previously utilized for the production CNT composite from master-batch [18]. Moreover, in the second step of filament extrusion, lower processing temperatures, 220 °C instead of 240 °C [18], were set in order to furtherly improve dispersion under high shear stresses.

3.1. Filament Extrusion and Melt Flow Index

The filament of neat ABS and of ABS/CNT composites were extruded with an orientation factor of about 1.0 at 220 °C, as evaluated by the ratio between the cross-sectional area of the extruder die hole (S_{DE}) and the cross-sectional area of the obtained filament (S_F) according to Equation (6)

$$OF_E = S_{DE}/S_F \tag{6}$$

Moreover, the orientation factor of fiber produced by 3D-printer as the cross-sectional area of filament (S_F) and the cross-sectional area of the obtained fiber (S_f), according to Equation (7)

$$OF_{3D} = S_F/S_f \tag{7}$$

The orientation factor is higher in the fiber (produced at 250 °C for ABS and at 280 °C for CNT6) than filament obtained at 220 °C due to the processing conditions. The higher the CNT content, the higher the orientation factor of fibers. Moreover, it is important to observe that linear density of fiber is progressively decreasing with CNT content, as shown in Table 3.

Table 3. Bulk density and linear density of ABS and ABS/CNT nanocomposite during filament extrusion and 3D fiber production. Extrusion and 3D printing draw ratio.

Samples	CNT Content (wt %)	Bulk Density (g/cm³)	Filament Linear Density (tex)	Filament Extrusion OF_E [1]	Fiber Linear Density (tex)	3D-Printing OF_{3D} [2]	Fiber Swelling DS [3]	Fiber OF_T [4]
ABS	0	1.042 ± 0.001	2389 ÷ 139	1.09	349 ± 17	7.1	2.6	7.7
CNT1	1	1.046 ± 0.001	2256 + 18	1.15	290 ± 7	8.1	2.2	9.3
CNT2	2	1.051 ± 0.001	2287 + 71	1.14	267 ± 5	8.8	2.0	10.1
CNT4	4	1.059 ± 0.002	2534 + 83	1.04	231 ± 3	11.2	1.7	11.6
CNT6	6	1.071 ± 0.002	2425 + 64	1.11	224 ± 3	11.1	1.7	12.2
CNT8	8	1.081 ± 0.002	2387 + 64	1.12	219 ± 2	11.3	1.6	12.7

[1] Draw ratio of filament (extrusion) see Equation (6). [2] Draw ratio in 3D printing see Equation (8). [3] Fiber swelling.
[4] Total Draw ratio of fiber (extrusion and 3D printing) see Equation (9).

This result could be explained by considering that the final diameter of fiber is decreasing with the nanofiller content (see detail in paragraph 3.6). Consequently, the free flow of the fibers from die of 3D-printer was used to evaluate the die-swelling (*DS*), according to Equation (8), where S_f is the cross-sectional area of fibers and S_{DP} is nozzle section of 3D-printer.

$$DS = S_F/S_{DP} \tag{8}$$

Table 3 shows that die-swelling of investigated composites is significantly reduced as the CNTs fraction increases; in particular, at 6 and 8 wt % of CNTs, die swelling in fiber is almost completely suppressed.

Moreover, the total orientation factor in fiber OF_T could be calculated combining Equations (6) and (7), as shown in Equation (9):

$$OF_T = S_{DE}/S_f \tag{9}$$

The total orientation factor in the fiber increased with CNT content in direct dependence on the first step of filament production at 220 °C and the subsequent extrusion from 3D printer at 250 °C (for ABS) or 280 °C (for nanocomposite), that is the most effective step. This cumulative effect could be a useful parameter for evaluating the processability of the various filaments.

The effect of CNT on the melt flow index (MFI) of extruded ABS filaments was also investigated. Figure 2 shows a strong decrease of MFI with the carbon nanotubes content, due to the increasing viscosity induced by the formation of a nanofiller network. This effect is also documented by a significant increase in the torque and internal pressure measured during the extrusion process after addition of CNT to ABS (see Table 1). Even though the MFI of nanocomposites with CNT content higher than 4 wt % is extremely low, it has been possible to produce feedstock filaments by using twin screw extruder up to 8 wt % of CNT, reaching maximum values of internal pressure of about 46 bar and 120 Nm of torque.

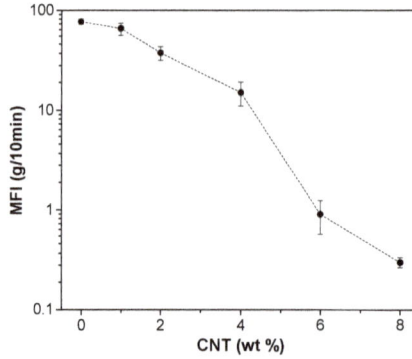

Figure 2. Melt flow index of ABS nanocomposites as a function of CNT fraction at 250 °C with applied load 10 kg.

3.2. Bulk Density

Density of CNTs was estimated to be 2.151 ± 0.033 g/cm^3 (see Supplementary Materials Figure S3). The bulk density of filaments is plotted in Figure 3 as a function of CNT volume fraction. The density of neat ABS filament is 1.042 g/cm^3, which is consistent with the reported value in the materials technical data sheet [19]. Density of ABS/CNT composites increases almost linearly with rising fraction of CNT up to 1.081 g/cm^3 at 8 wt % of CNT (corresponding to about 4 vol %). As it can be seen, the experimental density of ABS filled CNT nanocomposites is slightly lower than the theoretical density estimated by using the rule of mixture, which evidences the presence of microvoids, whose volume fraction (V_v) is reported in Figure 3. Details of voids determination are reported in Supplementary Materials Section 2.

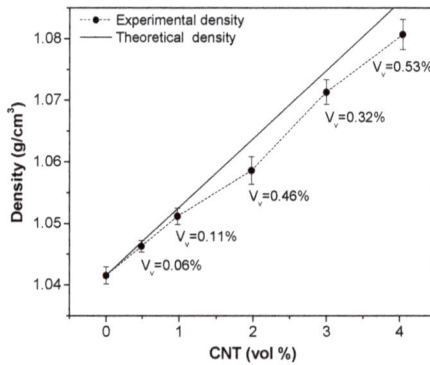

Figure 3. Experimental density values of ABS-CNT nanocomposite compared to theoretical density and voids fraction (V_v).

3.3. Morphological Analyses on Filaments and 3D-Printed Parts

The fracture surface of cryogenically broken filaments and 3D-printed specimens were analyzed by electron microscopy.

Figure 4 illustrates the SEM images of ABS/CNT filaments with a CNTs content of 6 and 8 wt % at increasing magnification. Regarding the CNTs dispersion in both compositions, a homogenous distribution of single nanotubes in ABS matrix can be observed (no aggregates of nanotubes were detected). This means that the adopted two-steps process, consisting of mixing in an internal mixer followed by twin-screw extrusion, was capable to avoid the formation of nanofiller aggregates and to properly disperse CNTs in the ABS matrix. In addition, at high magnifications, a good adhesion level between CNT and ABS can be observed.

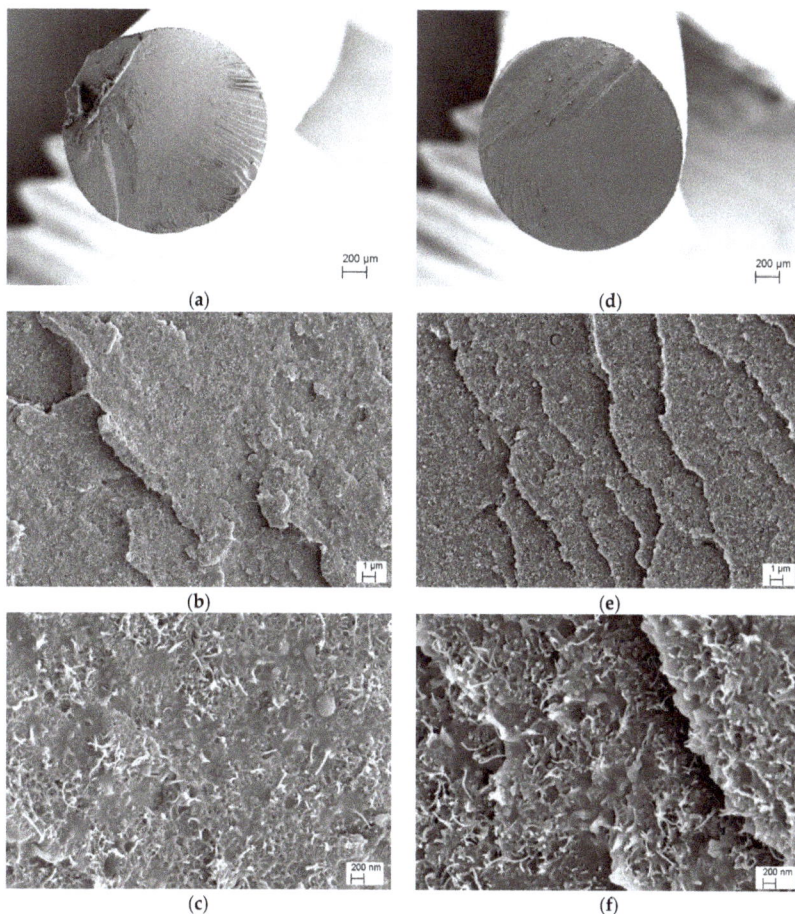

Figure 4. FESEM micrographs of CNT6 (**Left**) and CNT8 (**Right**) filaments at different magnifications (**a,d**) ×80, (**b,e**) ×10,000 and (**c,f**) ×50,000. Carbon nanotubes are identified in the form of small white lines in the highest magnification images.

In Figure 5a–f, the cross-sections of FDM nanocomposite specimens at low and high magnifications are visualized. Moreover, for FDM specimens the presence of voids (about 3 and 1 vol % as observed from Figure 5a,c respectively) is documented. Also, uniform dispersion of nanofillers can be observed in Figure 5b,d,f for all FDM specimens at different build orientations. By using the ImageJ software, the diameter of nanotubes was estimated to be about 33 ± 3 nm for all specimens (average of ten measurements).

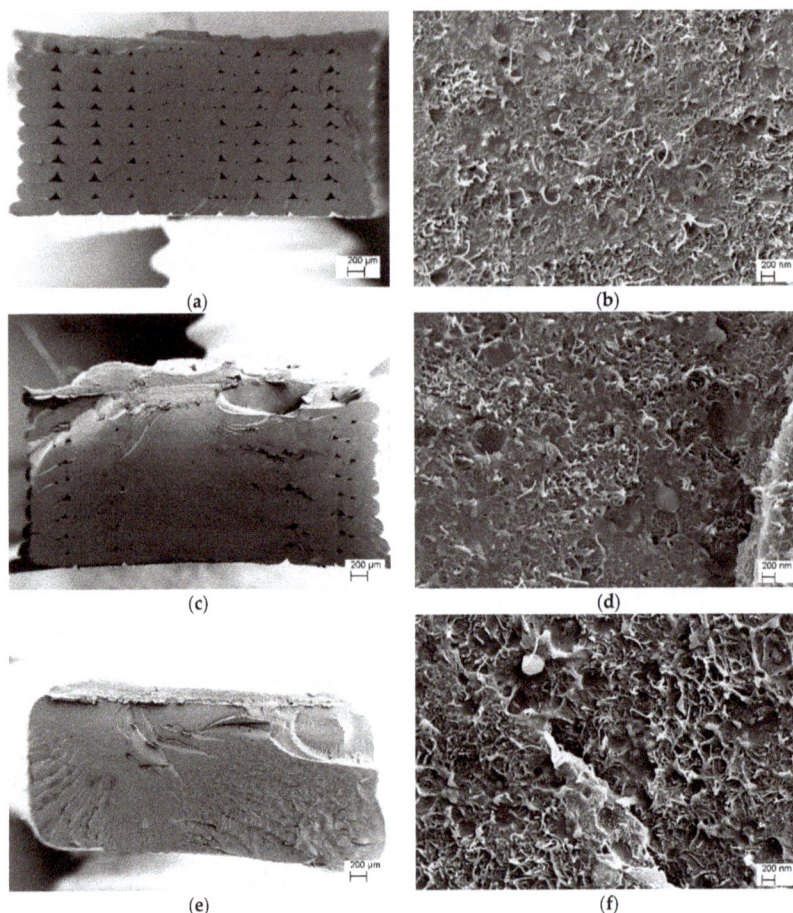

Figure 5. FESEM micrographs of 3D-printed dumbbell specimens printed from carbon nanotubes nanocomposites, CNT6-HC (**a**,**b**), CNT6-H45 (**c**,**d**) and CNT6-VC (**e**,**f**). Carbon nanotubes are identified in the form of small white lines in the high magnification images (see also Figure S7).

3.4. Thermal Degradation Behavior

Thermal stability of ABS matrix and prepared composites was investigated by using thermal gravimetric analysis (TGA). Figure 6a,b depicts the TGA thermogram of neat ABS and CNT-filled composite filaments, while the most important parameters are summarized in Table 4. For the neat ABS in air environment two main degradation steps can be clearly observed at 416 °C and 514 °C, that could be attributed to the molecular chain scission and the oxidation of residual species, respectively [26,27]. On the other hand, neat CNTs showed one single decomposition step at around 627 °C. The onset temperature (T_{onset}) and the maximum degradation temperature ($T_{d,max}$) of the composites slightly increase with rising CNTs fraction up to a maximum value for 2 wt % of CNTs; afterwards they decrease. Similar behavior was also observed for other systems, such as polylactic acid/CNT, where it was attributed to possible aggregation and breakage of CNTs at elevated concentrations [28].

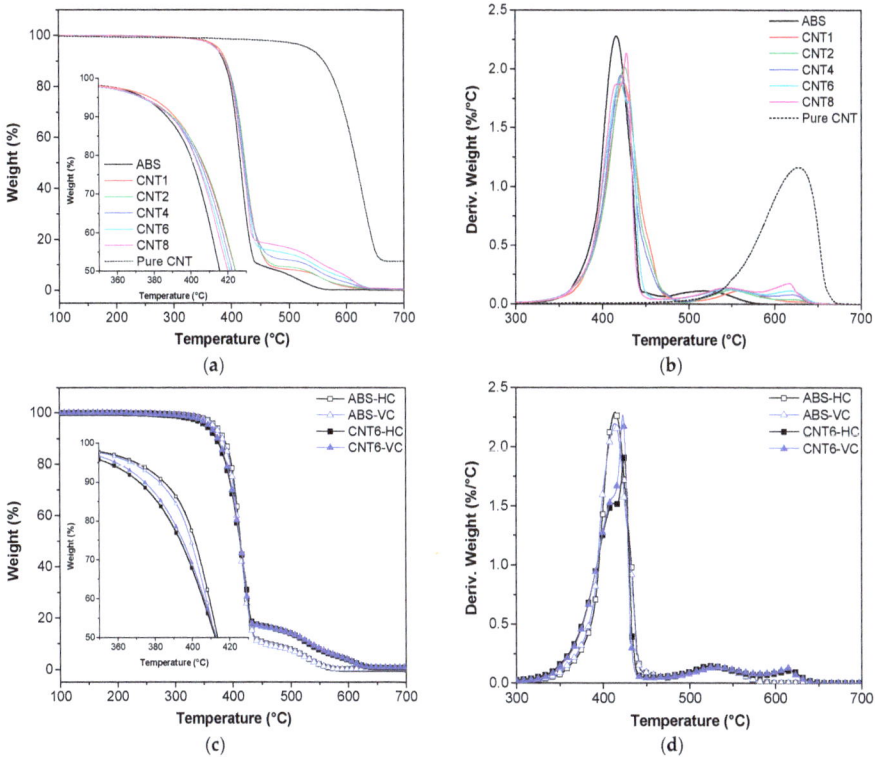

Figure 6. TGA curves of neat and nanofilled ABS filament and 3D-printed samples (HC and VC) under air atmosphere: (**a**,**c**) Residual mass as a function of temperature; (**b**,**d**) Derivative of the mass loss.

Table 4. TGA data of pure ABS and its nanocomposites in an air atmosphere.

Samples	T_{onset} (°C)	T_{max1} (°C)	T_{max2} (°C)	T_{CNT} (°C)	Residue at (wt %)			Relative Residue at (wt %) [1]	
					475 °C	575 °C	700 °C	475 °C	575 °C
ABS	394.0	416.5	514.4	/	7.9	0.1	0.0	0.0	0.0
CNT1	397.5	424.6	560.4	/	8.8	2.8	0.2	0.9	2.7
CNT2	399.4	426.2	548.6	/	9.7	3.0	0.2	1.8	2.9
CNT4	396.9	421.5	547.1	618.2	12.6	4.8	0.4	4.7	4.7
CNT6	394.8	420.6	542.2	617.5	15.4	6.6	0.6	7.5	6.5
CNT8	394.6	428.3	544.1	616.6	17.8	8.2	0.7	9.9	8.1
Pure CNT	576.3	/	/	627.3	97.7	84.6	11.3	/	/
ABS-HC	391.9	414.0	531.8	/	9.2	0.2	0.0	0.0	0.0
ABS-VC	391.0	415.2	535.6	/	8.9	0.2	0.0	0.0	0.0
CNT6-HC	382.3	424.5	525.3	618.0	15.8	5.4	0.6	6.6	5.2
CNT6-VC	388.0	423.1	529.0	613.6	15.6	5.7	0.8	6.7	5.5

[1] calculated according to Equation (1).

For CNT6 and CNT8 samples, it is possible to note that double peaks occurred between 420–430 °C. Moreover, an additional peak of nanocomposites with more than 4 wt % of CNT can be observed around 616–618 °C, which might be associated with the presence of CNT. The maximum mass loss rate (MMLR) in Figure 6b is progressively reduced by the presence of CNT since the nanofiller can hinder the diffusion of volatile products generated by polymer decomposition [27–29]. As reported in Table 4, the residue of tested composites at 700 °C increases with the CNT fraction. However, the residual mass is lower than the nominal amount of CNT because of the oxidation of CNT in the air in the course of experiments.

TGA thermograms reported in Figure 6c,d prove that 3D-printed specimens prepared at the different built orientations (HC and VC) exhibited a behavior similar to that observed for neat ABS filaments. However, as reported in Table 4, 3D-printed nanocomposite samples, i.e., CNT6-HC and CNT6-VC, showed a slightly lower T_{onset} than the corresponding neat ABS samples. The residue at 475 °C and 575 °C was considered to evaluate the CNTs content. In particular, the relative residue obtained after subtraction of ABS contribute fit quite well with the nominal CNT wt %.

3.5. Differential Scanning Calorimetry

Representative DSC thermograms are shown in Figure 7 for CNT6 filament. All DSC thermograms of neat matrix ABS and of CTN filled composites are depicted in Figure S4 and were used for the determination of the glass transition temperature T_g (Table S2). The T_g values found for SAN phase in neat ABS and in CNT-filled ABS filaments are about 106 °C and 108 °C at the first and the second heating run, respectively, which means that the presence of CNT has no significant effects on T_g of ABS/CNT composites. The glass transition temperature of 3D-printed neat ABS (ABS-HC and ABS-VC) is slightly higher than that of neat ABS filament at the first heating run but similar in the second heating run. Moreover, the presence of nanotubes does not have significant effects on the glass transition temperature of nanocomposites in all the three steps of the cycle (first heating-cooling second heating). Also, Yang et al. reported only a slight increase in T_g promoted by single wall carbon nanotubes (SWCNT) dispersed in ABS [27].

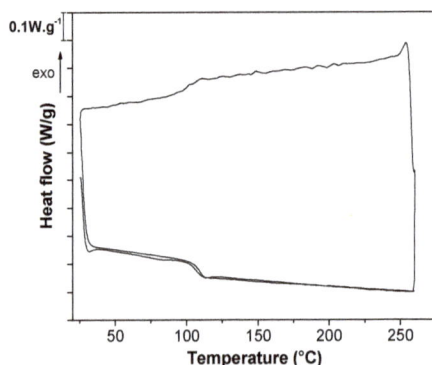

Figure 7. Representative DSC thermogram of CNT6 nanocomposite filament.

3.6. Mechanical Behavior

3.6.1. Quasi-Static Tensile Test

Tensile properties were measured for both filaments and fibers at various CNT contents. Table 5 shows an almost equivalent mechanical behavior of the different diameter extrudates (about 1.7 mm and 0.50–0.65 mm) with no direct dependence on the polymer orientation. Tensile energy to break (TEB) progressively decreases with CNT content and correspondingly the ductility factor for both filaments and fibers, especially above 4 wt % of nanofiller.

Table 5. Quasi-static tensile properties of ABS and its nanocomposite of filaments and single fiber (f) produced by twin screw and FDM extrusion, respectively.

Samples	Filament Diameter (mm)	E (MPa)	σ_y (MPa)	σ_b (MPa)	ε_b (%)	TEB ^{1} (MJ/m^3)	Ductility Factor ^{2} P/TEB
ABS	1.725 ± 0.049	2207 ± 65	42.8 ± 1.9	35.0 ± 0.4	25.6 ± 15.8	8.94 ± 5.61	0.907 ± 0.040
CNT1	1.679 ± 0.007	2132 ± 63	42.9 ± 0.4	35.1 ± 0.3	7.9 ± 2.4	2.59 ± 0.86	0.713 ± 0.104
CNT2	1.684 ± 0.025	2226 ± 48	43.3 ± 0.3	37.8 ± 1.8	4.4 ± 1.2	1.36 ± 0.45	0.464 ± 0.153
CNT4	1.765 ± 0.026	2320 ± 74	43.4 ± 0.9	41.9 ± 1.7	2.6 ± 0.3	0.65 ± 0.16	0.099 ± 0.083
CNT6	1.712 ± 0.035	2625 ± 55	47.1 ± 0.5	44.6 ± 1.0	3.2 ± 0.5	1.04 ± 0.24	0.273 ± 0.158
CNT8	1.702 ± 0.016	2650 ± 125	46.8 ± 1.2	46.5 ± 1.1	2.5 ± 0.2	0.73 ± 0.10	0.046 ± 0.065
f-ABS	0.648 ± 0.021	1918 ± 105	40.4 ± 0.9	33.6 ± 0.8	52.8 ± 27.2	18.5 ± 9.60	0.944 ± 0.023
f-CNT1	0.591 ± 0.012	1801 ± 122	39.3 ± 1.7	35.8 ± 1.5	6.4 ± 3.2	2.00 ± 1.20	0.463 ± 0.259
f-CNT2	0.567 ± 0.007	2033 ± 142	40.4 ± 0.5	33.6 ± 0.8	4.9 ± 1.2	1.53 ± 0.44	0.371 ± 0.168
f-CNT4	0.528 ± 0.001	2035 ± 58	42.9 ± 1.4	40.8 ± 2.4	4.9 ± 1.2	1.59 ± 0.48	0.335 ± 0.220
f-CNT6	0.515 ± 0.003	2099 ± 124	44.9 ± 1.3	44.1 ± 1.6	4.1 ± 0.6	1.29 ± 0.23	0.124 ± 0.098
f-CNT8	0.506 ± 0.005	2147 ± 80	47.1 ± 0.6	46.9 ± 0.9	4.0 ± 0.7	1.31 ± 0.31	0.096 ± 0.075

^{1} Total energy to break. ^{2} Ratio between the propagation energy (P) from the yield to break point, with respect to TEB.

Representative stress-strain curves of filaments of neat ABS and its nanocomposites are reported in Figure 8. It is worth noting that CNT enhances both tensile modulus (E) and yield strength (σ_y) of the composites (Table 5). At the highest concentration of nanotubes (8 wt %) the elastic modulus of ABS/CNT nanocomposites achieved a value 19% higher than that of ABS matrix. The highest σ_y was found for CNT6, while CNT8 shows a slight reduction in σ_y and almost brittle behavior. Therefore, ABS with 6 wt % of carbon nanotubes was an optimal compromise for FDM application.

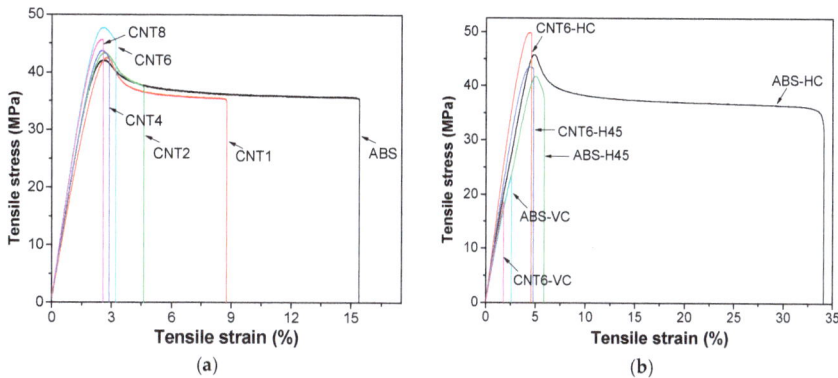

Figure 8. Tensile stress-strain curve of ABS and ABS-CNT filaments (**a**) and 3D-printed samples (**b**).

Stress-strain curves of 3D-printed specimens are shown in Figure 8b and the resulting mechanical parameters are summarized in Table 6. Tensile modulus of H45 sample is comparable to that of HC sample probably because of good contact between bead extruded microfilaments and a lower fraction of voids in H45, as documented by SEM images (Figure 5a,c). Similarly enough, the lower yield strength of H45 with respect to that of HC is most probably due to internal orientations of deposited filaments as shown in Figure 1a,b. H45 or HC samples are expected to behave almost as isotropic or transversally isotropic materials. On the other hand, ABS-VC samples manifest a brittle behavior due to the weakness of interlayer bonding and the same behavior is even clearer for CNT6-VC samples because interlayer bonding could be significantly reduced by the higher viscosity in the molten state. Correspondingly, the ductility factor is zero, due to the absence of any toughening mechanism in the fracture process. In addition, the presence of CNTs resulted in an enhancement of both tensile modulus and yield stress for all FDM samples. The elastic modulus of ABS/CNT nanocomposites continuously increased up to 22%, 18% and 5% above that of unfilled ABS at the orientation of HC, H45 and VC, respectively. The highest yield stress can be observed in CNT6-HC sample owing to deposited filaments parallel to the applied load and the reinforcing effect of carbon nanotubes. As a side effect, the elongation at break of FDM composites samples was significantly reduced proportionally to the CNT content.

Table 6. Quasi-static tensile properties of ABS and its nanocomposite of FDM samples.

Samples	E (MPa)	σ_y (MPa)	σ_b (MPa)	ε_b (%)	TEB [1] (MJ/m³)	Ductility Factor [2] P/TEB
ABS-HC	2235 ± 170	45.7 ± 0.5	31.9 ± 1.7	30.0 ± 10.4	10.7 ± 3.76	0.866 ± 0.077
ABS-H45	2308 ± 112	41.1 ± 0.9	37.9 ± 1.6	5.3 ± 0.5	1.30 ± 0.16	0.204 ± 0.120
ABS-VC	2077 ± 44	/	22.0 ± 4.4	2.4 ± 0.7	0.30 ± 0.10	0
CNT6-HC	2735 ± 158	49.6 ± 0.6	49.2 ± 0.6	4.5 ± 0.2	1.35 ± 0.10	0.048 ± 0.044
CNT6-H45	2739 ± 268	43.2 ± 0.3	42.6 ± 0.4	4.6 ± 0.3	1.19 ± 0.11	0.054 ± 0.056
CNT6-VC	2181 ± 51	/	18.7 ± 1.5	1.9 ± 0.1	0.18 ± 0.03	0

[1] Total energy to break. [2] Ratio between the propagation energy (P) from the yield to break point, with respect to TEB.

3.6.2. Fracture Mechanism

Figure S5 presents the fractured surface of ABS and CNT6 3D-printed specimens broken in liquid nitrogen. For both HC and H45 samples, it is easy to observe along the thickness in Z direction 10 flattened parallel deposited bead microfilaments with a dimension of about 420 microns in width and 210 microns in height for ABS-HC (and about 410 micron and 210 micron for CNT6-HC). Taking into consideration the initial diameter of freely extruded fibers (Table 5), a further orientation of about 3.7 and 2.4 could be calculated during 3D-printing of the microfilament for ABS-HC and CNT6-HC, respectively (see Supplementary Materials Table S3).

At the same time, along with the sample width in the Y direction, HC evidenced 10 deposited microfilaments, whereas only 4 deposited parallel microfilaments could be observed in the external contours of H45 samples (2 on the right and 2 on the left). The inner microfilaments oriented at +45°/−45° could not be easily distinguished and an almost homogeneous zone appeared. For this reason, the similar stiffness of HC and H45 can be attributed to the combined effect of both the larger number of voids and orientation of microfilaments. On the other hand, for ABS-VC and CNT6-VC no traces of voids and of deposited microfilaments were observed in the cryo-fractured surface. These results have been attributed to the higher temperature of interlayer overlapping that is dependent on two factors: (i) the lower deposition rate (16 mm/s of VC sample with respect to 40 mm/s of other samples) and consequently the lower viscosity of deposited microfilament; and (ii) the lower time of deposition of the layer in VC samples with respect to HC samples (23 s vs. 46 s, respectively) and hence the higher temperature of the last deposited layer in VC sample (surface of deposition).

Difference is the case of a fractured cross-section of 3D-printed samples derived from the tensile test, as shown in Figure S6. The clear shape and size of the triangle between the deposited microfilaments (see Figure S6a) were observed due to the plastic deformation under tensile load. Moreover, some traces of microfilaments were partially evidenced in VC sample of both neat ABS and its nanocomposites, as shown in Figure S6c,f, which suggests a weak adhesion of the inter-layer bonding between microfilaments.

3.7. Creep Stability

Figure 9a,b shows the creep compliance at 30 °C of neat ABS and composites found for (a) filaments and for (b) FDM samples. If no plastic deformation occurs, compliance of isothermal tensile creep, $D_{tot}(t)$, consists two components: elastic (instantaneous) D_{el} and viscoelastic (time-dependent) D_{ve}, as defined in Equation (10).

$$D_{tot}(t) = D_{el} + D_{ve}(t) \tag{10}$$

Incorporation of CNTs in ABS accounts for a pronounced reduction of both compliance components, as reported in Table 7. D_{el} is characterized by an almost linear decrease with CNTs fraction, which is in conformity with the inverse trend of tensile modulus (Tables 5 and 6). For example, the composite with 8 wt % of the nanofiller showed D_{el} or $D_{tot,3600s}$ by 21% or 26% lower than the neat matrix. For FDM samples, a similar effect of CNT on both elastic and viscoelastic creep compliance was observed: 6 wt % of the nanofiller in ABS matrix reduced the total compliance of nanocomposite by 16%, 12% and 10% for HC, H45 and VC respectively.

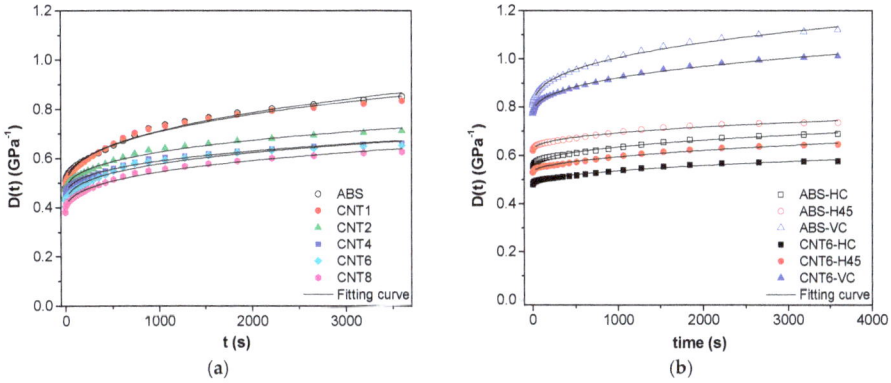

Figure 9. Creep compliance, D(t) at 30 °C, of neat ABS and nanocomposites as measured on (**a**) filaments at 3.9 MPa and (**b**) 3D-printed samples along different orientations at 3.0 MPa.

Table 7. Elastic (D_{el}), viscoelastic D_{ve} (t = 3600 s) and total D (t = 3600 s) creep compliance at 3600 s and fitting parameters (Equation (11)) of ABS and its nanocomposites as measured on filaments and FDM samples.

Samples	D_{el} (GPa^{-1})	$D_{ve,3600s}$ (GPa^{-1})	$D_{tot,3600s}$ (GPa^{-1})	D_e (GPa^{-1})	K (GPa^{-1} s^{-n})	n	R^2
ABS	0.482	0.369	0.851	0.488	0.012	0.419	0.9924
CNT1	0.471	0.362	0.833	0.460	0.020	0.364	0.9889
CNT2	0.450	0.261	0.710	0.447	0.016	0.345	0.9912
CNT4	0.436	0.225	0.660	0.436	0.014	0.342	0.9920
CNT6	0.405	0.246	0.652	0.393	0.020	0.319	0.9844
CNT8	0.380	0.246	0.626	0.374	0.016	0.347	0.9925
ABS-HC	0.521	0.169	0.689	0.547	0.005	0.402	0.9951
ABS-H45	0.587	0.148	0.735	0.616	0.005	0.392	0.9910
ABS-VC	0.756	0.364	1.120	0.783	0.019	0.355	0.9980
CNT6-HC	0.454	0.123	0.577	0.479	0.002	0.469	0.9905
CNT6-H45	0.501	0.145	0.645	0.531	0.002	0.503	0.9917
CNT6-VC	0.729	0.283	1.012	0.758	0.013	0.366	0.9981

The empirical Findley's model (power law), summarized in Equation (11) was used to describe the viscoelastic creep response [30–32]:

$$D(t) = D_e + kt^n \tag{11}$$

where D_e is the elastic (instantaneous) creep compliance, k is a coefficient related to the magnitude of the underlying retardation process and n is an exponent related to the time dependence of the creep process. The fitting parameters for experimental creep data are summarized in Table 7. The fitting model was satisfactory, as R^2 around 0.99 was found for all samples value. The addition of CNT reduced the creep compliance of composites; in particular, the value of parameter D_e, for both filaments and 3D-parts are in good agreement with the values of D_{el}. from Equation (10). The coefficient n reflects the kinetics of displacements of the segments of macromolecules in the viscous medium in the course of the creep and it was found to slightly decrease with the presence of CNT in ABS filaments.

3.8. Dynamic Mechanical Response and Coefficient of Thermal Expansion

Figure 10 documents that ABS matrix and all composites show two transitions which can be identified with the glass transition of butadiene phase (B-phase; T_{g1} = −84 °C) and the glass transition of styrene–acrylonitrile phase (SAN phase; T_{g2} = 125 °C). Incorporation of CNT accounts for enhancement of the storage modulus of composites above that of ABS matrix, which becomes more pronounced at higher temperatures. For instance, at the highest concentration of CNT (8 wt %), the storage modulus of composite filament exceeds that of ABS by about 16% at 30 °C and by 897% at 130 °C.

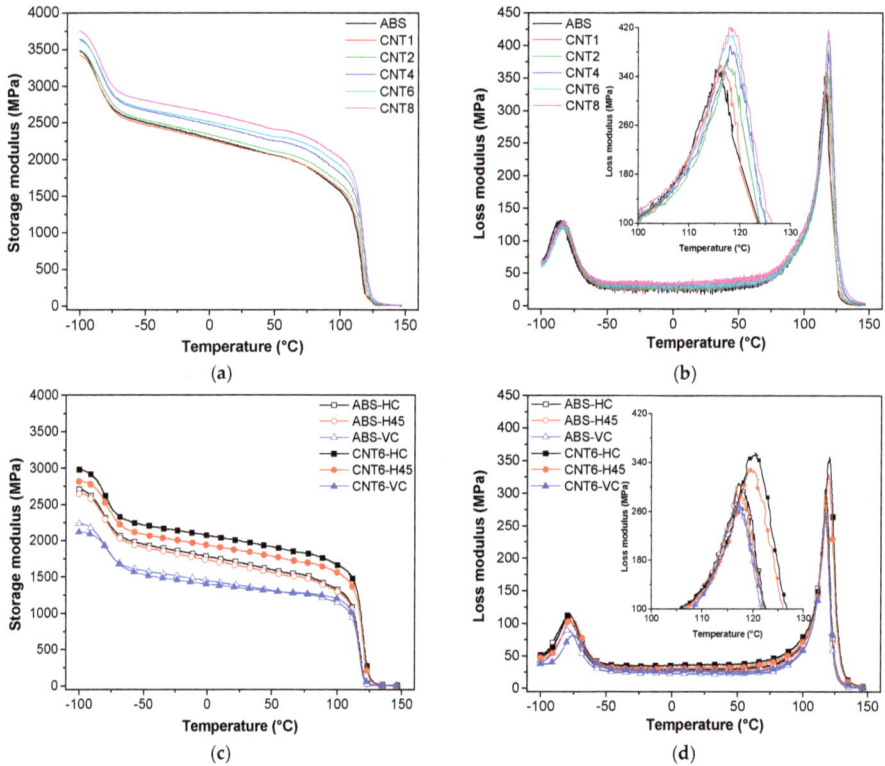

Figure 10. Dynamic mechanical thermograms of filament and 3D-printed samples (**a**,**c**) storage modulus (E') and (**b**,**d**) loss modulus (E''), of neat ABS and nanocomposite samples.

Incorporated CNT also contributes to enhancing the dissipation of mechanical energy, as represented by the dynamic loss modulus. Moreover, the nanofiller also increases the glass transition temperatures of both butadiene and styrene–acrylonitrile phases by about 3 °C due to the hindering of segmental motions at the interface. Similar observations were also reported in prior papers [23,33].

As expected, the storage modulus of 3D-printed specimens at build parallel and ±45° orientations (HC and H45) is higher than that measured on samples with the VC orientation. The behavior observed for HC and H45 samples is related to the direction of the deposited filaments preferentially aligned and isotropic materials inclined at ±45° along the tensile applied load respectively, while the deposited layers in VC specimens are mostly oriented transversally to the tensile force. In general, the 3D-printed samples show storage modulus lower than original filaments due to the presence of voids and specific orientation of extruded microfilaments in 3D-printed samples (HC and VC).

The data summarized in Table 8, clearly show that the storage modulus of HC or H45 at 30 °C is enhanced by about 15% or 12% due to the addition of carbon nanotubes. The observed effect is even more pronounced at higher temperatures: at 130 °C the storage modulus of CNT6-HC or CNT6-H45 is 5 times higher than that of neat ABS-HC and ABS-H45. On the other hand, CNT do not exhibit any stiffening effect on storage modulus along VC orientation in the temperature range −50 to 30 °C, while a three-fold increase in the storage modulus can be observed at 130 °C.

Table 8. Dynamic mechanical properties of neat ABS and its nanocomposites as measured on filaments and FDM samples.

Samples	Storage Modulus					Damping Peaks		Loss Modulus of SAN Peak		Stiffness Loss [1] at T_g	
	−100 °C (MPa)	−50 °C (MPa)	30 °C (MPa)	90 °C (MPa)	130 °C (MPa)	B-Phase T_{g1}	SAN-Phase T_{g2}	E''_{peak} (MPa)	T_{peak}	$SL\,T_{g1}$	$SL\,T_{g2}$
ABS	3474	2503	2145	1711	4.7	−84.9	122.7	347	115.6	0.453	0.795
CNT1	3417	2468	2129	1729	6.2	−83.6	123.2	355	116.3	0.446	0.809
CNT2	3487	2531	2197	1809	9.7	−82.4	125.5	356	117.6	0.435	0.819
CNT4	3636	2662	2342	1952	18.3	82.9	126.0	380	118.4	0.416	0.826
CNT6	3614	2685	2390	2044	28.0	−81.5	124.6	406	118.0	0.389	0.844
CNT8	3747	2799	2496	2139	42.2	−82.2	125.6	419	118.4	0.380	0.840
ABS-HC	2709	1948	1678	1415	5.4	−78.6	124.9	304	117.2	0.454	0.840
ABS-H45	2646	1904	1631	1384	4.7	−78.0	124.8	293	116.9	0.455	0.846
ABS-VC	2229	1583	1367	1207	3.9	−77.9	124.1	267	117.0	0.473	0.880
CNT6-HC	2980	2211	1977	1749	32.6	−77.5	127.6	354	120.3	0.389	0.868
CNT6-H45	2813	2080	1845	1632	25.9	−75.3	127.0	326	120.0	0.397	0.871
CNT6-VC	2114	1527	1338	1245	11.8	−73.9	123.4	270	117.1	0.439	0.922

[1] stiffness loss (*SL*) calculated according to Equation (12).

The presence of carbon nanotubes also increases the glass temperature of CNT6-HC and CNT6-H45 by about 3 °C, which is identical with previously reported an increase in T_g for nanocomposite filaments (see Supplementary Materials Table S2).

The stiffness loss (SL_{Tg}) at the glass transition temperature could be evaluated from the reduction of storage modulus before and after the transition ($\Delta E'$), according to Equation (12) as a function of storage modulus at 30 °C.

$$SL_{Tg} = (\Delta E')/E'_{30°C} \tag{12}$$

where $\Delta E'$ represents the modulus variation from −100 °C to −50 °C, or from 90 °C to 130 °C, in the case of transition of butadiene or SAN phase, respectively (see data in Table 8).

In the zone of butadiene transition, the parameter *SL* was found to progressively decrease from about 0.45 (ABS matrix) up to 0.38 for CNT8, in dependence on the content of CNT for all nanocomposite samples (both filaments and 3D-printed parts) as indication of the role of nanofiller in the relative stiffening at room temperature. On the other hand, the stiffness loss at the main glass transition (T_g of SAN phase) is almost linearly increasing with CNT content, from 0.80 (ABS filament) to about 0.84 for CNT8 filaments and it depends on the stiffening of rubbery phase above T_g.

Figure 11a,b shows the reduction of the main transition of storage modulus (*R*) and *F*-factor [23,24] which are plotted as functions of the CTN fraction. In Figure 11a, the stiffening effect of CNTs in the rubbery phase above T_g of SAN is well documented. In particular, this effect seems to be more pronounced for FDM samples (HC and H45) with respect to filaments CNT6, probably owing to the higher orientation and adhesion/dispersion of carbon nanotubes in FDM process.

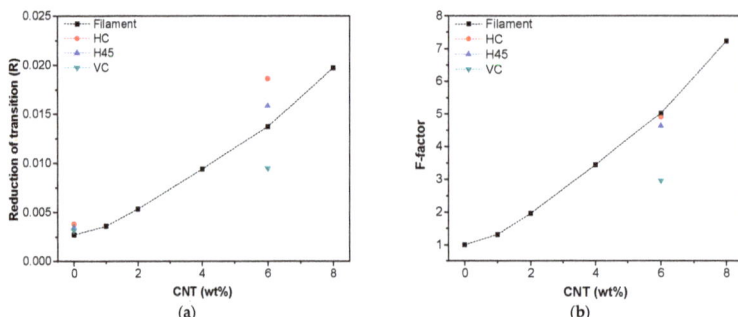

Figure 11. Reduction of main transition of storage modulus-*R* (**a**) and *F*-factor (**b**) as function of CNT nanofiller loading measured on filaments and 3D-printed samples (HC, H45 and VC).

Moreover, the *F*-factor represents a relative measure of modulus in the temperature interval of the glass transition, assuming that modulus at glassy state is dominated by the strength of intermolecular forces when polymer chains and nanofillers are packed [23]. Thus, the higher *F*-factor, the higher the effectiveness of the filler. Figure 11b presents the increase in the *F*-factor of filaments with the fraction of CNT and it confirms the relative effectiveness of CNT nanofiller with its fraction in composites in the rubbery phase.

For FDM sample (HC and H45), the reinforcing efficiency is slightly lower than filament at 6 wt % of CNT, maintaining almost the same adhesion level of nanofiller and matrix during FDM process. On the other hand, VC sample shows a different effect since the properties of these specimens are mainly dependent on the inter-layer matrix adhesion and mostly independent on the compatibility of the polymer chains and nanofiller.

Thermal strain of ABS/CNT filaments is plotted in Figure 12a and the coefficient of thermal expansion of all samples is reported in Table 9. The thermal strain of composite filaments exhibited the linear trend up to 100 °C, i.e., approximately to the glass transition temperature. The steep increment of thermal strain indicates the transition from the glassy state to the rubbery state with the much higher mobility of polymer chains. Above 120 °C, the thermal strain showed negative slope with respect to the temperature scale due to some shrinkage of the polymer chains orientated during extrusion. Incorporated CNT markedly reduced the coefficient of thermal expansion (see Table 9). As expected, the composite with the highest concentration of CNT shows the largest drop of the coefficient of thermal expansion, i.e., 79.6 for ABS to 52.3×10^{-6}/K for CNT8 at the room temperature (20~50 °C) and from -891 for ABS to -51×10^{-6}/K for CNT8 at the temperature (130~150 °C).

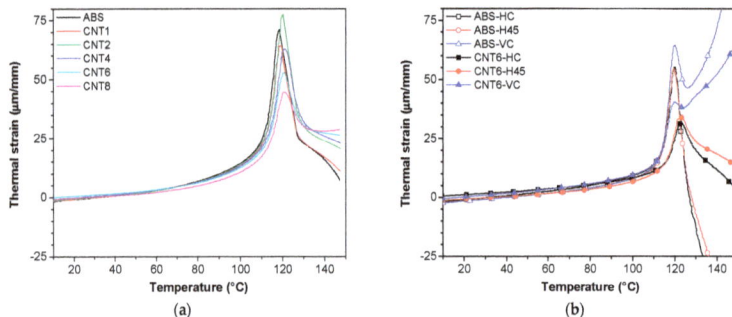

Figure 12. Thermal strain of neat ABS and nanocomposite samples as measured on filaments (**a**) and 3D-printed samples (**b**) along different orientations (HC, H45 and VC).

Table 9. Coefficients of linear thermal expansion (CLTE) and linear thermal deformation (CLTD) of ABS and its nanocomposites as measured on filament and FDM samples (see Equation (4) for detail).

Sample	CLTE ($\times 10^{-6}$/K)				CLTD ($\times 10^{-6}$/K)
	$\Delta T_1 = -50/-20\,°C$	$\Delta T_2 = 20/50\,°C$	$\Delta T_3 = 70/90\,°C$	$\Delta T_4 = 108/113\,°C$	$\Delta T_5 = 130/150\,°C$
ABS	49.7 ± 0.2	79.6 ± 0.4	262.9 ± 2.1	2350 ± 68	-891 ± 11
CNT1	51.6 ± 0.3	78.5 ± 0.5	246.1 ± 1.8	1880 ± 55	-678 ± 55
CNT2	48.4 ± 0.1	70.3 ± 0.4	250.8 ± 2.0	1770 ± 54	-491 ± 7
CNT4	43.2 ± 0.1	67.5 ± 0.5	223.8 ± 2.1	1450 ± 44	-465 ± 7
CNT6	38.1 ± 0.1	55.9 ± 0.3	207.0 ± 2.1	1290 ± 26	-206 ± 7
CNT8	33.7 ± 0.1	52.3 ± 0.2	191.3 ± 1.9	1150 ± 21	-51 ± 7
ABS-HC	61.0 ± 0.1	85.8 ± 0.3	156.6 ± 1.2	1040 ± 41	-4860 ± 50
ABS-H45	58.0 ± 0.2	74.5 ± 0.2	146.5 ± 1.1	1210 ± 36	-3620 ± 32
ABS-VC	61.0 ± 0.2	79.1 ± 0.2	147.3 ± 0.9	1330 ± 36	3310 ± 67
CNT6-HC	40.2 ± 0.1	59.0 ± 0.2	106.7 ± 0.9	479 ± 11	-805 ± 4
CNT6-H45	41.1 ± 0.1	54.0 ± 0.2	114.3 ± 1.0	587 ± 16	-506 ± 2
CNT6-VC	57.8 ± 0.1	79.4 ± 0.3	139.7 ± 0.8	1010 ± 28	1090 ± 12

In the temperature interval 20/50 °C, FDM specimens (HC, H45 and VC) printed from neat ABS, exhibit CLTE values of 85.8, 74.5 and 79.1 $\times 10^{-6}$/K, respectively (see Table 9). The presence of CNTs accounts for a reduction of the CLTE of FDM specimens by 31% or 27% for HC or H45 but no effect was observed for VC build orientation.

3.9. Electrical Behavior

3.9.1. Electrical Resistivity

The previous text has illustrated how CNT affects mechanical properties of prepared composites but most important effects of CNT can be expected in the field of electrical properties. Improvements of conductivity and electrical properties by incorporated CNT in different polymer, such as polyamide [34], polypropylene [35], polylactide [28] and ABS [18,29,36] have been documented in literature According to the technical data sheet [19], the volume resistivity of neat ABS bulk materials is 10^{15} Ω·cm. Our measurements reveal that the volume resistivity of the composites significantly decreases with at least 4 wt % of nanofiller (see Figure 13), whereas at CNTs fractions up to 2 wt %, the materials still exhibit an insulating behavior and filaments could not be tested by means of the four probes configuration.

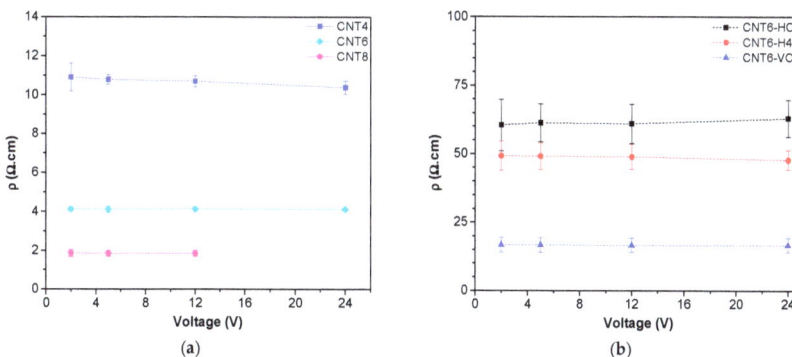

Figure 13. Electrical resistivity of ABS nanocomposites: (**a**) filament and (**b**) 6 wt % filled nanocomposites with different 3D printing as a function of the applied voltage.

The incorporation of CNTs decreases the electrical resistivity of filaments to about 11 Ω·cm, 4.1 Ω·cm and 1.8 Ω·cm for CNT4, CNT6 and CNT8, respectively. The volume resistivity of all samples is directly dependent on the CNTs content and independent on the applied voltages (Figure 13a), which suggests that these nanocomposites behave as ohmic conductors. It is also worth noting that CNT8 filaments could not be tested at 24 V due to the high resistive heating effect.

The resistivity of 3D-printed samples CNT6-H45 and CNT6-VC shown in (Figure 13b) is independent of applied voltages and is higher than the correspondent filament. This partial reduction of conductivity not only in comparison with single filaments but also with compression molded specimens at the same composition (see Supplementary Materials Figure S9) could be attributed to the internal features of FDM samples. Moreover, it should be noted that CNT6-HC shows the highest resistivity, whereas CNT-VC the lowest. These results could be related to the better contact between deposited bead microfilaments, resulting in higher conductivity of samples; these findings are in good conformity with the documentation of SEM images (Figure 5e), where VC-CNT6 specimens exhibit better and extensive contacts between the layer of deposited microfilaments in the direction of electrical measurements.

Similarly, in literature, Zhang et al. reported that the resistivity of 3D-printed components was found lower than the pristine 3D-printing fibers and the results were also confirmed to be highly dependent on the contact resistivity by numerical simulation method [10].

In order to understand the electrical behavior of composite filaments and to evaluate the effect of CNT orientation in ABS, the most conductive filaments (i.e., CNT6 and CNT8) were compression-molded for the production of homogeneous plates (resistivity results are shown in Figure S9). It worth noting that the electrical resistivity of CNT8-0, CNT8-45 and CNT8-90 was found to directly depend on the angles of filament orientation in the plate (Figure S8c). The resistivity of CNT8-0 is similar to that of CNT8 filament owing to the almost identical filaments orientation, whereas CNT8-90 leads to the lower level of filaments alignment with respect to the electrical field. The higher the angle, the higher the resistivity. And the same for specimens CNT6-0, CNT6-45 and CNT6-90, at resistivity even higher. From these findings, the electrical resistivity of filaments could be considered a quasi-isotropic behavior of materials with partial random oriented CNT.

However, after FDM process the conductivity of 3D-printed fibers (see Table S4) slightly increases, so that is comparable to that of plate samples. The results suggest that the orientation CNT during the extrusion contributes to the reduction of the resistivity of the composites. A similar effect was also observed for the composites with graphene oxide [8] and carbon black [10].

The beneficial effect of CNTs could be summarized in the double results to increase the stiffness and to reduce the electrical resistivity of ABS nanocomposites, considering the experimental ratio modulus/resistivity, expressed as MPa/Ω·cm. Figure 14 depicts this double effect, revealing that filaments and fibers exhibited the best behavior especially with a CNT of 6–8 wt %. The relative lower values of 3D-printed specimens directly depend on the specific FDM process.

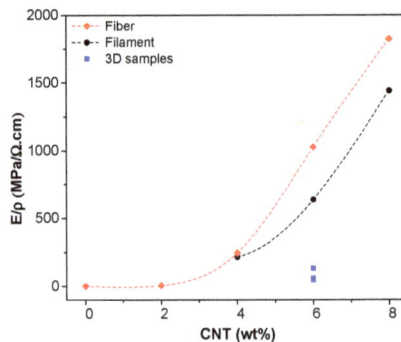

Figure 14. The ratio modulus/resistivity reference at 5 V as a function of CNT % for filament (●), fiber (♦) and 3D samples (■).

3.9.2. Surface Temperature under Applied Voltage

The measurement of Joule's heating upon voltage application of the samples with different fractions of CNTs was performed for 12 and 24 V which are commonly reached by batteries for automotive applications. In Figure 15, we can see the representative images of the evolution of surface temperature upon voltage application to CNT6 filament and FDM samples. The highest temperature was obviously concentrated in the center of samples due to the cooling effect at the border of samples due to the relatively low thermal conductivity. We monitored the evolution of the surface temperature as a function of the voltages, the time and the composition of nanocomposite materials. As shown in Figure 16a,b the increment of the temperature of all samples under both voltages (12 V and 24 V) seems to reach the plateau after 60 s. Obviously, the higher the applied voltage, the higher the increase in temperature. Besides, the resistivity of composite materials of ABS/CNT filaments evidences a good correlation with the increase of the temperature. The higher the conductivity, the higher the increase in the surface temperature of samples due to the dissipation of thermal energy. For example, CNT4 sample does not show any significant increase in temperature, whereas a rather high increase in temperature can be seen for CNT8. It is worth noting that, at applied voltage of 24 V for 120 s (Figure 16b), the generated surface temperature of CNT8 sample exceeds the glass transition temperature of ABS. Therefore, in order to avoid thermal degradation of materials during prolonged voltage application, between the various ABS materials studied in this research, the CNT6 samples appeared the most convenient nanocomposite materials for electro-conductive applications.

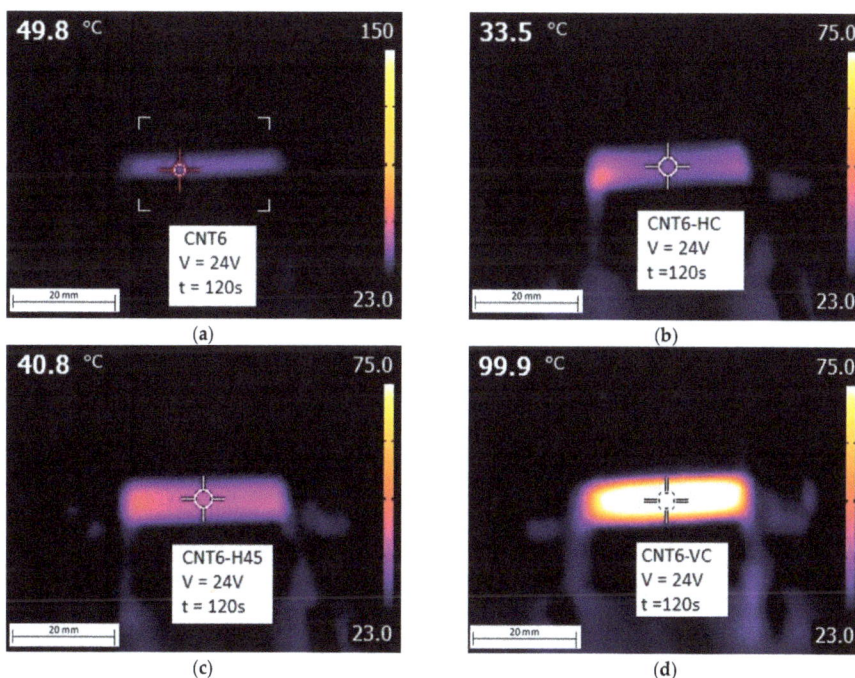

Figure 15. Results of thermal imaging upon voltage application at 24 V at 120 s: CNT6 filament (**a**); CNT6-HC (**b**); CNT6-H45 (**c**) and CNT6-VC (**d**).

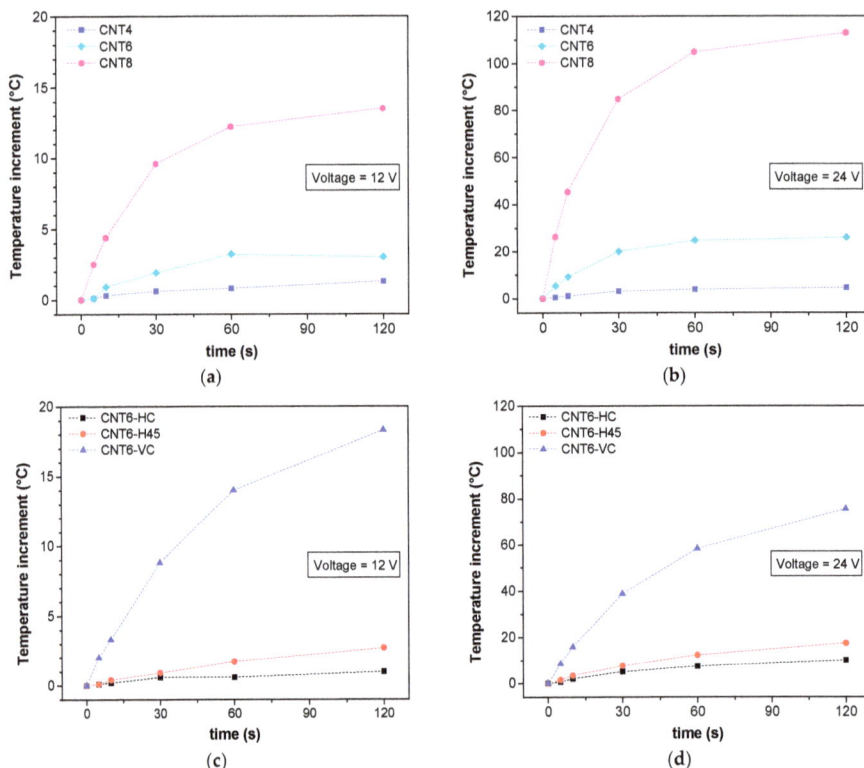

Figure 16. Increment of surface temperature upon a voltage of 12 V (**a,c**) and 24 V (**b,d**) for ABS nanocomposites filaments and 3D-printed samples with different CNT loading at room temperature of 23 °C.

The electrical measurements of 3D-printed samples with CNT contents of 6 wt % built with different orientations were performed by using the same two voltages applied to the filaments (12 V and 24 V). Surface temperature under applied voltage shows good correlation with resistivity measurements. Lower resistivity resulted in a higher increment of temperature, e.g., CNT6-VC reached the highest temperature of about 100 °C after 120 s. However, the local temperature of all FDM samples achieved via the Joule's effect remains below the glass transition temperature of ABS, which allows us to presume good thermal stability of produced nanocomposite materials in electrical applications.

4. Conclusions

Carbon nanotubes (in fractions up to 8 wt %) were directly melt compounded with relatively high viscosity ABS matrix by using a completely solvent-free process. Subsequently, by using a twin-screw extruder, composite filaments were appositely extruded for application in 3D printing with fused deposition modelling.

The optimum CNT fraction for fused deposition modelling process was found to be 6 wt %. Thermal, mechanical and electrical properties of neat ABS and ABS/CNT composites have been investigated on produced filaments and 3D-printed parts. CNT has the positive effect on the resistance to long-lasting loads due to the reduction of creep compliance. Besides, the enhancement of both tensile modulus and strength was found for filaments and FDM products, except for vertical 3D built specimens. On the other hand, elongation at break of the composites was reduced in proportion to the

Nanomaterials **2018**, *8*, 49

CNT fraction. The presence of CNT also promoted the thermal stability of 3D-printed parts due to the reduction in coefficient of thermal expansion.

Electrical conductivity of 3D-printed samples was markedly incremented but a partial loss in conductivity with respect to filament nanocomposite was also observed. Moreover, the resistivity of 3D-printed parts is highly dependent on the build microfilaments orientation, which consequently leads to different surface temperature increment under applied voltages. For FDM-printed parts, the carbon nanotubes in playing the best reinforcement in thermal mechanical behavior for HC and H45 orientation but less effective in electrical properties.

Supplementary Materials: The following are available online at www.mdpi.com/2079-4991/8/1/49/s1, Figure S1: Schematic of 3D-printed Parallelepiped: (a) horizontal concentric (HC); (b) horizontal 45° angle (H45) and (c) vertical concentric (VC), Figure S2: 3D-printed dumbbells, resistivity and resistivity heating specimens of ABS and ABS-CNT nanocomposites, Figure S3: Density of carbon nanotube measured through a Micromeritics® Accupyc 1330 helium pycnometry (23 °C) with 10 cm^3 chamber, Figure S4: DSC thermogram of neat ABS and its ABS/CNT nanocomposites: (a) filaments; (b) 3D-printed samples, Figure S5: Frozen fracture of cross-section of 3D-printed dumbbells: (a) ABS-HC; (b) ABS-H45; (c) ABS-VC; (d) CNT6-HC; (e) CNT6-H45 and (f) CNT6-VC, Figure S6: Tensile fracture of cross-section of 3D-printed dumbbells: (a) ABS-HC; (b) ABS-H45; (c) ABS-VC; (d) CNT6-HC; (e) CNT6-H45 and (f) CNT6-VC, Figure S7: SEM micrographs of 3D-printed dumbbell specimens of CNT6-HC with indicating CNTs (red arrow). Figure S8: Summary of preparation of filament plate with the mold 50 mm × 50 mm × 1.0 mm starting with filaments at 6 and 8 wt % of CNT: (a) before compression and (b) after compression; (c) Schematic of samples at the different angles (0, 45 and 90°) for measuring electrical resistivity (see Figure S8), Figure S9: Electrical volume resistivity of ABS 6 wt % and 8 wt % filled nanocomposites of filament plates at different angles (0, 45 and 90°) as a function of the applied voltage, Table S1: Dimensions and processing parameters of FDM specimens, Table S2: Glass transition temperatures (T_g) of styrene–acrylonitrile phase in ABS and in nanocomposite (from inflection point of DSC thermogram). Table S3: Evaluation of orientation factor in microfilament during 3D printing (according Equation (S4)). Table S4: Volume resistivity of different kinds of ABS-CNT samples at an applied voltage of 5 V.

Acknowledgments: Authors wish to thank Versalis S.p.A. (Mantova, Italy) for donating ABS pellet polymer for this work. Authors are also thankful to Sharebot S.r.l. (Nibionno, LC, Italy) for providing the prototype of the HT Next Generation desktop 3D-printer. One author (S.D.) is grateful to AREAS + EU Project of Erasmus Mundus Action 2 Programme for financial support.

Author Contributions: S.D., L.F. and A.P. conceived and designed the experiments; S.D. performed the experiments; S.D., L.F. and A.P. analyzed the data and wrote the paper.

Conflicts of Interest: The authors declare no conflict of interest.

References

1. Campbell, T.A.; Ivanova, O.S. 3D printing of multifunctional nanocomposites. *Nano Today* **2013**, *8*, 119–120. [CrossRef]
2. Kalsoom, U.; Nesterenko, P.N.; Paull, B. Recent developments in 3D printable composite materials. *RSC Adv.* **2016**, *6*, 60355–60371. [CrossRef]
3. Ghoshal, S. Polymer/carbon nanotubes (CNT) nanocomposites processing using additive manufacturing (three-dimensional printing) technique: An overview. *Fibers* **2017**, *5*, 40. [CrossRef]
4. Leigh, S.J.; Bradley, R.J.; Purssell, C.P.; Billson, D.R.; Hutchins, D.A. A simple, low-cost conductive composite material for 3D printing of electronic sensors. *PLoS ONE* **2012**, *7*, e49365. [CrossRef] [PubMed]
5. Farahani, R.D.; Dalir, H.; Le Borgne, V.; Gautier, L.A.; El Khakani, M.A.; Lévesque, M.; Therriault, D. Direct-write fabrication of freestanding nanocomposite strain sensors. *Nanotechnology* **2012**, *23*, 085502. [CrossRef] [PubMed]
6. Muth, J.T.; Vogt, D.M.; Truby, R.L.; Mengüç, Y.; Kolesky, D.B.; Wood, R.J.; Lewis, J.A. Embedded 3D printing of strain sensors within highly stretchable elastomers. *Adv. Mater.* **2014**, *26*, 6307–6312. [CrossRef] [PubMed]
7. Chizari, K.; Arjmand, M.; Liu, Z.; Sundararaj, U.; Therriault, D. Three-dimensional printing of highly conductive polymer nanocomposites for EMI shielding applications. *Mater. Today Commun.* **2017**, *11*, 112–118. [CrossRef]
8. Zhang, D.; Chi, B.; Li, B.; Gao, Z.; Du, Y.; Guo, J.; Wei, J. Fabrication of highly conductive graphene flexible circuits by 3D printing. *Synth. Met.* **2016**, *217*, 79–86. [CrossRef]
9. Sun, K.; Wei, T.-S.; Ahn, B.Y.; Seo, J.Y.; Dillon, S.J.; Lewis, J.A. 3D printing of interdigitated Li-Ion microbattery architectures. *Adv. Mater.* **2013**, *25*, 4539–4543. [CrossRef] [PubMed]

10. Zhang, J.; Yang, B.; Fu, F.; You, F.; Dong, X.; Dai, M. Resistivity and its anisotropy characterization of 3D-printed acrylonitrile butadiene styrene copolymer (ABS)/carbon black (CB) composites. *Appl. Sci.* **2017**, *7*, 20. [CrossRef]

11. Wei, X.; Li, D.; Jiang, W.; Gu, Z.; Wang, X.; Zhang, Z.; Sun, Z. 3D printable graphene composite. *Sci. Rep.* **2015**, *5*, 11181. [CrossRef] [PubMed]

12. Zhang, Q.; Zhang, F.; Medarametla, S.P.; Li, H.; Zhou, C.; Lin, D. 3D printing of graphene aerogels. *Small* **2016**, *12*, 1702–1708. [CrossRef] [PubMed]

13. Gnanasekaran, K.; Heijmans, T.; van Bennekom, S.; Woldhuis, H.; Wijnia, S.; de With, G.; Friedrich, H. 3D printing of CNT- and graphene-based conductive polymer nanocomposites by fused deposition modeling. *Appl. Mater. Today* **2017**, *9*, 21–28. [CrossRef]

14. Dul, S.; Fambri, L.; Pegoretti, A. Fused deposition modelling with ABS–graphene nanocomposites. *Compos. Part A Appl. Sci. Manuf.* **2016**, *85*, 181–191. [CrossRef]

15. Guo, S.-Z.; Yang, X.; Heuzey, M.-C.; Therriault, D. 3D printing of a multifunctional nanocomposite helical liquid sensor. *Nanoscale* **2015**, *7*, 6451–6456. [CrossRef] [PubMed]

16. Postiglione, G.; Natale, G.; Griffini, G.; Levi, M.; Turri, S. Conductive 3D microstructures by direct 3D printing of polymer/carbon nanotube nanocomposites via liquid deposition modeling. *Compos. Part A Appl. Sci. Manuf.* **2015**, *76*, 110–114. [CrossRef]

17. Wu, C.S.; Liao, H.T. Interface design of environmentally friendly carbon nanotube-filled polyester composites: Fabrication, characterisation, functionality and application. *Express Polym. Lett.* **2017**, *11*, 187–198. [CrossRef]

18. Dorigato, A.; Moretti, V.; Dul, S.; Unterberger, S.H.; Pegoretti, A. Electrically conductive nanocomposites for fused deposition modelling. *Synth. Met.* **2017**, *226*, 7–14. [CrossRef]

19. Sinkral® F 322—ABS—Versalis S.p.A Material Data. Available online: https://www.materialdatacenter. com/ms/en/Sinkral/Versalis+S%252ep%252ea/SINKRAL%C2%AE+F+332/c6da6726/1895 (accessed on 18 October 2017).

20. Nanocyl SA. Nanocyl® NC7000™ Technical Data Sheet. Available online: http://www.Nanocyl.com/ product/nc7000 (accessed on 18 October 2017).

21. TA Instruments DMA Q800 Product Data. Available online: http://www.tainstruments.com/pdf/literature/ TA284.pdf (accessed on 10 January 2018).

22. Fambri, L.; Kesenci, K.; Migliaresi, C. Characterization of modulus and glass transition phenomena in poly(l-lactide)/hydroxyapatite composites. *Polym. Compos.* **2003**, *24*, 100–108. [CrossRef]

23. Jyoti, J.; Singh, B.P.; Arya, A.K.; Dhakate, S.R. Dynamic mechanical properties of multiwall carbon nanotube reinforced ABS composites and their correlation with entanglement density, adhesion, reinforcement and C factor. *RSC Adv.* **2016**, *6*, 3997–4006. [CrossRef]

24. Pothan, L.A.; Oommen, Z.; Thomas, S. Dynamic mechanical analysis of banana fiber reinforced polyester composites. *Compos. Sci. Technol.* **2003**, *63*, 283–293. [CrossRef]

25. Dul, S.; Mahmood, H.; Fambri, L.; Pegoretti, A. Graphene-abs nanocomposites for fused deposition modelling. In Proceedings of the 17th European Conference on Composite Materials, Munich, Germany, 26–30 June 2016.

26. Hong, N.; Zhan, J.; Wang, X.; Stec, A.A.; Richard Hull, T.; Ge, H.; Xing, W.; Song, L.; Hu, Y. Enhanced mechanical, thermal and flame retardant properties by combining graphene nanosheets and metal hydroxide nanorods for acrylonitrile–butadiene–styrene copolymer composite. *Compos. Part A Appl. Sci. Manuf.* **2014**, *64*, 203–210. [CrossRef]

27. Yang, S.; Rafael Castilleja, J.; Barrera, E.V.; Lozano, K. Thermal analysis of an acrylonitrile–butadiene–styrene/SWNT composite. *Polym. Degrad. Stab.* **2004**, *83*, 383–388. [CrossRef]

28. Wang, L.; Qiu, J.; Sakai, E.; Wei, X. The relationship between microstructure and mechanical properties of carbon nanotubes/polylactic acid nanocomposites prepared by twin-screw extrusion. *Compos. Part A Appl. Sci. Manuf.* **2016**, *89*, 18–25. [CrossRef]

29. Al-Saleh, M.H.; Al-Saidi, B.A.; Al-Zoubi, R.M. Experimental and theoretical analysis of the mechanical and thermal properties of carbon nanotube/acrylonitrile–styrene–butadiene nanocomposites. *Polymer* **2016**, *89*, 12–17. [CrossRef]

30. Pegoretti, A. Creep and fatigue behaviour of polymer nanocomposites. In *Nano- and Micromechanics of Polymer Blends and Composites*; Karger-Kocsis, J., Fakirov, S., Eds.; Carl Hanser Verlag GmbH & Co. KG.: Munich, Germany, 2009; pp. 301–339.

31. Findley, W.N. 26-year creep and recovery of poly(vinyl chloride) and polyethylene. *Polym. Eng. Sci.* **1987**, *27*, 582–585. [CrossRef]
32. Williams, G.; Watts, D.C. Non-symmetrical dielectric relaxation behaviour arising from a simple empirical decay function. *Trans. Faraday Soc.* **1970**, *66*, 80–85. [CrossRef]
33. Pandey, A.K.; Kumar, R.; Kachhavah, V.S.; Kar, K.K. Mechanical and thermal behaviours of graphite flake-reinforced acrylonitrile-butadiene-styrene composites and their correlation with entanglement density, adhesion, reinforcement and C factor. *RSC Adv.* **2016**, *6*, 50559–50571. [CrossRef]
34. Krause, B.; Pötschke, P.; Häußler, L. Influence of small scale melt mixing conditions on electrical resistivity of carbon nanotube-polyamide composites. *Compos. Sci. Technol.* **2009**, *69*, 1505–1515. [CrossRef]
35. Müller, M.T.; Krause, B.; Kretzschmar, B.; Pötschke, P. Influence of feeding conditions in twin-screw extrusion of PP/MWCNT composites on electrical and mechanical properties. *Compos. Sci. Technol.* **2011**, *71*, 1535–1542. [CrossRef]
36. Jyoti, J.; Basu, S.; Singh, B.P.; Dhakate, S.R. Superior mechanical and electrical properties of multiwall carbon nanotube reinforced acrylonitrile butadiene styrene high performance composites. *Compos. Part B Eng.* **2015**, *83*, 58–65. [CrossRef]

nanomaterials

MDPI

Article

Morphological and Optical Characteristics of Chitosan$_{(1-x)}$:Cu$^0{}_x$ (4 ≤ x ≤ 12) Based Polymer Nano-Composites: Optical Dielectric Loss as an Alternative Method for Tauc's Model

Shujahadeen B. Aziz [1,2]

[1] Advanced Polymeric Materials Research Laboratory, Department of Physics, College of Science, University of Sulaimani, Sulaimani 46001, Kurdistan Regional Government, Iraq; shujahadeenaziz@gmail.com or shujahadeen.aziz@univsul.edu.iq

[2] Komar Research Center (KRC), Komar University of Science and Technology, Sulaimani 46001, Kurdistan Regional Government, Iraq

Received: 26 October 2017; Accepted: 8 December 2017; Published: 13 December 2017

Abstract: In this work, copper (Cu) nanoparticles with observable surface plasmonic resonance (SPR) peaks were synthesized by an in-situ method. Chitosan host polymer was used as a reduction medium and a capping agent for the Cu nanoparticles. The surface morphology of the samples was investigated through the use of scanning electron micrograph (SEM) technique. Copper nanoparticles appeared as chains and white specks in the SEM images. The strong peaks due to the Cu element observed in the spectrum of energy dispersive analysis of X-rays. For the nanocomposite samples, obvious peaks due to the SPR phenomena were obtained in the Ultraviolet-visible (UV-vis) spectra. The effect of Cu nanoparticles on the host band gap was understood from absorption edges shifting of absorption edges to lower photon energy. The optical dielectric loss parameter obtained from the measurable quantities was used as an alternative method to study the band structure of the samples. Quantum mechanical models drawbacks, in the study of band gap, were explained based on the optical dielectric loss. A clear dispersion region was able to be observed in refractive indices spectra of the composite samples. A linear relationship with a regression value of 0.99 was achieved between the refractive index and volume fractions of CuI content. Cu nanoparticles with various sizes and homogenous dispersions were also determined from transmission electron microscope (TEM) images.

Keywords: biopolymer; Cu nanoparticles; SEM and EDAX analysis; TEM analysis; optical properties

---

1. Introduction

A recent study revealed that nano-scale materials exhibit unique electronic and optical properties, which are unlike those in their bulk state [1]. Metallic nanoparticles have attracted attention of many researchers, because of the local field enhancement at nano interfaces, which is beneficial for a number of applications, from sensors to nonlinear optics [2]. Plasmonic metal nanoparticles are described by their strong interactions with UV-visible radiation through the localized surface plasmon resonance (LSPR) excitation [3]. Copper (Cu) is the most commonly used metal in electrical/electronic applications, because of its high conductivity and low cost. The development of nano-devices that combine electronic, photonic, chemical and/or biological features is crucial for future electronic and sensing devices [4]. The importance of copper nanoparticles (CuNPs) arises from the advantageous properties of this metal, such as its good thermal and electrical conductivities at a cost much less than that of silver. This leads to potential applications in cooling fluids for electronic systems and conductive inks [5]. Copper "incorporated into" or "supported on" solid matrices is broadly utilized in the catalysts and nanocomposites preparations with unusual optical, electrical and magnetic properties [6]. The photosensitivity of noble metal nanostructures makes them promising platforms for highly sensitive optical nanosensors, photonic components and surface-enhanced spectroscopies [4]. Nanoparticles have been synthesized through several methods, such as the polyol, reverse micelles, electron beam irradiation, micro-emulsion and wire explosion techniques and in-situ chemical synthesis. Among all the procedures, a compound that has the ability to form a complex with metal ions, such as soluble polymers, is an important method for CuNPs synthesis, since it prevents the nanoparticles aggregation [7,8]. Previous work indicated that physical and chemical methods have been used to synthesize polymer nanocomposites, depending on the nanoparticle-polymer interactions [9]. The usage of synthetic polymers in general and polymer composites in particular appears to be ever increasing. However, despite the fact that synthetic polymer and their composite are cost effective; they are non-biodegradable materials and produced from petroleum sources. On the other hand, biodegradable polymers produced from renewable sources are cheap and easy to treat without any hazardous chemicals [10]. Recently, the use of natural bio-polymers as stabilizers for the synthesis of CuNPs has been gaining momentum because of their availability, biocompatibility and low toxicity [11]. Chitosan (CS) has attracted significant attention as functional, nontoxic and biodegradable natural biopolymer for many applications [12]. Chitosan is a cationic polysaccharide and obtained by alkaline N-acetylation of chitin, which is the second-most abundant natural polymer after cellulose [13]. The amine (NH_2) and hydroxyl (OH) functional groups on the CS backbone structure explain its ability to form complexes with inorganic salts [12]. Previous studies confirmed that the existences of lone pair electrons on chitosan functional groups are found to be responsible for complexation as well as reduction of silver ions to silver nanoparticles [14–18]. However, such problem can be overcome by the use of in-situ technique [19]. Thus, the direct use of chitosan can solve the problems of aggregation. Earlier study revealed that hybrid (organic-inorganic) materials represents an intrinsic interdisciplinary field of research and development because it includes a variety of communities such as organometallics, colloids, soft matter, polymers, nanocomposites, biomaterials and biochemistry [20]. This is related to the fact that organic-inorganic materials lie at the interface of the organic and inorganic areas. These materials present outstanding chance not only to combine the fundamental properties from both worlds but to create entirely new compositions with exclusive properties [21]. The optical properties of hybrid materials are currently of considerable interest, due to their wide applications in sensors, single-molecule detection, optical data storage and light-emitting diodes (LED) [22,23]. Moreover, recent study indicated that hybrid materials are crucial in the development of numerous types of organic transistors, organic light-emitting diodes and organic solar cells [24]. The intensive and extensive survey of literature reveals that band gap study of polymer composites are not studied in detail. The objective of this work was to study the optical properties of synthesized Cu nanoparticles via the in-situ method inside the chitosan host polymer at room temperature. The results shown in this paper reveals that a Cu nanoparticle significantly reduces the optical band gap of chitosan host polymer. The Cu-chitosan polymer composites were characterized using scanning electron micrograph (SEM), energy dispersive analysis of X-rays (EDAX), transmission electron microscopy (TEM) and UV-visible techniques. The effect of CuNPs on the optical properties of the chitosan host polymer was clarified.

2. Experimental Details

2.1. Materials and Sample Preparation

Chitosan ($\geq$75% deacetylated, molecular weight $M_w = 1.1 \times 10^5$) and copper iodide (CuI) were supplied by Sigma Aldrich (Sigma Aldrich, Warrington, PA, USA). Acetic acid (1%) was prepared using glacial acetic acid solution and used as a solvent to prepare the nanocomposite solid polymer electrolytes (SPEs). The standard solution cast technique was used to prepare the SPE films. Here, 1 g of chitosan was dissolved in 100 mL of 1% acetic acid solution. The mixture was stirred continuously with a magnetic stirrer for several hours at room temperature until the chitosan powder was completely dissolved in the 1% acetic acid solution. Different amounts (4 to 12 wt. %) of CuI were dissolved in 20 mL of acetonitrile (CH$_3$CN) solvent. To prepare nanocomposite samples, the CuI solution was added to the chitosan solutions separately and the mixture was stirred continuously for several hours. The solutions were then cast into different clean and dry Petri dishes and allowed to evaporate at room temperature until solvent-free films were obtained. The films were kept in desiccators with blue silica gel desiccant for further drying. The greenish colors of the chitosan nanocomposite films are evidence of the formation of Cu nanoparticles. The thickness of the films ranged from 121–123 μm was controlled by casting the same amount of CS. The chitosan nanocomposite (CSN) samples were coded as CSN0, CSN1, CSN2 and CSN3 for CS incorporated with 0 wt. %, 4 wt. %, 8 wt. % and 12 wt. % CuI, respectively.

2.2. Characterization Techniques

The UV-visible spectra of the chitosan-silver triflate membrane films and their nanocomposites were recorded, using a Jasco V-570 UV-Vis-NIR spectrophotometer (Jasco SLM-468, Tokyo, Japan) in the absorbance mode. Scanning electron microscopy (SEM) was taken to study the morphological appearance of the samples, using the FEI Quanta 200 Field Emission Scanning Electron Microscopy (FESEM) (FEI Company, Hillsboro, OR, USA). The microscope was fitted with an Oxford instruments INCA Energy 200 energy-dispersive X-ray microanalysis (EDXA) system (Abingdon, UK) [Detector: Si (Li) crystal] to detect the overall chemical composition of solid chitosan nanocomposites. Transmission electron microscope (TEM) image was also obtained for the Cu nanoparticles (CSN3 sample), using a LEO LIBRA instrument (Carl Zeiss, Oberkochen, Germany)with accelerating voltage of 120 kV. For TEM measurement, a drop of chitosan:CuI solution containing Cu nanoparticles was placed on a carbon-coated copper grid. The excess solution was then removed by filter paper and the grid was left at room temperature to dry prior being imaged.

3. Results and Discussion

3.1. Morphological Studies

Figure 1 shows the SEM images and EDAX results for the CS nanocomposite samples. It is clear from the figures that, at high CuI concentration, white chains and spots with different sizes can be observed on the surface of the CS nanocomposite samples. Earlier studies have confirmed that the use of SEM and EDAX techniques are sufficient to detect the formation of plasmonic metal nanoparticles in polymer composites [13–15,17,18]. The EDAX taken for the CSN2 sample focusing on the aggregated white spots is illustrated in Figure 1D, which shows the existence of significant amounts of metallic Cu particles. The electron image was taken at 100× magnifications and shows very small white specs for CSN1 sample, while many white aggregates can be seen for images of CSN2 and CSN3 samples. The SEM-EDAX results presented here confirm the successful formation of Cu nanoparticles via the in-situ method inside the chitosan host polymer. Such results can be further examined by the use of UV-visible and TEM techniques. The results of the present work suggest that chitosan can be used as a novel polar polymer for synthesis of plasmonic metallic nanoparticles. The distinguishable intense peaks of Cu nanoparticles appeared in the EDAX spectrum at approximately 1 and 8 keV confirms the formation of Cu nanoparticles. Similar peaks for Cu nanoparticles in EDAX spectra have been observed by other researchers [25–27]. From this discussion, it is understood that morphological (SEM and EDAX) study is significant for nanoparticle characterization.

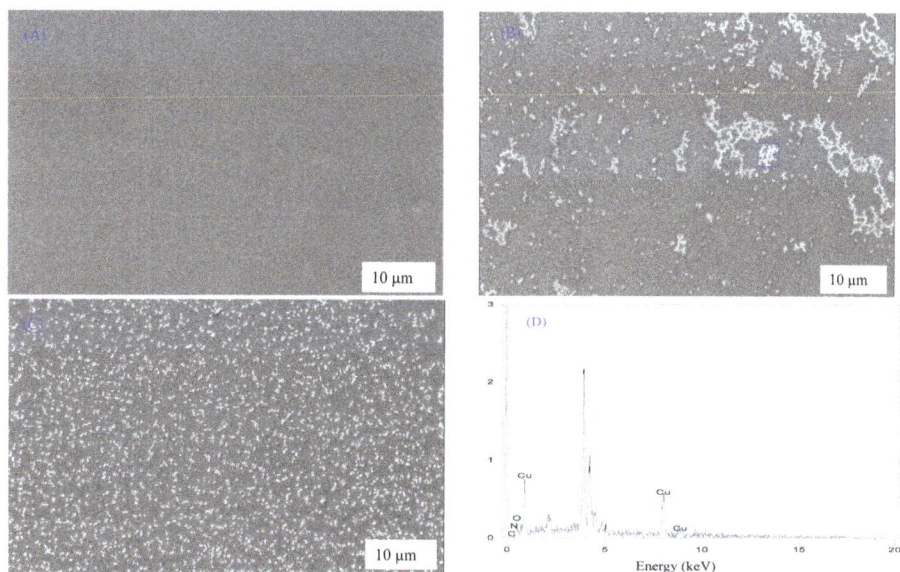

Figure 1. Scanning electron microscopy (SEM) image for (**A**) CSN1; (**B**) CSN2; (**C**) CSN3 and (**D**) EDAX for white specks inside the blue box. The white spots appearing on the film surface are attributable to Cu metallic particles.

3.2. Optical Properties

3.2.1. Absorption and Absorption Edge Study

Figure 2 shows the absorption spectra of pure CS and CS nanocomposite samples. It can be seen that the pure CS does not indicate any absorption peak at 644 nm, whereas the samples incorporated with CuI exhibit distinguishable peaks at wavelength ranges from 550 to 750 nm. These peaks can only be attributable to the LSPR excitation that occurs due to the nanoscale-size metal particles [12,28]. An earlier study established that the LSPR band occurring near 620–640 nm is related to the formation of copper and copper oxide nanoparticles [29]. On the other hand, in our previous works, we have observed distinguishable and enhanced LSPR peaks for CuO- and CuS-based nanoparticles [29–31]. Therefore, the SPR peaks presented in this work can be attributed to the Cu nanoparticles alone. It is interesting to note that the results of absorption spectra achieved in this work are close to those reported by other researchers [2,8]. The shift of absorption spectra to the visible ranges for the nanocomposite samples reveals the role of Cu nanoparticles on optical properties of CS host polymer. More insights about the change in the band structure of chitosan can be grasped from the band gap study. Earlier researchers used ion implantation to fabricate gold (Au)-polyimide hybrid. They achieved SPR peaks with low intensity and they do not study the effect of Au nanoparticles on optical properties of polyimide [32]. Significant shifting of absorption spectra to visible region reveals that Cu nanoparticles forms charge transfer complexes through the CS host polymer. The results of the current work show that polymer composites incorporated with Cu nanoparticles possess great changes in their optical properties compared to hybrid materials containing silver nanoparticles [33–35]. It is obvious that around 750 nm an artifact or a step rise occurs. The appearance of artifact around 750 nm may be related to the change of the range of measurement from visible to infrared IR region [12].

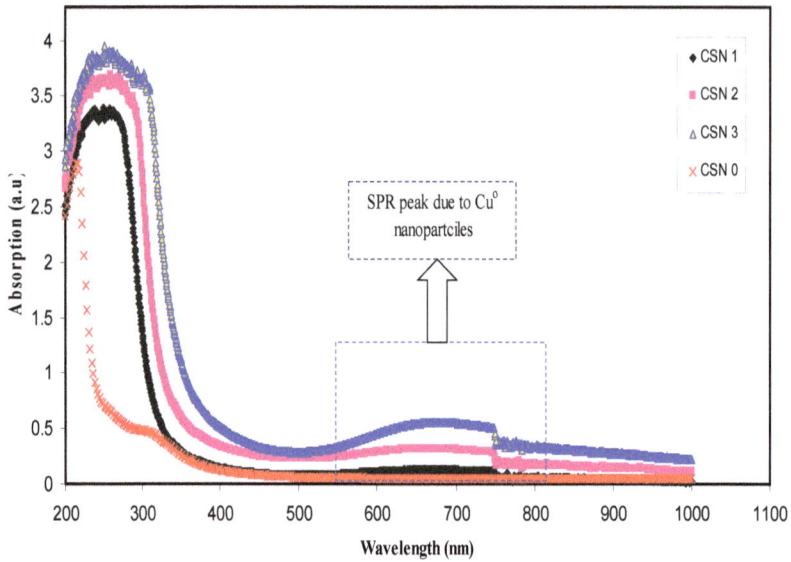

Figure 2. Absorption spectra of pure CS and CS:CuI solid films. The surface plasmonic resonance (SPR) peak appearing at approximately 667 nm for CS:CuI samples is related to the existence of Cu metallic nanoparticles.

The absorption of light by an optical medium is quantified by its absorption coefficient (α) [36]. The absorption coefficient is calculated by,

$$\alpha_{(\lambda)} = (2.303) \times [\frac{A}{d}] \tag{1}$$

where d is the thickness of the sample and A is the absorption data. Figures 3 and 4 illustrate the absorption coefficient as a function of photon energy for pure CS and CS nanocomposite samples, respectively. Absorption coefficient is defined as the fraction of the power absorbed in a unit length of the medium [36]. It can be clearly seen that the absorption edge of the CS shifts toward the lower photon energy for the samples containing Cu nanoparticles. Essential information about the band structure and the energy band gap in the crystalline and non-crystalline materials can be obtained from the optical absorption spectra determination [37]. The shift towards the lower photon energy reveals the reduction in the optical band gap of the CS nanocomposite samples. Absorption edge values were estimated from the intersection of the extrapolation of the linear relationship. The obtained absorption edge values are presented in Table 1. It is clear that the absorption edge decreases from 4.7 eV for pure CS to 3.51 eV for CS incorporated with 12 wt. % CuI. Such significant changes in absorption edge values are evidence of the occurrence of changes in the band structure of the CS nanocomposite samples.

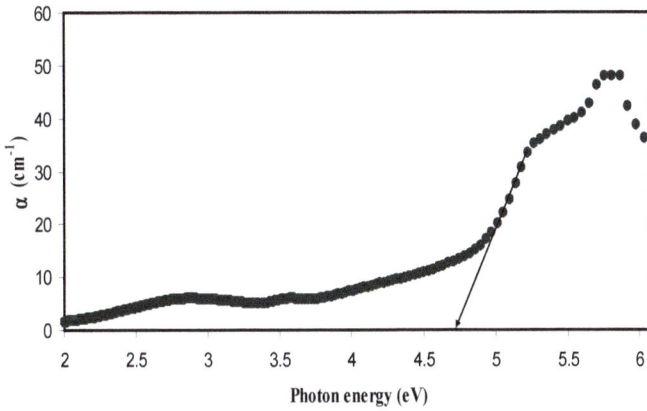

Figure 3. Variation of absorption coefficient (α) versus photon energy ($h\upsilon$) for pure CS sample.

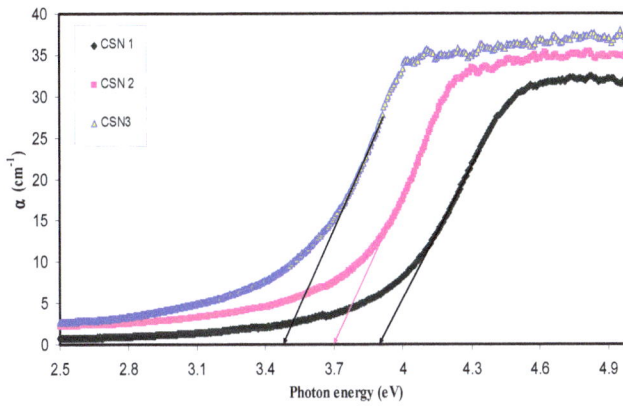

Figure 4. Variation of absorption coefficient (α) versus photon energy ($h\upsilon$) for doped samples.

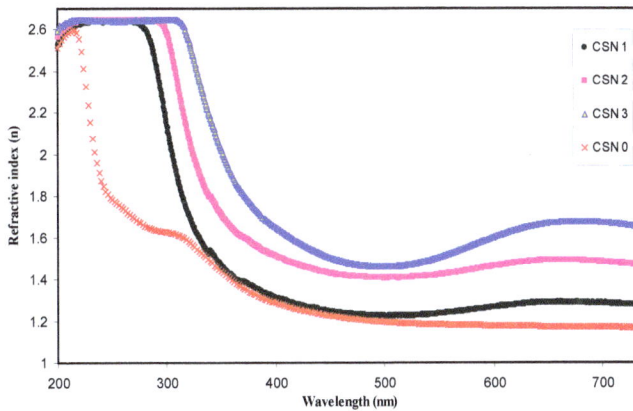

Figure 5. Refractive index (n) as a function of wavelength for pure CS and doped CS samples.

3.2.2. Refractive Index and Optical Dielectric Constant Study

Changes in the refractive index of composite films are crucial for controlling optical properties of materials. An earlier study revealed that in refractive index analysis, the real part (ε_1) and imaginary part (ε_2) of the optical dielectric constant are important for designing new materials. Refractive index is an important optical parameter for designing prisms, optical windows and optical fibers [38]. From the reflection coefficient R and the optical extinction data, the refractive indexes (n) can be estimated, using the Fresnel formulae as follows [28]:

$$n = \left(\frac{1+R}{1-R}\right) + \left[\frac{4R}{(1-R)^2} - K^2\right]^{1/2} \tag{2}$$

where R is reflection coefficient and $K = \alpha\lambda/4\pi$ is the extinction coefficient. Figure 5 illustrates the refractive index of pure CS and CS nanocomposite samples. A significant change in refractive index has been occurred for the doped samples. It is obvious that the refractive index increased from 1.16 for pure CS to 1.68 for CS incorporated with 12 wt. % CuI. The synthesis of Cu nanoparticles through the host polymer makes the polymer matrix denser in nature and thus exhibits high refractive index according to the well-known Lorentz–Lorenz formula [31,39]. Materials with high refractive index play an important role in many top end advanced optical and optoelectronic equipment, including waveguides, antireflective coatings and light-emitting diodes [40]. It is evident from Figure 5 that the dispersion curve becomes steeper for samples containing a higher concentration of Cu nanoparticles. The extension of plateau region of Figure 5 at high wavelengths to the Y-axis was used to estimate the index of refraction. It is clear that the refractive index increases linearly with Cu concentration, as shown in Figure 6. The linear nature of the refractive index with filler dopant has been reported experimentally and theoretically [28,30,36,41,42]. Earlier studies established that the linear dependence of the refractive index indicates the homogeneous dispersion of fillers within a polymer matrix [28,30,36,42]. This is further illustrated in the transmission electron microscopy (TEM) image Figure 7, which taken for the CSN3 sample. Here, the homogeneous dispersion of Cu nanoparticles with various sizes can be seen. The linear behavior of the refractive index and TEM result indicates that the in-situ method is an excellent technique for the preparation of polymer nanocomposites with homogeneous dispersion of fillers in the matrices.

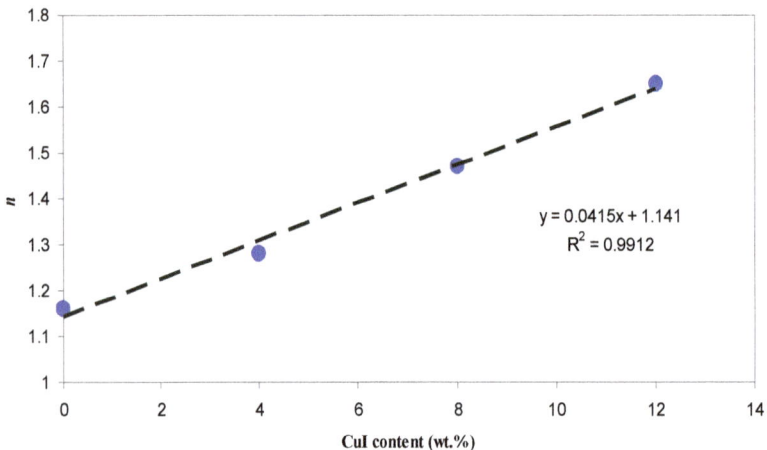

Figure 6. Refractive index (n) as a function of volume fraction of CuI.

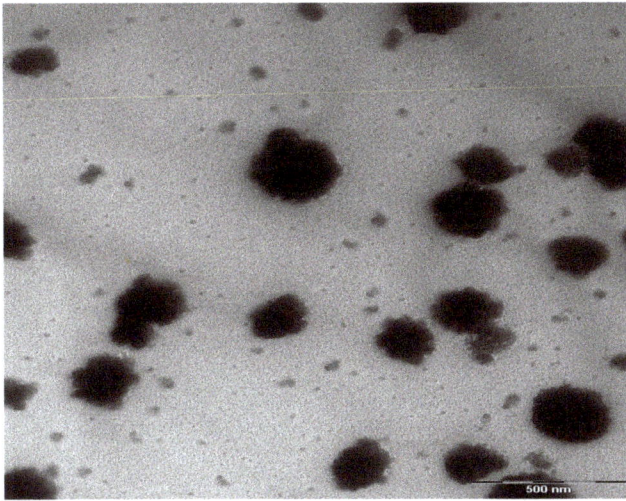

Figure 7. TEM image for CSN3 samples.

Figure 8 shows the optical dielectric constant (ε_1) versus wavelength. From the graph, an obvious increase in dielectric constant upon the incorporation of CuI content can be seen. Such an increase is related to the increment of the density of states since ε_1 can be directly associated with the density of states inside the forbidden gap of the solid polymer films [43]. Previous studies established that ε_1 is related to the electronic part and depends strongly on the optical bandgap [28,44]. This can be better understood from the well-known Penn model [45], as given by:

$$\varepsilon_1(0) \approx 1 + \left(\hbar\omega_p / E_g\right)^2 \tag{3}$$

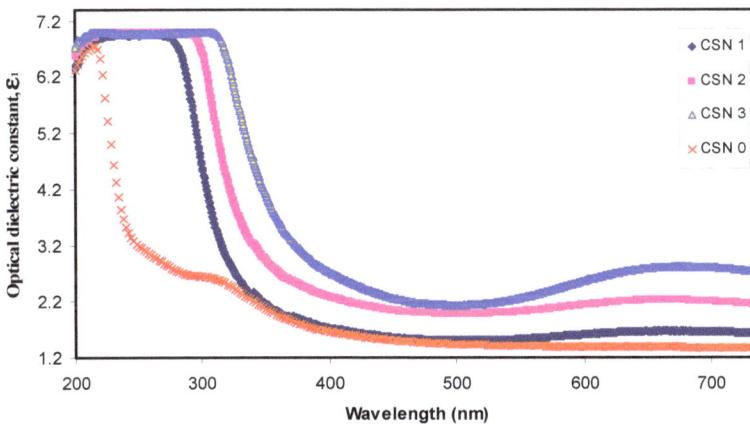

Figure 8. Optical dielectric constant spectra versus wavelength for pure CS and doped CS samples.

It is evident from Equation (3) that a smaller energy gap (E_g) yields a larger ε_1 value. The ε_1 and n values are presented in Table 1. Here, it is clear that the reduction in optical band gap is associated with the increase in refractive index. The decrease in the optical band gap (see Table 1) can be related to an increase in optical dielectric constant. An increase in optical dielectric constant means the introduction of more charge carriers to the host material and thus an increase in the density of states [46]. The results obtained in this work reveal the validity of the Penn model, in which the increment of the density of states within the band gap causes the energy band gap to be increased. It can be observed that the values of ε_1 (see Table 1) increase with the square of refractive index, which exactly meets the relation of $\varepsilon_1 = n^2$.

Table 1. Refractive index and optical dielectric constant for all the samples.

Sample Designation	Estimated Refractive Index from Figure 5	Estimated Optical Dielectric Constant (ε_1) from Figure 8	Estimated Refractive Index from $n = (\varepsilon_1)^{1/2}$
CSN0	1.16	1.36	1.16619
CSN1	1.28	1.63	1.276715
CSN2	1.47	2.17	1.473092
CSN3	1.65	2.76	1.661325

3.2.3. Band Gap Study

In the present work, based on experimental and theoretical approaches, two methods were performed for the band-gap study. Tauc's method has been developed from the theory of optical absorption. It is a familiar method for the study of band-gap, which relates the absorption coefficient to the photon energy. On the other hand, theoretical physicists have developed various models based on quantum approaches for band-gap study. Theoretically, all intrinsic effects corresponding to light-matter interaction processes are contained in the optical dielectric function [47]. It is recognized from the previous studies that the main peak of optical dielectric loss (ε_2) is directly related to the electron transitions from valence band to conduction band [28,46,48]. Therefore, the optical dielectric loss parameter versus photon energy can be considered for band gap study.

It was reported that the band edges, in the amorphous materials, have contributions from the different orbital types of the metallic complex and ligand. Therefore, it is difficult to predict whether the band will be a direct or an indirect type [15,49]. In this regard, it should be pointed out that the Tauc's equation alone is not enough to specify the type of transition, since four figures is needed to be plotted depending on the values that the exponent takes [15]. From a quantum mechanics viewpoint, the existence of strong correlation between optical dielectric loss (ε_2) and band structure of the materials is a subject of various theoretical studies [46,50–52]. The use of ε_2 to estimate the optical band gap, is related to the fact that ε_2 is a function of the electronic band-structure, density of filled and empty states and magnitude of optical transition probability between the filled and empty states [53,54], as follows:

$$\varepsilon_2(\omega) = \frac{4\pi^2}{\Omega\omega^2} \sum_{i\in VB, j\in CB} \sum_K W_K \left|p_{ij}^a\right|^2 \delta(\epsilon_{kj} - \epsilon_{ki} - \omega) \tag{4}$$

where Ω is the unit-cell volume and ω is photon frequency, VB and CB denote the valence and conduction bands, W_k is the weight associated with a k-point, p_{ij}^a is the transition probability and a denotes a particular direction. The delta function (δ) is used to ensure the conservation of energy in electronic transitions. The transition can occur only when the photon energy matches the energy difference between the valence and conduction state. Thus, from Equation (4) it is clear that optical dielectric loss is directly related to the material band structure and can be used to estimate the optical band gap.

The optical band gap has been studied by other researchers both experimentally and theoretically [50,54]. They observed that the experimental values for the band gap energy are found to be larger than the theoretical values. They have attributed this discrepancy to neglecting excitons and to poor description of strong Coulomb exchange interaction among electrons. The measurable optical dielectric loss has been used experimentally due to the shortcoming of the quantum models. In our previous works, the optical dielectric loss from the measured quantities, such as refractive index (n) and extinction (K) coefficient ($\varepsilon_2 = 2nK$), has been successfully used to estimate the optical band gap. From the quantum mechanical viewpoint, Equation (4) was found to be a microscopic approach for the study of band-gap and required lengthy numerical methods. By contrast, it is found that using the experimental value of optical dielectric loss ($\varepsilon_2 = 2nK$), which is a macroscopic approach, is easier and more accurate method to estimate the optical band gap [28,46].

Figures 9 and 10 show the optical dielectric loss versus photon energy for pure- and doped-CS samples. It is well established that crystalline materials have sharp structures in the fundamental absorption region, whereas amorphous materials exhibit broad peaks [55]. Compared to the doped samples pure CS has exhibited a sharp peak. This is related to the existence of large amount of crystalline fraction in pure CS [56–58]. In our previous work, broad peaks in the plot of optical dielectric loss have been observed due to the amorphous nature of the composite samples [28,46]. The broad peaks depicted in Figure 10 may be ascribed to the amorphous structure of the samples. The estimated band gaps from the optical dielectric loss plots are presented in Table 2. To specify the type of electronic transition, Tauc's method must be studied.

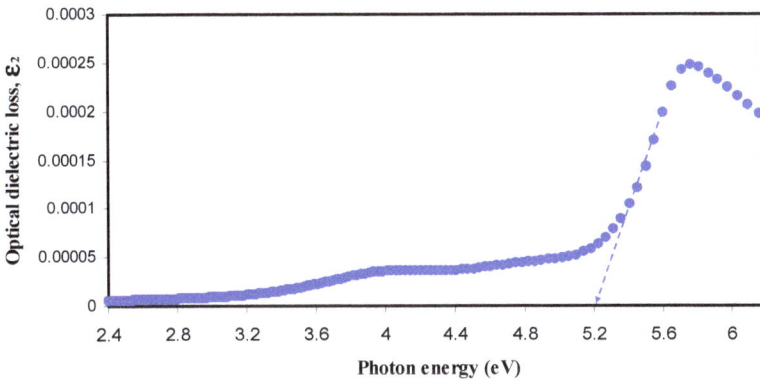

Figure 9. Optical dielectric loss versus photon energy (hv) for pure CS sample.

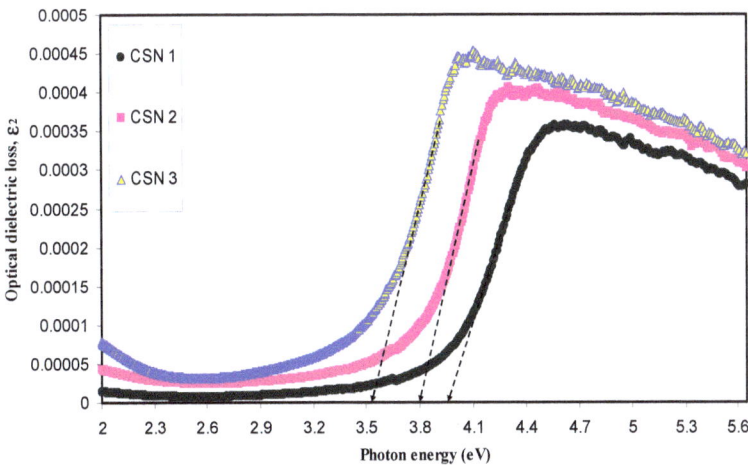

Figure 10. Optical dielectric loss versus photon energy (hv) for doped CS samples.

Table 2. Estimated energy band gap (E_g) from optical dielectric loss and Tauc method for all samples.

Sample Designation	Estimated Bandgap (eV) from ε_2 Plot	E_g (eV) from Tauc Method ($\gamma = 1/2$)
CSN0	5.2	5.24
CSN1	3.94	4.14
CSN2	3.8	3.95
CSN3	3.53	3.72

The theory of optical absorption provides the relationship between the absorption coefficient (α) and photon energy ($h\upsilon$) for direct allowed transition [59]. The absorption coefficient for non-crystalline materials related to the incident photon energy can be determined as [28,60]:

$$\alpha = \frac{\beta}{h\upsilon}(h\upsilon - E_g)^{\gamma} \tag{5}$$

where β is a constant and E_g is the optical energy bandgap. The exponent γ may have values of 1/2, 2, 3/2 and 3, which correspond to the allowed direct, allowed indirect, forbidden direct and forbidden indirect excitations, respectively [46]. Thus, from Tauc's relation, four figures can be plotted to specify the type of transition. In this work, the plot corresponding to allowed direct excitation ($\gamma = 1/2$) is only presented due to that the plots corresponding to other values of γ exhibit optical band gaps that cannot be compared with those achieved from the optical dielectric loss plots (Figures 9 and 10).

Table 2 shows the band gap values estimated from the linear parts of Figures 11 and 12. In the table, one can see that the band gap decreases with increasing CuI concentration. Previous studies concluded that an electronic interaction occurs between the nanoparticles and the host polar polymer and results in an increase in the absorption intensity [28,31,61]. This has been related to the fact that the embedded nanoparticles inside the host polymer may set up many localized charge carrier levels called trapping sites [31]. It is clear from Table 2 that the achieved optical band gaps estimated from Tauc's model are close enough to those from the optical dielectric loss spectra. Thus, it is understood that optical dielectric loss is essential for band-gap study and Tauc's method is an important way to determine the types of transition, by which the electron can cross the forbidden gap from the valence band to the conduction band. The band gap achieved from Tauc's relation must be quite close to the expected values from optical dielectric loss. This is related to the main peak appearing in ε_2 versus photon energy corresponds to the strong interband transitions [28].

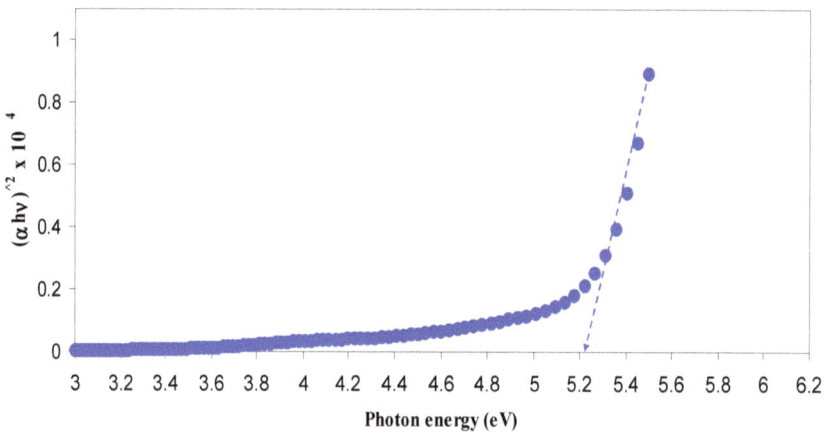

Figure 11. Plot of $(\alpha h\upsilon)^2$ versus photon energy ($h\upsilon$) for pure CS sample.

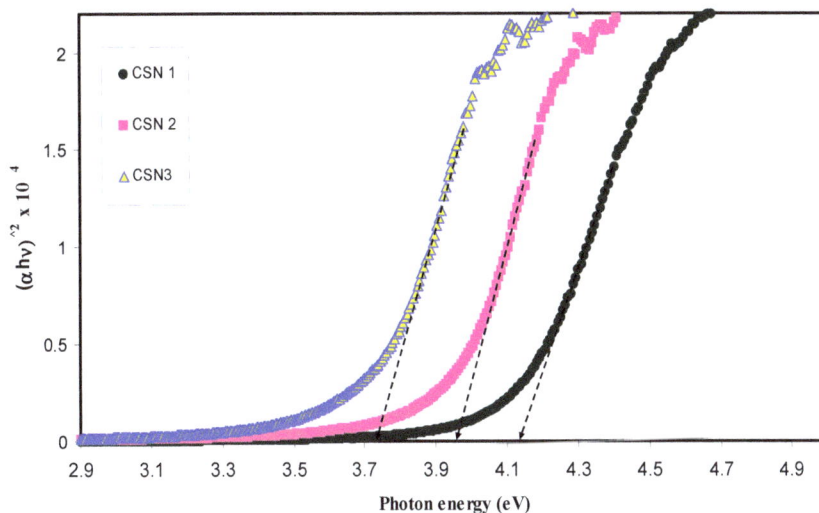

Figure 12. Plot of $(\alpha h \upsilon)^2$ versus photon energy $(h\upsilon)$ for doped CS samples.

4. Conclusions

The results show that copper nanoparticles with observable SPR peaks can be synthesized by the in-situ method inside the chitosan host polymer. The white specks and chains of Cu nanoparticles were observed through SEM images. Strong peaks due to the Cu element appeared at approximately 1 and 8 keV in EDAX. Obvious peaks due to SPR phenomena were obtained in the UV-visible spectra of the nanocomposite samples. The effect of Cu nanoparticles on band gap of the host polymer could be expected from shifting of absorption edges to lower photon energy. The obtained optical dielectric loss parameter from measurable quantities was used to study the band structure of the samples. Tauc's model was used to specify the electronic transition type. The drawbacks of the quantum models based on optical dielectric loss for band-gap study were explained. A clear dispersion region was observed in the spectra of the refractive indices. A linear relationship with a regression value of 0.99 between the refractive index and volume fractions of CuI content reveals the homogeneous distribution of Cu nanoparticles. The TEM image shows Cu nanoparticles with various sizes and homogeneous dispersions.

Acknowledgments: The authors gratefully acknowledge the financial support for this study from the Ministry of Higher Education and Scientific Research-Kurdistan Regional Government; the Department of Physics, College of Science, University of Sulaimani, Sulaimani and the Komar Research Center (KRC), Komar University of Science and Technology.

Conflicts of Interest: The authors declare no conflict of interest.

References

1. Rahman, A.; Ismail, A.; Jumbianti, D.; Magdalena, S.; Sudrajat, H. Synthesis of copper oxide nano particles by using phormidium cyanobacterium. *Indones. J. Chem.* **2009**, *9*, 355–360.
2. Almeida, J.M.P.; de Boni, L.; Avansi, W.; Ribeiro, C.; Longo, E.; Hernandes, A.C.; Mendonca, C.R. Generation of copper nanoparticles induced by fs-laser irradiation in borosilicate glass. *Opt. Express* **2012**, *20*, 15106. [CrossRef] [PubMed]
3. Linic, S.; Aslam, U.; Boerigter, C.; Morabito, M. Photochemical transformations on plasmonic metal nanoparticles. *Nat. Mater.* **2015**, *14*, 567–576. [CrossRef] [PubMed]

4. Chan, G.H.; Zhao, J.; Hicks, E.M.; Schatz, G.C.; Duyne, R.P.V. Plasmonic Properties of Copper Nanoparticles Fabricated by Nanosphere Lithography. *Nano Lett.* **2007**, *7*, 1947–1952. [CrossRef]

5. Dang, T.M.D.; Le, T.T.T.; Fribourg-Blanc, E.; Dang, M.C. Synthesis and optical properties of copper nanoparticles prepared by a chemical reduction method. *Adv. Nat. Sci.* **2011**, *2*, 015009. [CrossRef]

6. Pestryakov, A.N.; Petranovskii, V.P.; Kryazhov, A.; Ozhereliev, O.; Pfander, N.; Knop-Gericke, A. Study of copper nanoparticles formation on supports of different nature by UV-Vis diffuse reflectance spectroscopy. *Chem. Phys. Lett.* **2004**, *385*, 173–176. [CrossRef]

7. Alzahrani, E.; Ahmed, R.A. Synthesis of Copper Nanoparticles with Various Sizes and Shapes: Application as a Superior Non-Enzymatic Sensor and Antibacterial Agent. *Int. J. Electrochem. Sci.* **2016**, *11*, 4712–4723. [CrossRef]

8. Cuevas, R.; Durán, N.; Diez, M.C.; Tortella, G.R.; Rubilar, O. Extracellular Biosynthesis of Copper and Copper Oxide Nanoparticles by Stereum hirsutum, a Native White-Rot Fungus from Chilean Forests. *J. Nanomater.* **2015**, *2015*, 789089. [CrossRef]

9. Zhang, D.; Gökce, B. Perspective of laser-prototyping nanoparticle-polymer composites. *Appl. Surf. Sci.* **2017**, *392*, 991–1003. [CrossRef]

10. Sadananda, V.; Rajinib, N.; Rajiluc, A.V.; Satyanarayana, B. Preparation of cellulose composites with in-situ generated copper nanoparticles using leaf extract and their properties. *Carbohydr. Polym.* **2016**, *150*, 32–39. [CrossRef] [PubMed]

11. Usman, M.S.; Ibrahim, N.A.; Shameli, K.; Zainuddin, N.; Yunus, W.M.Z.W. Copper Nanoparticles Mediated by Chitosan: Synthesis and Characterization via Chemical Methods. *Molecules* **2012**, *17*, 14928–14936. [CrossRef] [PubMed]

12. Aziz, S.B.; Rasheed, M.A.; Abidin, Z.H.Z. Optical and Electrical Characteristics of Silver Ion Conducting Nanocomposite Solid Polymer Electrolytes Based on Chitosan. *J. Electron. Mater.* **2017**. [CrossRef]

13. Aziz, S.B.; Abidin, Z.H.Z.; Kadir, M.F.Z. Innovative method to avoid the reduction of silver ions to silver nanoparticles in silver ion conducting based polymer electrolytes. *Phys. Scr.* **2015**, *90*, 035808. [CrossRef]

14. Aziz, S.B.; Abdullah, O.G.; Saber, D.R.; Rasheed, M.A.; Ahmed, H.M. Investigation of Metallic Silver Nanoparticles through UV-Vis and Optical Micrograph Techniques. *Int. J. Electrochem. Sci.* **2017**, *12*, 363–373. [CrossRef]

15. Aziz, S.B.; Abdullah, O.G.; Rasheed, M.A. A novel polymer composite with a small optical band gap: New approaches for photonics and optoelectronics. *J. Appl. Polym. Sci.* **2017**, *134*, 44847. [CrossRef]

16. Aziz, S.B.; Abidin, Z.H.Z.; Arof, A.K. Influence of silver ion reduction on electrical modulus parameters of solid polymer electrolyte based on chitosan-silver triflate electrolyte membrane. *Express Polym. Lett.* **2010**, *5*, 300–310. [CrossRef]

17. Aziz, S.B.; Abidin, Z.H.Z. Electrical and morphological analysis of chitosan: AgTf solid electrolyte. *Mater. Chem. Phys.* **2014**, *144*, 280–286. [CrossRef]

18. Wei, D.; Sun, W.; Qian, W.; Ye, Y.; Ma, X. The synthesis of chitosan-based silver nanoparticles and their antibacterial activity. *Carbohydr. Res.* **2009**, *344*, 2375–2382. [CrossRef] [PubMed]

19. Vodnik, V.V.; Božanic, D.K.; Džunuzovic, E.; Vukovic, J.; Nedeljkovic, J.M. Thermal and optical properties of silver–poly(methylmethacrylate) nanocomposites prepared by in-situ radical polymerization. *Eur. Polym. J.* **2010**, *46*, 137–144. [CrossRef]

20. Sanchez, C.; Belleville, P.; Popalld, M.; Nicole, L. Applications of advanced hybrid organic–inorganic nanomaterials: From laboratory to market. *Chem. Soc. Rev.* **2011**, *40*, 696–753. [CrossRef] [PubMed]

21. Shea, K.J.; Loy, D.A. Bridged Polysilsesquioxanes. Molecular-Engineered Hybrid Organic-Inorganic Materials. *Chem. Mater.* **2001**, *13*, 3306–3319. [CrossRef]

22. Sanchez, C.; Lebeau, B.; Chaput, F.; Boilot, J.-P. Optical Properties of Functional Hybrid Organic-Inorganic Nanocomposite. *Adv. Mater.* **2003**, *15*, 1969–1994. [CrossRef]

23. Faupel, F.; Zaporojtchenko, V.; Strunskus, T.; Elbahri, M. Metal-Polymer Nanocomposites for Functional Applications. *Adv. Eng. Mater.* **2010**, *12*, 1117–1190. [CrossRef]

24. Ruffino, F.; Torrisi, V.; Marletta, G.; Grimaldi, M.G. Effects of the embedding kinetics on the surface nano-morphology of nano-grained Au and Ag films on PS and PMMA layers annealed above the glass transition temperature. *Appl. Phys. A* **2012**, *107*, 669–683. [CrossRef]

25. Tamaekong, N.; Liewhiran, C.; Phanichphant, S. Synthesis of Thermally Spherical CuO Nanoparticles. *J. Nanomater.* **2014**, *2014*, 507978. [CrossRef]

26. Kozak, D.S.; Sergiienko, R.A.; Shibata, E.; Iizuka, A.; Nakamura, T. Non-electrolytic synthesis of copper oxide/carbon nanocomposite by surface plasma in super-dehydrated ethanol. *Sci. Rep.* **2016**, *6*, 21178. [CrossRef] [PubMed]

27. Shah, M.A.; Al-Ghamdi, M.S. Preparation of Copper (Cu) and Copper Oxide (Cu$_2$O) Nanoparticles under Supercritical Conditions. *Mater. Sci. Appl.* **2011**, *2*, 977–980.

28. Aziz, S.B.; Rasheed, M.A.; Ahmed, H.M. Synthesis of Polymer Nanocomposites Based on [Methyl Cellulose]$_{(1-x)}$:(CuS)$_x$ (0.02 M ≤ x ≤ 0.08 M) with Desired Optical Band Gaps. *Polymers* **2017**, *9*, 194. [CrossRef]

29. Muniz-Miranda, M.; Gellini, C.; Simonelli, A.; Tiberi, M.; Giammanco, F.; Giorgetti, E. Characterization of Copper nanoparticles obtained by laser ablation in liquids. *Appl. Phys. A* **2013**, *110*, 829–833. [CrossRef]

30. Aziz, S.B.; Abdulwahid, R.T.; Rsaul, H.A.; Ahmed, H.M. In-situ synthesis of CuS nanoparticle with a distinguishable SPR peak in NIR region. *J. Mater. Sci.* **2016**, *27*, 4163–4171. [CrossRef]

31. Abdullah, O.G.; Aziz, S.B.; Omer, K.M.; Salih, Y.M. Reducing the optical band gap of polyvinyl alcohol (PVA) based nanocomposite. *J. Mater. Sci.* **2015**, *26*, 5303–5309. [CrossRef]

32. Maggioni, G.; Vomiero, A.; Carturan, S.; Scian, C. Structure and optical properties of Au-polyimide nanocomposite films prepared by ion implantation. *Appl. Phys. Lett.* **2004**, *85*, 5712. [CrossRef]

33. Takele, H.; Kulkarni, A.; Jebril, S.; Chakravadhanula, V.S.K.; Hanisch, C.; Strunskus, T.; Zaporojtchenko, V.; Faupel, F. Plasmonic properties of vapour-deposited polymer composites containing Ag nanoparticles and their changes upon annealing. *J. Phys. D* **2008**, *41*, 12540. [CrossRef]

34. Biswas, A.; Aktas, O.C.; Schürmann, U.; Saeed, U.; Zaporojtchenko, V.; Faupel, F. Tunable multiple plasmon resonance wavelengths response from multicomponent polymer-metal nanocomposite systems. *Appl. Phys. Lett.* **2004**, *84*, 2655–2657. [CrossRef]

35. Avasthi, D.K.; Mishra, Y.K.; Kabiraj, D.; Lalla, N.P.; Pivin, J.C. Synthesis of metal–polymer nanocomposite for optical applications. *Nanotechnology* **2007**, *18*, 125604. [CrossRef]

36. Aziz, S.B.; Ahmed, H.M.; Hussein, A.M.; Fathulla, A.B.; Wsw, R.M.; Hussein, R.T. Tuning the absorption of ultraviolet spectra and optical parameters of aluminum doped PVA based solid polymer composites. *J. Mater. Sci.* **2015**, *26*, 8022–8028. [CrossRef]

37. Deshmukh, S.H.; Burghate, D.K.; Shilaskar, S.N.; Chaudhari, G.N.; Deshmukh, P.T. Optical properties of polyaniline doped PVC-PMMA thin films. *Indian J. Pure Appl. Phys.* **2008**, *46*, 344–348.

38. Bhavsar, V.; Tripathi, D. Study of refractive index dispersion and optical conductivity of PPy doped PVC films. *Indian J. Pure Appl. Phys.* **2016**, *54*, 105–110.

39. Urs, T.G.; Gowtham, G.K.; Nandaprakash, M.B.; Mahadevaiah, D.; Sangappa, Y.; Somashekar, R. Determination of force constant and refractive index of a semiconducting polymer composite using UV/visible spectroscopy: A new approach. *Indian J. Phys.* **2017**, *91*, 53–56. [CrossRef]

40. Potzsch, R.T.; Stahl, B.C.; Komber, H.; Voit, C.J.H.B.I. High refractive index polyvinylsulfide materials prepared by selective radical mono-addition thiol–yne chemistry. *Polym. Chem.* **2014**, *5*, 2911–2921. [CrossRef]

41. Tao, P.; Li, Y.; Rungta, A.; Viswanath, A.; Gao, J.; Benicewicz, B.C.; Siegel, R.W.; Schadler, L.S. TiO$_2$ nanocomposites with high refractive index and transparency. *J. Mater. Chem.* **2011**, *21*, 18623. [CrossRef]

42. Abdulwahid, R.T.; Abdullah, O.G.; Aziz, S.B.; Hussein, S.A.; Muhammad, F.F.; Yahya, M.Y. The study of structural and optical properties of PVA: PbO$_2$ based solid polymer nanocomposites. *J. Mater. Sci.* **2016**, *27*, 12112–12118.

43. Aziz, S.B. Modifying poly (vinyl alcohol) (PVA) from insulator to small-bandgap polymer: A novel approach for organic solar cells and optoelectronic devices. *J. Electron. Mater.* **2016**, *45*, 736–745. [CrossRef]

44. Biskri, Z.E.; Rached, H.; Bouchear, M.; Rached, D.; Aida, M.S. A Comparative Study of Structural Stability and Mechanical and Optical Properties of Fluorapatite (Ca$_5$(PO$_4$)$_3$F) and Lithium Disilicate (Li$_2$Si$_2$O$_5$) Components Forming Dental Glass–Ceramics: First Principles Study. *J. Electron. Mater.* **2016**, *45*, 5082–5095. [CrossRef]

45. Ravindra, N.M.; Ganapathy, P.; Choi, J. Energy gap–refractive index relations in semiconductors—An overview. *Infrared Phys. Technol.* **2007**, *50*, 21–29. [CrossRef]

46. Aziz, S.B.; Rasheed, M.A.; Hussein, A.M.; Ahmed, H.M. Fabrication of polymer blend composites based on [PVA-PVP]$_{(1-x)}$:(Ag$_2$S)$_x$ (0.01 ≤ x ≤ 0.03) with small optical band gaps: Structural and optical properties. *Mater. Sci. Semicond. Process.* **2017**, *71*, 197–203. [CrossRef]

47. Alaya, R.; Slama, S.; Hashassi, M.; Mbarki, M.; Rebey, A.; Alaya, S. Theoretical Predictions of Structural, Electronic and Optical Properties of Dilute Bismide $AlN_{1-x}Bi_x$ in Zinc-Blend Structures. *J. Electron. Mater.* **2017**, *46*, 1977–1983. [CrossRef]

48. Wu, M.; Sun, D.; Tan, C.; Tian, X.; Huang, Y. Al-Doped ZnO Monolayer as a Promising Transparent Electrode Material: A First-Principles Study. *Materials* **2017**, *10*, 359. [CrossRef] [PubMed]

49. Rodrguez, A.; Vergara, M.E.S.; Montalvo, V.G.; Ortiz, A.; Alvarez, J.R. Thin films of molecular materials synthesized from $C_{32}H_{20}N_{10}M$ (M = Co, Pb, Fe): Film formation, electrical and optical properties. *Appl. Surf. Sci.* **2010**, *256*, 3374–3379. [CrossRef]

50. Bouzidi, C.; Horchani-Naifer, K.; Khadraoui, Z.; Elhouichet, H.; Ferid, M. Synthesis, characterization and DFT calculations of electronic and optical properties of $CaMoO_4$. *Phys. B* **2016**, *497*, 34–38. [CrossRef]

51. Feng, J.; Xiao, B.; Chen, J.C.; Zhou, C.T.; Du, Y.P.; Zhou, R. Optical properties of new photovoltaic materials: $AgCuO_2$ and $Ag_2Cu_2O_3$. *Solid State Commun.* **2009**, *149*, 1569–1573. [CrossRef]

52. Logothetidis, S. Optical and electronic properties of amorphous carbon materials. *Diam. Relat. Mater.* **2003**, *12*, 141–150. [CrossRef]

53. Nematollahin, M.; Yang, X.; Aas, L.M.S.; Ghadyani, Z.; Kildemo, M.; Gibson, U.J.; Reenaas, T.W. Molecular beam and pulsed laser deposition of ZnS:Cr for intermediate band solar cells. *Sol. Energy Mater. Sol. Cells* **2015**, *141*, 322–330. [CrossRef]

54. Bao, H.; Ruan, X. Ab initio calculations of thermal radiative properties: The semiconductor GaAs. *Int. J. Heat Mass Transf.* **2010**, *53*, 1308–1312. [CrossRef]

55. Ortenburger, I.B.; Rudge, W.E.; Herman, F. Electronic denstity of states and optical properties of polytypes of germanium and silicon. *J. Non-Cryst. Solids* **1972**, *8–10*, 653–658. [CrossRef]

56. Aziz, S.B.; Abidin, Z.H.Z.; Arof, A.K. Effect of silver nanoparticles on the DC conductivity in chitosan-silver triflate polymer electrolyte. *Phys. B Condens. Matter* **2010**, *405*, 4429–4433. [CrossRef]

57. Aziz, S.B.; Kadir, M.F.Z.; Abidin, Z.H.Z. Structural, morphological and electrochemical impedance study of CS: LiTf based solid polymer electrolyte: Reformulated Arrhenius equation for ion transport study. *Int. J. Electrochem. Sci.* **2016**, *11*, 9228–9244. [CrossRef]

58. Aziz, S.B.; Abidin, Z.H.Z. Ion-transport study in nanocomposite solid polymer electrolytes based on chitosan: Electrical and dielectric analysis. *J. Appl. Polym. Sci.* **2015**, *132*, 41774. [CrossRef]

59. Kumar, S.; Koh, J.; Tiwari, D.K.; Dutta, P.K. Optical Study of Chitosan-Ofloxacin Complex for Biomedical Applications. *J. Macromol. Sci. Part A* **2011**, *48*, 789–795. [CrossRef]

60. Kumar, K.K.; Ravi, M.; Pavani, Y.; Bhavani, S.; Sharma, A.K.; Rao, V.V.R.N. Investigations on the effect of complexation of NaF salt with polymer blend (PEO/PVP) electrolytes on ionic conductivity and optical energy band gaps. *Phys. B Condens. Matter* **2011**, *406*, 1706–1712. [CrossRef]

61. Al-Osaimi, J.; Al-Hosiny, N.; Abdallah, S.; Badawi, A. Characterization of optical, thermal and electrical properties of SWCNTs/PMMA nanocomposite films. *Iran. Polym. J.* **2014**, *23*, 437–443. [CrossRef]

nanomaterials

MDPI

Article

Molecular Mechanics of the Moisture Effect on Epoxy/Carbon Nanotube Nanocomposites

Lik-ho Tam and Chao Wu *

School of Transportation Science and Engineering, Beihang University, 37 Xueyuan Road, Beijing 100191, China; leo_tam@buaa.edu.cn
* Correspondence: wuchao@buaa.edu.cn; Tel.: +86-010-8233-9923

Received: 14 August 2017; Accepted: 9 October 2017; Published: 13 October 2017

Abstract: The strong structural integrity of polymer nanocomposite is influenced in the moist environment; but the fundamental mechanism is unclear, including the basis for the interactions between the absorbed water molecules and the structure, which prevents us from predicting the durability of its applications across multiple scales. In this research, a molecular dynamics model of the epoxy/single-walled carbon nanotube (SWCNT) nanocomposite is constructed to explore the mechanism of the moisture effect, and an analysis of the molecular interactions is provided by focusing on the hydrogen bond (H-bond) network inside the nanocomposite structure. The simulations show that at low moisture concentration, the water molecules affect the molecular interactions by favorably forming the water-nanocomposite H-bonds and the small cluster, while at high concentration the water molecules predominantly form the water-water H-bonds and the large cluster. The water molecules in the epoxy matrix and the epoxy-SWCNT interface disrupt the molecular interactions and deteriorate the mechanical properties. Through identifying the link between the water molecules and the nanocomposite structure and properties, it is shown that the free volume in the nanocomposite is crucial for its structural integrity, which facilitates the moisture accumulation and the distinct material deteriorations. This study provides insights into the moisture-affected structure and properties of the nanocomposite from the nanoscale perspective, which contributes to the understanding of the nanocomposite long-term performance under the moisture effect.

Keywords: carbon nanotube; composite; moisture; mechanical property; molecular dynamics simulation

---

1. Introduction

Polymer nanocomposites comprise the reinforcements distributed in a matrix, which have attracted substantial interest in both academic and industrial communities. The nanocomposites reinforced by the nanoparticles in the form of the sphere, cylinder, and plate possess enhanced material properties, such as light weight, high stiffness, increased fatigue resistance, good chemical stability, and improved heat and thermal resistance [1]. In recent years, these nanocomposites have been increasingly used in a wide array of engineering fields across multiple length scales, ranging from the microscopic sensors and actuators in the micro-electromechanical systems, to the macroscopic composite materials in the aerospace, automotive, and construction industries [2–4]. In practice, a wide variety of the polymer matrix and the reinforcements have been used for fabricating the nanocomposite. Particularly, the epoxy is one of the most commonly used polymers, which possesses a three-dimensional cross-linked network formed by the cross-linking process occurred in the molecular level [5]. Meanwhile, due to the remarkable mechanical properties including the high elastic modulus and tensile strength, the carbon nanotube (CNT) is regarded as one of the most promising reinforcement materials [1]. The incorporation of a small amount of the CNT can substantially improve the mechanical properties of the resulting polymer nanocomposites, which can sustain the large degree of the elastic deformation, and ensure the durability of the engineering applications. However, as the epoxy is sensitive to the environment humidity, the moisture diffuses into the nanocomposite and interacts with the structure, which influences the mechanical performance of the nanocomposite [6,7]. For these reasons, the structural integrity of the epoxy nanocomposites is significantly affected in the moist environment.

As the moisture-affected structural integrity of the polymer composites is a matter of concern during the intended service life, several experimental studies have been carried out to study the composite material under the influence of the moisture [8–13]. The diffusion and absorption of the moisture into the composite have been successfully obtained through the gravimetric measurements at different moisture conditioning durations [8–12]. With the incorporation of the reinforcement, it is reported that the saturated moisture concentration of the composite is either lower than, or close to, that of the neat polymer. The moisture absorption of the composite material is affected by the reinforcement, which also has an influential role in the moisture-affected material properties. Particularly, a recent experimental investigation of the epoxy/CNT composite has shown that the storage moduli decreases with increasing moisture ingression, and the value for the composite is higher when compared to the neat polymer after the same conditioning time [11]. Similar moisture-affected deterioration has been observed in the glass transition temperature of the epoxy/organoclay composite [8] and the shear stress of the epoxy/CNT composite adhesive obtained from the lap shear tests [13]. The lower degree of the degradation in case of the composite is attributed to the existence of the stiff reinforcement, which might affect the structure and the molecular interaction inside the composite under the wet condition. So far, the molecular origin of the moisture-affected property variations of the polymer composite is still unclear, such as the distribution, structure and molecular interactions of the water molecules inside the structure.

The polymer composite features the hierarchical structure ranging from the atomistic building blocks, including the polymer monomer and the CNT segment, to the molecular model with the CNT segment embedded in the polymer matrix, and to the mesoscale structure consisting of the CNT cluster, as shown in Figure 1. To understand the origin of the moisture effect, it is important to study the material behavior in the molecular level, so as to predict the system performance at the larger length and time scales. Molecular dynamics (MD) simulation has been considered as a fundamental and powerful technique for studying the molecular interaction inside the structure [14–18], and it has been widely used to investigate the structure and properties of the polymer, the carbon material, and the polymer composite at the nanoscale [19–23]. Microscopic information about the moisture diffusion and absorption inside the polymer and the nanocomposite has been obtained through observing the molecular motions [24–26]. The mechanism of the water molecule diffusion in the polymer is reported recently, where the water molecules absorbed in the structure can move between the microcavities and form hydrogen bonds (H-bonds) with the network polar groups [24]. Further simulation investigations have provided the information about the interactions among the polymer, the CNT, and the water molecules [27–30]. Notably, an atomistic simulation of the epoxy network has quantitatively shown that at the low moisture concentration, most of the water molecules tend to form H-bonds with the structure functional groups, while forming bonds with other water molecules at the high concentration, but the detailed H-bond network inside the structure is not provided [27]. Furthermore, the MD simulation has been adopted for investigating the moisture effect on the properties of the epoxy and epoxy-based composite [20,26]. Particularly, the adhesion energy of the epoxy/nanoclay composite to the substrate is significantly reduced when the water molecule is added at the interface close to the substrate, but the mechanism of the moisture effect is not clear [26]. These MD studies have advanced the understanding of the moisture effect on the polymer, but until now, very few investigations have been focused on the relationship between the molecular interaction among the polymer nanocomposite and the absorbed water molecules and the structure and properties of the nanocomposite. Such knowledge enables one to understand more in-depth about the molecular origin of the moisture-affected variation in the structure and properties of the epoxy/CNT nanocomposite.

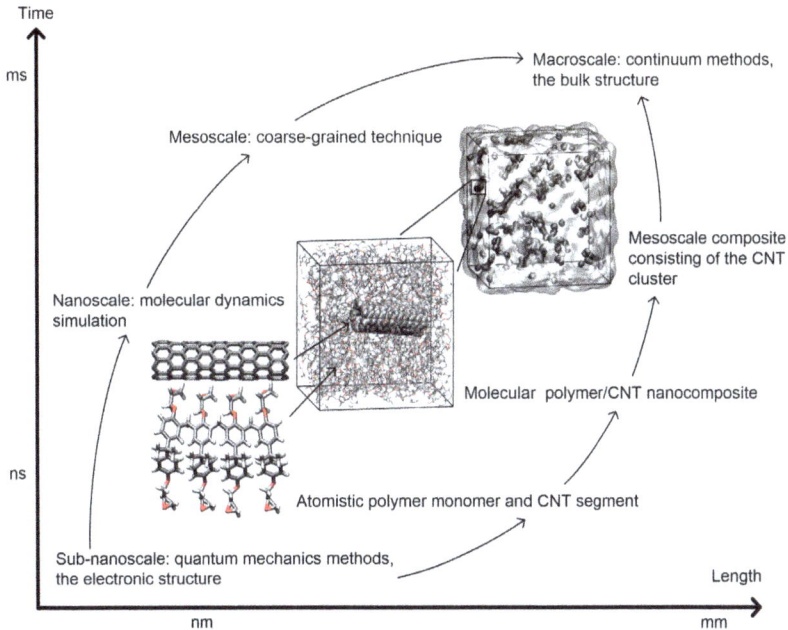

Figure 1. Overview of the polymer composite at different scales: from the atomistic building blocks of the polymer monomer and the carbon nanotube (CNT) segment, to the molecular model of the polymer/CNT nanocomposite, and to the mesoscale structure consisted of the CNT cluster. To understand the origin of the moisture effect, it is important to study the material behavior in the molecular level, so as to predict the system performance at the larger levels.

The objectives of this study are to understand the molecular interactions between the water molecules and the epoxy/CNT nanocomposite in the molecular level, so as to evaluate the effect of the moisture on the structure and properties of the nanocomposite, and to predict the durability of the related engineering applications at the larger scales. The epoxy nanocomposite reinforced by a CNT segment is modeled by using a dynamic cross-linking algorithm, which is applicable to the construction of the epoxy-based materials involving the highly cross-linked network [19]. To mimic the different moisture absorption states in reality, the constructed model is solvated with a successive amount of the water molecules. Through examining the local structure and the H-bond network in the systems, the molecular interactions between the nanocomposite structure and the added water molecule are characterized, which are correlated with the mechanical behavior of the solvated structures. This paper provides the knowledge on the structure and properties of the epoxy/CNT nanocomposite with various moisture concentrations, which forms the fundamental basis for predicting the structural integrity of the nanocomposite during the intended service life, as well as for developing better products with improved moisture barrier properties.

2. Results and Discussion

The polymer matrix used in this study is the cross-linked epoxy, and its cross-linking process and physical properties have been investigated in our previous atomistic investigation [19]. The epoxy monomer consists of four components of the diglycidyl ether of bisphenol A, which are connected by the methylene group, as shown in Figure 1. For the CNT filler, it is reported that the chirality and diameter of the single-walled CNT (SWCNT) have limited effects on its Young's modulus [31]. In this study, an armchair (5,5) SWCNT segment with a diameter of 6.8 Å is selected as a representative model of the reinforcement, and the SWCNTs with the similar diameter have been synthesized in the experiment [32,33]. According to various experimental and simulation studies, the mass fraction of the CNT in the composite is usually designed to be less than 10.0 wt% [10–13,34,35]. Accordingly, the SWCNT segment with a length of 24.6 Å is used in the simulation, as shown in Figure 1, and its mass fraction in the composite equals 4.3 wt%. Meanwhile, it should be noted that the SWCNT length is less than that of the periodic simulation cell edge of around 4.5 nm. Though the length of the CNT is several orders of magnitude shorter than the bulk material, the CNT reinforcing effect on the nanocomposite has been observed as equal to that in experiments, which indicates that the short aspect ratio SWCNT segment is reasonable as the filler in the nanocomposite modeling [35,36]. The MD simulation results based on the molecular model of the epoxy/SWCNT nanocomposite as shown in Figure 1 are presented here, including the validation of the nanocomposite model and the discussions about the moisture effect on the structure and properties of the epoxy/SWCNT nanocomposite from the nanoscale perspective.

2.1. Structure and Properties of the Epoxy/SWCNT Nanocomposite

The epoxy/SWCNT nanocomposite model with no moisture is validated first by determining its physical properties, i.e., the density (ρ) and the Young's modulus (E), and comparing these against the experimental data. During the structural equilibration, the system is equilibrated under the atmospheric pressure for 5 ns and the three orthogonal directions are adjusted independently. The ρ of the modeled nanocomposite structure is sampled at a 2 ps interval, and the recorded data from the last 1 ns equilibration run is used for the calculation, such that the statistical error is minimized. With the addition of the SWCNT, the ρ of the nanocomposite model is measured to be 1.07 ± 0.002 g·cm^{-3}, which is higher than the corresponding simulated value of the neat cross-linked epoxy in the range of 1.04 g·cm^{-3} to 1.05 g·cm^{-3} as reported in our previous work [19]. The increase of the composite density due to the addition of the CNT has been observed in various experimental and simulation studies, which can be attributed to the change in the cross-linked epoxy matrix, especially in the interfacial layer surrounding the CNT [37–39]. Specifically, previous simulation studies have demonstrated that a closely packed epoxy structure can be formed close to the epoxy-CNT interface, which leads to the increase of the nanocomposite density [38,39]. Another reason to the higher density is that the free volume of the epoxy matrix is decreased when the SWCNT is added. By using the Connolly surface algorithm, the free volume of the modeled structure is analyzed [40]. During the last 1 ns equilibration process, the fractional free volume of the model is sampled at a 200 ps interval. The measured fractional free volume of the epoxy and the nanocomposite is $23.85 \pm 0.26\%$ and $21.08 \pm 0.41\%$, respectively. From these observations, it is found that the nanocomposite with the addition of the SWCNT possesses the structural characteristics close to the reported observation, i.e., the densified structure and the decrement in the structure free volume compared to the neat epoxy.

After the structural characterization, the Young's modulus of the nanocomposite model is quantified. The stress-strain data obtained from the tensile deformation of the nanocomposite model are shown in the Supplementary Materials (Figure S1), in comparison to those of the neat epoxy obtained from the experiment and simulation [41,42]. During the dynamic deformation, the stress of the nanocomposite model along the stretched direction shows a linear relation to the applied strain, which indicates that the structure is elongated elastically within this small-strain deformation. Based on the obtained stress-strain data, the E is measured by carrying out a linear regression analysis on the data, and the coefficient of determination for the fitted curve is over 0.99. The E of the nanocomposite model is measured to be 4.88 ± 0.07 GPa, revealing a significant increase in the stiffness in comparison to the neat cross-linked epoxy, with the experimental value ranging from 2.70 GPa to 4.02 GPa and the simulated results in the range of 2.67–4.43 GPa [19,41,42]. The enhancement of the nanocomposite

Young's modulus by the embedded CNT agrees closely with various measurements [1,7,22,23], which is due to the stiff CNT reinforcement and the efficient stress transfer from the epoxy matrix to the CNT. In view of the good agreement of the structural and mechanical properties of the modeled nanocomposite with the available data, the generated model of the epoxy/SWCNT nanocomposite is regarded as the reasonable structure close to those found in the real system, which is used as the basis in the following discussion.

2.2. Moisture Effect on the Structure and Properties of the Epoxy/SWCNT Nanocomposite

In order to study the moisture effect on the epoxy/SWCNT nanocomposite, the equilibrated structure is solvated with a successive amount of the water molecule, such that the different states of the moisture absorption are mimicked. The maximum moisture concentration of the epoxy nanocomposite is reported to be less than 4.0 wt% [10–12]. In this study, the moisture concentration of the nanocomposite model ranges from 1.0, 2.0, 3.0, to 4.0 wt% of the molar mass of the structure, and the concentration of 4.0 wt% is regarded as the saturated condition. Based on the modeled systems, the moisture effect is firstly characterized by examining the local structural changes in the nanocomposite with various moisture concentrations. The epoxy matrix of the nanocomposite includes two oxygen-containing functional groups, i.e., the hydroxyl groups and the ether oxygen atoms, which are important potential hydrogen bonding sites and tend to form the H-bonds among the functional groups and with the absorbed water molecules in the structure. Due to the polar interactions with the functional groups, the distribution of the water molecules in the nanocomposite structure can be influenced. In order to determine the water distribution, the arrangement of the functional groups around the SWCNT segment is firstly characterized. In general, the radial distribution function (RDF) is utilized to determine the normalized probability of finding an object at a distance r away from another object, which provides the useful information about the arrangement of the functional group. Here, the intermolecular RDFs between the oxygen of the functional groups and the carbon of the SWCNT are calculated and shown in Figure S2, with respect to the moisture concentration. The RDFs with similar shapes are reported for the polymer distribution in various nanocomposites [43,44]. For the RDFs of the hydroxyl groups, the small peaks at the distance of 2.5–5.0 Å indicate the favorable distribution of these functional groups in the interfacial layer surrounding the SWCNT. As the quantity of the hydroxyl groups in the structure is smaller than that of the ether groups, the sparse hydroxyl groups in the interfacial layer can aggregate at a certain distance due to the H-bond interaction and lead to the small peaks. With various moisture concentrations, the RDFs for the same functional group overlap, showing that the arrangement of the epoxy matrix are relatively the same for the different nanocomposite models, which is also observed from examining the orientation of the SU-8 monomers against the SWCNT axis.

After characterizing the arrangement of the epoxy matrix around the SWCNT segment, the distribution of the added water molecules in relation to the functional groups is quantified. Specifically, the RDFs between the oxygen of the functional groups and the oxygen of the water molecule are calculated, as shown in Figure 2, with respect to the moisture concentration. All of the RDFs display a sharp peak at the distance between 2.5 Å to 3.5 Å, which features the presence of the H-bond interaction between the added water molecules and the functional groups. The higher peak of the water-hydroxyl RDFs indicates that the water molecules preferentially locate in the vicinity of the hydroxyl groups, as they are highly electrophilic in nature. Meanwhile, it can be seen that the intensity of the peak decreases monotonically with the increasing moisture concentration, which is due to the fact that the number of the functional groups in the system is constant, and the fraction of the added water molecules forming the H-bonds with the nanocomposite decreases with the increasing concentration, leading to the decrease in the RDF peak intensity. For comparison, the intermolecular RDFs for the neat epoxy structure are shown in the Supplementary Materials (Figure S3) [42]. The RDFs for the water-hydroxyl and water-ether interactions show similar shapes and positions as those for the nanocomposite structure, while the peak of the RDFs for the nanocomposite structure is relatively higher, indicating the water aggregation in the vicinity of the functional groups is more pronounced in the nanocomposite structure, which could be due to the existence of the hydrophobic SWCNT segment in the system, as discussed subsequently.

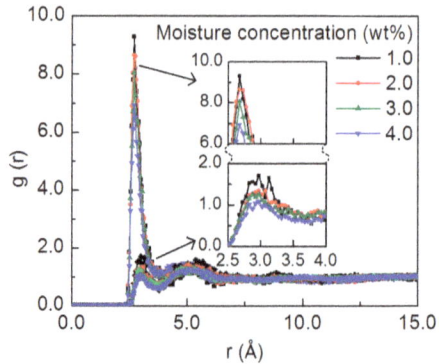

Figure 2. The radial distribution function (RDF) between the oxygen of the functional groups and the oxygen of the water molecule with respect to the moisture concentration. The RDFs for the water-hydroxyl interactions display higher peaks than those for the water-ether interactions, as demonstrated in the enlarged picture. Meanwhile, the intensity of the peak decreases monotonically with the increasing moisture concentration for both water-functional group interactions.

Apart from the effect of the epoxy functional groups, the distribution of the water molecules can be affected by the SWCNT segment, which is characterized by measuring the water-SWCNT RDFs for the systems with different moisture concentrations, as shown in Figure 3. Meanwhile, the two-dimensional distribution map of the water molecules in the plane perpendicular to the SWCNT axis after the equilibration is shown in Figure S4, with respect to the moisture concentration. At the concentration of 1.0 wt%, the RDF shows a well-defined first peak around the distance of 3.8 Å and a second peak at the distance of about 10.3 Å. With the existence of the SWCNT segment, some water molecules can diffuse to the free volume in the epoxy-SWCNT interface, which leads to the first peak. Meanwhile, due to the repulsive effect of the hydrophobic SWCNT, the other water molecules are repulsed by the SWCNT through the short-range van der Waals (vdW) interaction and some of them are accumulated at the cutoff distance from the SWCNT surface, which leads to the second peak. With the increasing concentration, due to the limited available space in the interface, only a small portion of the added water molecules diffuse to the interface region, as demonstrated by the two-dimensional distribution map. Therefore, the peak close to the interface is not obvious at the high moisture concentrations. Specifically, at the 2.0 wt% concentration, the newly-added water molecules repulsed by the SWCNT tend to accumulate at the cutoff distance, leading to the noticeable peak intensity. When the moisture concentration continuously increases, as there is limited available space at the cutoff distance, the added water molecules tend to diffuse to other available space in the structure, and the peak intensity drops gradually. The simulation results suggest that the distribution of the added water molecules is affected by the functional groups and the embedded SWCNT in the structure, which varies with the moisture concentration. Furthermore, it is noted that there is a relatively high correlation among the distribution map of the water molecules with various concentrations, which is resulted from the successive addition of the water molecules. As the amount of 1.0 wt% water molecules is added successively to the equilibrated model for the condition with various concentrations, the newly-added water molecules are also affected by the water molecules existing in the structure, which can diffuse to the vicinity of the existing water molecules. Therefore, it leads to the relatively high correlation among the final distribution of the water molecules with various concentrations.

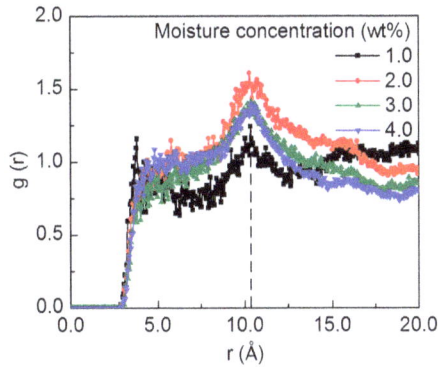

Figure 3. The RDF between the oxygen of the water molecules and the carbon of the SWCNT with respect to the moisture concentration. The peak at the distance of about 10.3 Å is resulted from the accumulation of the water molecules at the distance where the short-range van der Waals (vdW) interaction is cutoff, which is caused by the repulsive effect of the hydrophobic SWCNT.

With the knowledge of the preferential distribution of the water molecules in the nanocomposite, the structure of the water molecules in the system is characterized, by focusing on the water-water RDF for the systems containing various amounts of the water molecules, as shown in Figure 4. The water-water RDFs show the sharp peak at a distance of around 2.7 Å for all the systems, with the highest peak height for the system with the lowest concentration in this study, i.e., 1.0 wt%. The relationship between the water-water RDF peak and the moisture concentration is in good accord with previous MD simulation results, where the peak height of the water-water RDF decreases with the moisture concentration [45–47], and the RDF peak of the oxygen atoms in the bulk water is lower than that in the solvated polymer structures at any moisture concentration [47,48]. This observation demonstrates that the water molecules are more likely to be hydrogen bonded with each other and form the cluster at high moisture concentrations.

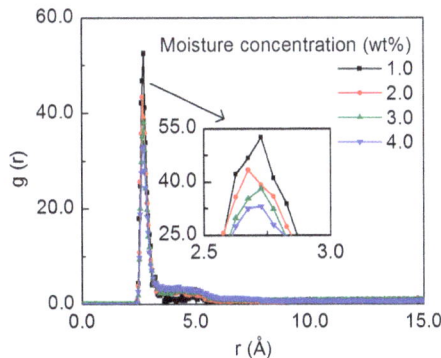

Figure 4. The RDF of the oxygen atoms in the water molecules with respect to the moisture concentration, with the highest peak height for the system with the concentration of 1.0 wt%.

Further information about the structure of the water molecules is obtained by quantifying the size probability of the water cluster. When the oxygen of the water molecules is less than a cutoff distance of 3.5 Å and is also within a cutoff angle of 50°, the water molecules are considered to be in the same cluster [49]. The probability of the cluster size is shown in Figure 5, with respect to the moisture concentration. It should be mentioned that the ordinate indicates the probability that a randomly-chosen water molecule will be in the cluster of a provided size. When the moisture concentration is low, a large portion of the water molecules exist as single molecules in the structure, with about 35% existing as pairs. When the moisture concentration increases, the water molecules tend to form clusters. The observation of the water clustering behaviors with different moisture concentrations is consistent with the recent observation in the polymer materials [28,50]. A close examination of the MD simulation process shows that some water molecules can diffuse to the free volume in the epoxy-SWCNT interface, as shown in Figure 6. With the increasing moisture concentration, more water molecules are involved in forming the large clusters in the system.

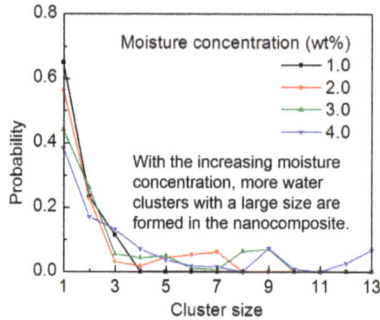

Figure 5. The cluster size distribution of the water molecules in the epoxy/SWCNT nanocomposite with respect to the moisture concentration: with the increasing moisture concentration, more water molecules are involved in forming the large clusters in the system.

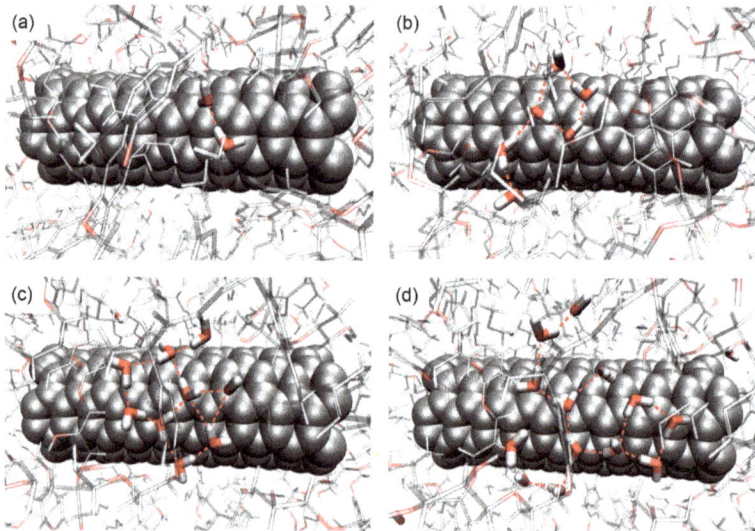

Figure 6. Simulation snapshots of the water cluster in the epoxy-SWCNT interface in the nanocomposite: the water molecules diffuse to the free volume in the epoxy-SWCNT interface, and form the clusters in the system: (**a**) 1.0 wt%; (**b**) 2.0 wt%; (**c**) 3.0 wt%; and (**d**) 4.0 wt%. For clarity, the epoxy structure is faded and the H-bond is indicated by the dash line.

Due to existence of the epoxy functional groups and the hydrophobic SWCNT, the preferential distribution and the structure of the water molecules in the nanocomposite vary with different moisture concentrations, which can affect the molecular interactions inside the structure. To gain insight into the interactions between the nanocomposite structure and the water molecules, the H-bond network is characterized, with the focus on the probability of forming the water-nanocomposite and water-water H-bond for a given water molecule, and the probability of forming the H-bond between the water cluster and the nanocomposite. The H-bond is defined based on two geometric criteria: the distance between the donor and the acceptor oxygen is less than 3.5 Å, and the donor-hydrogen-acceptor angle is above 130° [49]. The cutoff distance value corresponds to the position of the first minimum of the water-functional group RDFs as shown in Figure 2. Meanwhile, all of the potential hydrogen bonding sites in the nanocomposite structure are considered for determining the water-nanocomposite H-bond, including the oxygen and hydrogen of the hydroxyl groups and the oxygen of the ether groups. During the equilibration run, the H-bond network is monitored at a 100 ps interval, and the recorded data from the last 1 ns equilibration run is used for the calculation. The measured H-bond probability distribution for a given water molecule and for a given cluster is shown in Figure 7 and Figure S5, respectively, for different moisture concentration conditions.

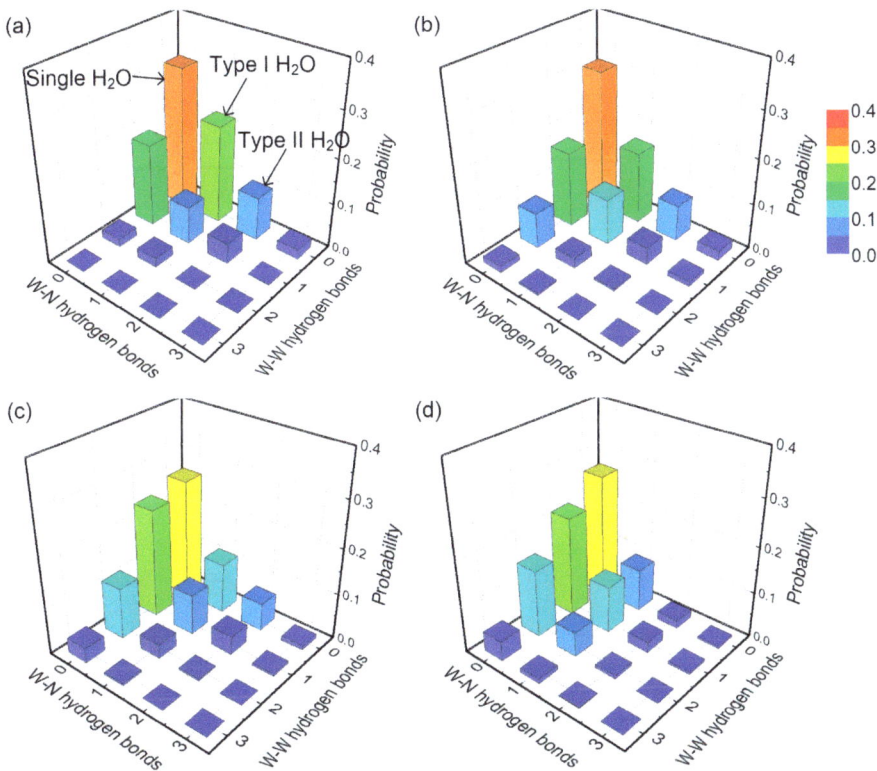

Figure 7. The distribution of the hydrogen bond (H-bond) probability for a given water molecule in the epoxy/SWCNT nanocomposite with respect to the moisture concentration: (**a**) 1.0 wt%; (**b**) 2.0 wt%; (**c**) 3.0 wt%; and (**d**) 4.0 wt%. The single water molecules do not form the H-bond, the Type I water molecules form one water-nanocomposite H-bond, and the Type II water molecules form more than one water-nanocomposite H-bond. The W-N and W-W denote the water-nanocomposite and water-water, respectively.

From these figures, it is observed that a large portion of the water molecules do not form H-bonds with the nanocomposite structure and other water molecules, existing as single water molecules. These single water molecules neither show strong interactions with the nanocomposite structure nor with other water molecules, and they can only weakly affect the mechanical properties of the nanocomposite with various moisture concentrations. Meanwhile, there are a certain number of water molecules forming one to three H-bonds with the nanocomposite, which have no H-bond with other water molecules. These water molecules can be classified into two types, as indicated in Figures 7 and 8. The Type I water molecules forming one water-nanocomposite H-bond can interfere with the molecular interactions established by the vdW and Coulombic force inside the structure, leading to the material plasticization, as demonstrated in Figure 8a [6,51]. Comparatively, the Type II water molecules forming more than one H-bond with the nanocomposite can effectively form the bridge between the functional groups in the structure, leading to the secondary cross-linking process, as demonstrated in Figure 8b. When the moisture concentration increases, the added water molecules tend to form the H-bonds, especially with other existing water molecules. The water molecules interacting by the H-bonds form the water clusters, which can affect the molecular interactions inside the system and cause a deviation in the nanocomposite properties. Accordingly, these water clusters can be classified into two types, as indicated in Figure S5 and Figure 8. Among these water clusters, a portion of the water clusters form the H-bonds with the nanocomposite, being the Type I water cluster with one water-nanocomposite H-bond, or the Type II water cluster forming more than one water-nanocomposite H-bond, as demonstrated in Figure 8c,d, respectively. From the simulation results, it is observed that when there is a scarcity of water, the water molecules predominantly form H-bonds with the functional groups in the structure rather than with other water molecules. When the moisture concentration increases, the water molecules tend to form a larger number of the H-bonds with the existing water molecules rather than with the nanocomposite structure.

The formation of the water cluster and its interactions with the nanocomposite structure vary with different moisture concentrations, which affect the movements of the water molecules. The diffusion of the water molecules is determined by examining the mean squared displacement (MSD) of the water molecules, as shown in Figure S6. It is observed that with the increasing concentration, the diffusion of the water molecules inside the structure is restricted, which is because more and more water molecules are involved in the H-bond interactions either with other water molecules or with the functional groups, and the dynamics of the water molecules reach the equilibrium state. The simulation results support the observation for the cluster formation and the H-bond analysis as discussed previously.

The structural variation under the moisture effect plays an influential role in the material properties of the epoxy/SWCNT nanocomposite. During the equilibration process, the ρ and volume of the nanocomposite model is recorded at a 2 ps interval, and the recorded data from the last 1 ns equilibration run is averaged. The measured ρ of the nanocomposite model with various moisture concentrations is shown in Figure 9a, with the error bar showing the standard derivation of the averaged recorded data. For the solvated nanocomposite systems, the ρ shows a monotonic increase with the moisture concentration, and reaches the maximum of 1.10 ± 0.002 g·cm^{-3} at the moisture concentration of 4.0 wt%. In order to understand the reason for such behavior, the measured volume of the nanocomposite is shown in Figure 9b. For comparison, the volume of the dry model and the corresponding water molecules at various concentrations are added together and regarded as the reference value. With the increasing moisture concentration, the volume increment of the nanocomposite model is smaller than that of the added water molecules, which indicates that the added water molecules mainly diffuse to the free volume and the vicinity of the existing water molecules in the structure, leading to the decrease of the fractional free volume, as shown in Figure 9a. With the limited expansion of the system and the weight gain by the added water molecules, it results in the monotonic increase of the nanocomposite density with the increasing moisture concentration. Furthermore, the simulation results show the moisture induced swelling of the nanocomposite, which can affect the structural integrity of the nanocomposite.

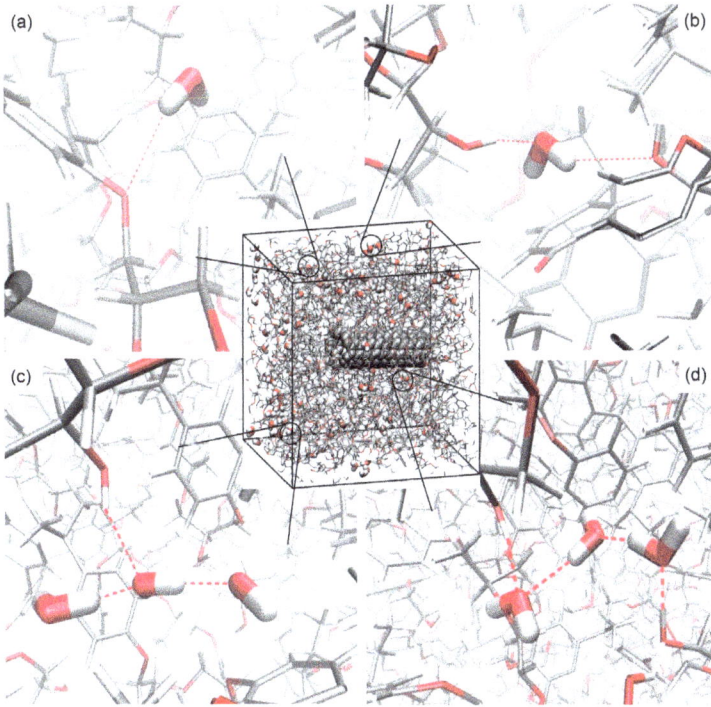

Figure 8. The schematic diagram of (**a**) the Type I water molecule forming one water-nanocomposite H-bond; (**b**) the Type II water molecule forming more than one water-nanocomposite H-bond; (**c**) the Type I water cluster; and (**d**) the Type II water cluster. For clarity, the epoxy structure is faded, the water cluster only consists of three water molecules, and the H-bond is indicated by the dashed line.

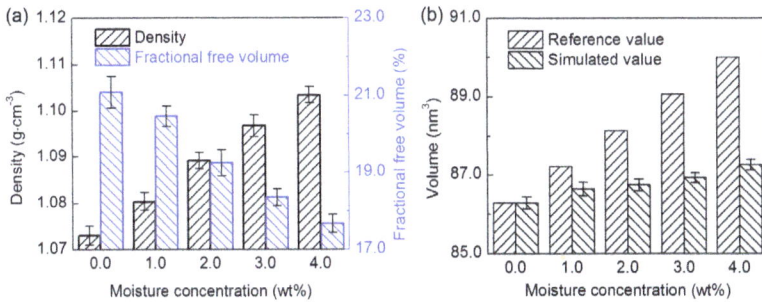

Figure 9. The (**a**) density and fractional free volume and (**b**) volume of the epoxy/SWCNT nanocomposite with respect to the moisture concentration at the temperature of 300 K and the pressure of 1 atm.

After determining the moisture effect on the volumetric properties, the moisture-affected elastic properties are now quantified by using the Young's modulus as an indicator. The stress-strain data of the nanocomposite model with various moisture concentrations from the tensile deformations are averaged and shown in Figure 10, together with the calculated E and the error bar obtained from the linear regression of the averaged simulated data. For the solvated structures, the E decreases gradually to 4.39 ± 0.04 GPa for the structure with the maximum moisture concentration of 4.0 wt%, revealing a

9.9% reduction compared to the dry sample. Such moisture-affected deterioration in the mechanical properties is consistent with the experimental observations [8,11,13], which can be related to the variation in the structure and the molecular interactions in the nanocomposite due to the moisture absorption. When the moisture concentration is low, i.e., 1.0 wt% in this study, the water molecules are located in the vicinity of the functional groups and expand the nanocomposite structure. The water molecules and the small clusters affect the molecular interaction inside the structure by forming a significant number of water-nanocomposite H-bonds. The existence of the Type I water molecules and clusters plays a vital role in the plasticization, while the bridging of the Type II water molecules and clusters leads to a secondary cross-linking process, and the stiffening of the nanocomposite to a certain extent. Under the influence of these various factors, the E of the nanocomposite shows a marginally decrease of 2.5% at this low moisture concentration. When the concentration continuously increases, the water molecules preferably form a large number of the water-water H-bonds and the large water clusters. The water clusters are located in the vicinity of the epoxy functional groups and the available space in the epoxy-SWCNT interface, which can significantly disrupt the molecular interaction, and lead to the continuous decrease of the elastic modulus. The inferred trend of the mechanical properties with various moisture concentrations reported here is valid for the CNTs with a larger length at the nanometer scale.

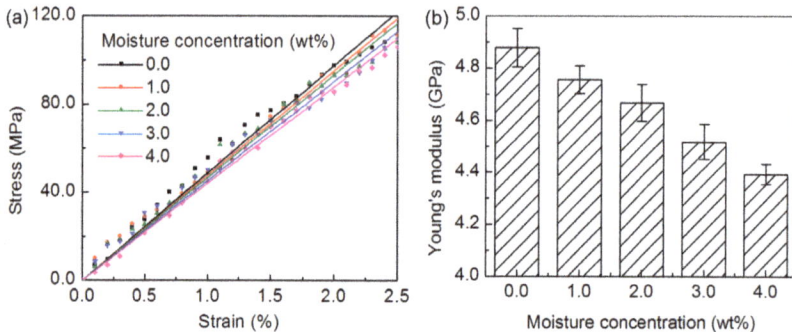

Figure 10. The (**a**) stress-strain data, and the (**b**) Young's modulus of the epoxy/SWCNT nanocomposite obtained from the small-strain uniaxial tensile deformation at the temperature of 300 K and the pressure of 1 atm.

To further quantify the moisture-affected mechanical properties, the nanocomposite models with moisture concentrations of 0 and 4 wt% are subjected to the computational tensile deformation to the plastic regime. The stress-strain curves for the models are shown in Figure 11. After the initial elastic deformation stage, the structure yields and undergoes the plastic deformation, as similar to the stress-strain responses observed for different polymer/CNT composites [52–54]. During the plastic regime, the stress of the nanocomposite with no moisture fluctuates around a relatively constant level, which indicates that the nanocomposite structure is strained adequately. However, for the nanocomposite with the 4 wt% moisture concentration, the stress drops occasionally, implying that the existence of the water molecules interferes with the molecular interactions inside the structure and, thus, weakens the material performance. From the stress-strain curves, it is inferred that the ultimate strength and the toughness of the nanocomposite decreases with the addition of the water molecules, which is in accordance with the trend of the Young's modulus. To understand the mechanism of the effect of the water molecules on the nanocomposite plastic deformation, the configuration of the structure is examined and shown in Figure 12, by selecting the SWCNT segment and the same epoxy chain for demonstration. It is observed that, with no water molecules in the structure, the epoxy chain can stick more firmly with the SWCNT segment during the tensile deformation, thus ensuring the efficient stress transfer from the epoxy matrix to the CNT, as shown in Figure 12a–c.

With the addition of the water molecules, they can diffuse to the interface region and interact with the epoxy chain, and the strained epoxy matrix slides along the SWCNT surface more easily, as shown in Figure 12d–f. The slippage event occurs continuously and leads to the drop in stress. This observation can be further confirmed by examining the energy evolution of the nanocomposite structure during the deformation, as shown in Figure 11b–d. The value of the energy in the nanocomposite model is slightly different with the addition of the water molecules. With no water molecules, the bonded energy and non-bonded energy of the nanocomposite, as well as the interaction energy between the epoxy and the SWCNT change steadily, which indicates that the stress dissipation is mainly through the deformation within the epoxy matrix. With the addition of the 4 wt% moisture concentration, the evolution of the interaction energy between the epoxy and the SWCNT experiences relatively large fluctuations, during both the elastic and plastic deformation regime, as demonstrated in Figure 11b. This phenomenon further demonstrates that with the existence of the water molecules, the epoxy chain can slide along the SWCNT more easily, and it affects the non-bonded interaction between the epoxy matrix and the SWCNT, and the local high stress dissipation. The finding is in good accordance with recent observations, where the non-bonded interaction between the constituents in the nanocomposite plays an important role in dissipating the local high stress in the plastic deformation regime [55,56]. In addition, the bonded energy of the nanocomposite with 4 wt% moisture concentration evolves steadily at the beginning, which can be due to the reason that as the energy dissipation is mainly through the slippage between the epoxy matrix and the CNT, the strained epoxy matrix after the slippage reaches the local energy minimum state, thus maintaining the bonded energy.

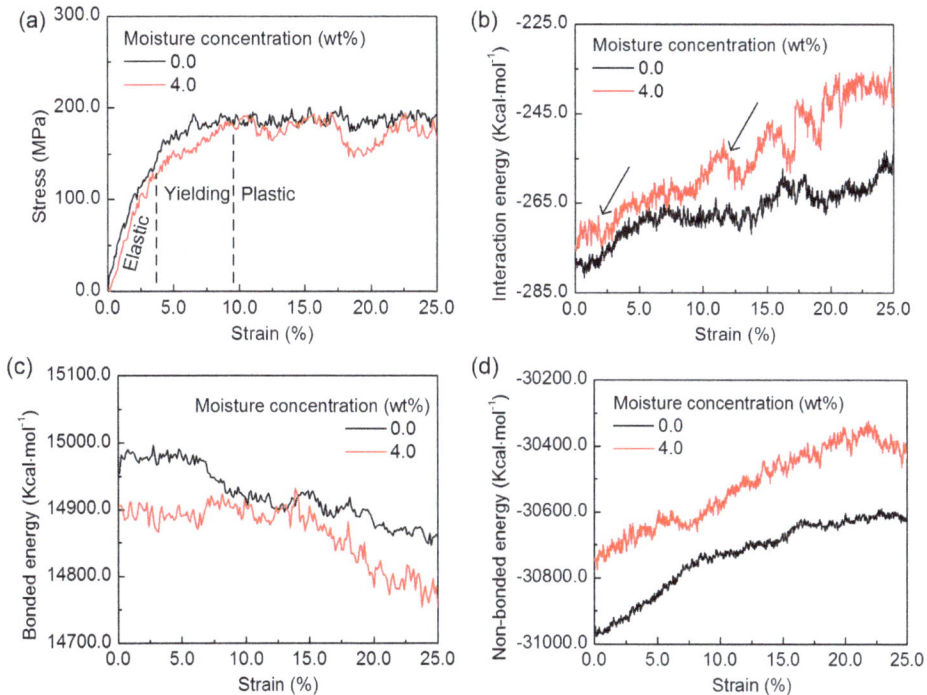

Figure 11. The (**a**) stress-strain curve and (**b–d**) the energy evolution of the epoxy/SWCNT nanocomposite during the tensile deformation to the plastic regime: (**b**) the interaction energy between the epoxy and the SWCNT (the arrows show the fluctuation during the elastic and plastic deformation regime); (**c**) the bonded energy, and (**d**) non-bonded energy of the nanocomposite.

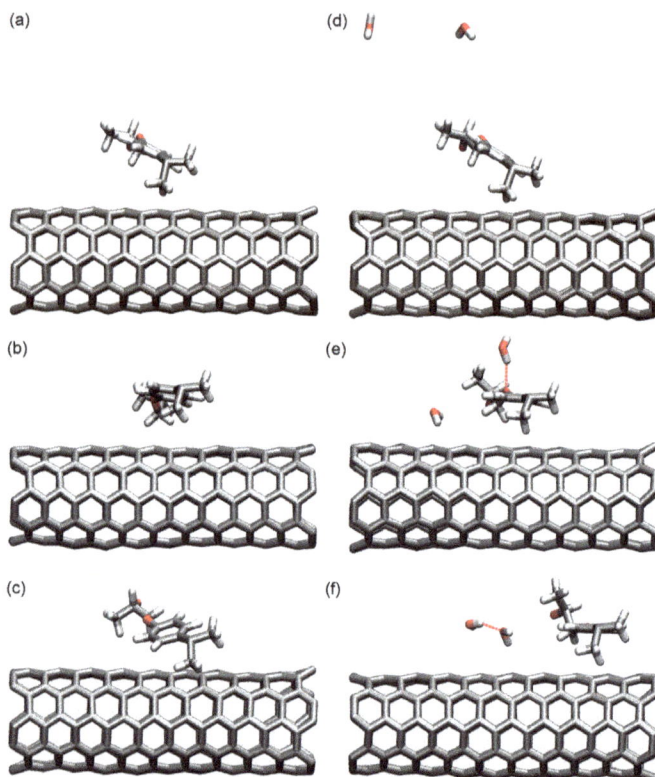

Figure 12. The configuration of the SWCNT segment and the selected epoxy chain in the nanocomposite with no moisture concentration when the strain is (**a**) 0.0%, (**b**) 13.0%, and (**c**) 25.0%; and with 4 wt% moisture concentration when the strain is (**d**) 0.0%, (**e**) 13.0%, and (**f**) 25.0%. The H-bond is indicated by the dashed line.

From these simulation results, it is observed that, due to the existence of the functional groups and the hydrophobic SWCNT, the distribution and structure of the water molecules, as well as the interactions between the water molecules and the nanocomposite structure, are altered with various moisture concentrations, which significantly affect the structural integrity of the epoxy/SWCNT nanocomposite, such as the swelling and plasticization to a certain degree. The material degradation under the moisture effect agrees very well with the recent experimental studies of various epoxy-based nanocomposites [8,11,13]. When focusing at the nanoscale, the molecular model can mimic the mechanism of the moisture effect by capturing the variation in the interactions between the added water molecules and the nanocomposite structure, and the formation of the water cluster in the epoxy matrix and the epoxy-SWCNT interface. The MD simulation results indicate that in the humid environment, the water molecules are absorbed into the free volume in the cross-linked structure and the epoxy-bonded interface, which interfere with the molecular interaction inside the nanocomposite structure, resulting in the degeneration of the material performance. The observations at the nanoscale provide the structural basis for the global moisture-affected structural deterioration obtained from the macroscale experiments, such as the dynamic thermal mechanical analysis of the epoxy/organoclay and epoxy/CNT nanocomposites, as well as the lap shear tests of the epoxy/CNT composite adhesive [8,11,13]. This paper uses the nanoscale simulation to explain the global structural degradation of the epoxy/CNT nanocomposite, which leads to the understanding of the relationship between the molecular interactions and the material structure and properties with various moisture concentrations.

3. Materials and Methods

In this study, the simulation approaches include the modeling of the epoxy/SWCNT nanocomposite by using Materials Studio software [57], and the structural relaxation and the dynamic deformation of the nanocomposite model in the open source code LAMMPS [58]. The interactions inside the epoxy matrix and between the epoxy and the SWCNT are described by the consistent valence force field (CVFF) [59,60], as it has been extensively used in investigating the epoxy and the epoxy/CNT nanocomposite, which yields good agreements with the theoretical and experimental results [19,20,39,61]. Meanwhile, the interactions between the carbon atoms in the SWCNT are defined with the adaptive intermolecular reactive empirical bond-order (AIREBO) potential, which is parameterized against the density functional theory calculations and experimental results [62], and its applicability in the investigation of the CNT in the polymer nanocomposite has been validated in recent studies [63,64]. The parameters of the TIP3P model are used to simulate the water molecule [65], and its bond length and angle are kept constant during the simulation by using the SHAKE algorithm [66]. The van der Waals (vdW) and the short-range Coulombic interactions between the non-bonded atoms are calculated with a cutoff distance of 10 Å, which is normally used in the investigation of the epoxy and epoxy/CNT composites [19,20,34,38,61], and the long-range Coulombic interaction is treated with the particle-particle particle-mesh (PPPM) solver [67]. Partial charges of the atoms are determined by using the bond increment method [68]. The detailed simulation method used in this work is provided in the subsequent sections.

3.1. Atomistic Models

In the modeling process, the initial epoxy/SWCNT composite structure is constructed by placing a SWCNT segment into the simulation cell, and by randomly distributing the epoxy monomers in the available space, as shown in Figure 1. As a single SWCNT segment is the basic unit of the agglomeration of the CNTs, it is chosen as the filler in the nanocomposite structure. Similar configuration of a single CNT embedded in the polymer matrix is considered as the representative model of the nanocomposite used in the simulation studies [35,36]. A total of thirty-eight epoxy monomers (7562 atoms) are packed in the periodic simulation cell with a density of 1.07 g·cm^{-3}, which is based on the commercially-available data. After that, the initial equilibration is carried out on the uncross-linked structure. Subsequently, the cross-linking process of the epoxy matrix is performed by adopting a cross-linking algorithm, which involves the structural relaxation after each cross-linking reaction, so that the epoxy matrix can get rid of the serious geometrical distortion. The detailed steps of the cross-linking process are described in previous studies and in the Supplementary Materials (Figure S7) [19,20]. The constructed epoxy matrix possesses the maximum cross-linking density of 81%, which is in the range of the synthesized polymer in the experiment [69,70]. After the cross-linking process, the nanocomposite model is equilibrated by adopting an equilibration scheme, which consists of the pressure control to significantly improve the accuracy of the achieved density [19,20]. A final equilibration process is carried out in the isothermal and isobaric (NPT) ensemble at a constant temperature of 300 K and a constant pressure of 1 atm for 8 ns. Before the equilibration is completed, the root mean squared displacement (RMSD) of the atoms remains at a constant level, which implies that the model is equilibrated properly.

3.2. Equilibration and Tensile Deformation

After the model construction, the amount of 1.0 wt% water molecules is firstly added in the model, which is then equilibrated to reach the equilibrium state. Accordingly, the model is equilibrated in the NPT ensemble for 5 ns. The fully equilibrium state is confirmed by examining the RMSD of the atoms. Based on the equilibrated structure with 1.0 wt% moisture concentration, another of

1.0 wt% water molecules is added to the model for the condition with the concentration of 2.0 wt%. A 5 ns equilibration process is carried out on the nanocomposite model. Similarly, the amount of 1.0 wt% water molecules is added to the equilibrated model successively for the condition with the concentration of 3.0 and 4.0 wt%. After the structural equilibration, the models with various moisture concentrations are subjected to the computational tensile deformation to characterize the moisture-affected mechanical behavior. During the tensile deformation, one direction of the simulation cell containing the model is deformed in a step-wise manner and the atmospheric pressure in the two transverse directions is maintained. The deformation is applied at a strain rate of $1 \times 10^8 \text{ s}^{-1}$, which is typically used for obtaining the system response within a reasonable timespan [19,20]. Each orthogonal direction is subjected to the tensile test for three times, and for all the deformation, the nanocomposite model is deformed by 3.0% in total. The reported stress-strain data are the averaged results from all the simulation runs. Furthermore, the nanocomposite models with the moisture concentrations of 0.0 and 4.0 wt% are deformed to a strain of 25.0% to quantify the moisture effect on the nanocomposite plastic deformation. The tensile deformation is carried out along the direction parallel to the SWCNT axis for three times, and the recorded data are averaged for the reported simulation results.

4. Conclusions

In this work, the moisture effect on the structure and properties of the epoxy/SWCNT nanocomposite is explored at the nanoscale by using MD simulations. The nanocomposite model is constructed with a SWCNT segment embedded in the highly cross-linked epoxy matrix, which possesses the structural characteristic in good accord with various experimental and simulation observables, and its structure and properties are affected significantly by the addition of water molecules. Due to the electrophilic nature, the added water molecule has a strong affinity for the epoxy hydroxyl groups, while its affinity for the hydrophobic SWCNT is low, and the difference in the distribution and the structure of the water molecules affects the molecular interactions inside the structure. At the low moisture concentration, the water molecules predominantly form the water-nanocomposite H-bonds and small clusters, i.e., the Type I water molecule and cluster acting as the important plasticizer and the Type II water molecule and cluster bridging several functional groups and, thus, the elastic properties of the nanocomposite only deteriorate slightly. As the moisture concentration increases, the added water molecules begin to form a large number of the water-water H-bonds and the large clusters located in the vicinity of the epoxy functional groups and in the available space close to the epoxy-SWCNT interface, which significantly disrupt the molecular interactions inside the structure and lead to the softening of the network. Furthermore, with the existence of the water molecules, they affect the interaction between the epoxy and the SWCNT, which deteriorates the nanocomposite performance at the plastic deformation regime.

The interplay between the water molecules and the structure and properties of the epoxy/SWCNT nanocomposite that is investigated in this work is useful for predicting the durability of the nanocomposite, as the polymer composite tends to absorb the moisture during the intended service life. The simulation results suggest that the water clusters at the high moisture concentration, especially those close to the epoxy-SWCNT interface, lead to the local structural swelling and the plasticization of the mechanical properties, which can deteriorate the global behavior of the nanocomposite material, and reduce its service life. Therefore, the environment humidity level should be under a constant control to prevent the excess moisture diffusion to the epoxy-based nanocomposite, especially for those used in the load-bearing applications, where the moisture-induced ageing can be more dangerous. This work shows the applicability of the MD simulations in understanding the mechanisms of the moisture-affected behaviors of the nanocomposite which has been commonly observed in the applications at larger length scales. It is envisioned that this work will be beneficial to the design, manufacturing, and engineering applications of the polymer composite with an enhanced moisture-resistant property.

Supplementary Materials: The following are available online at http://www.mdpi.com/2079-4991/7/10/324/s1, Figure S1: The stress-strain data of the nanocomposite model obtained in this study, in comparison to those of the neat epoxy obtained from the experiment and simulation [41,42], Figure S2: The radial distribution function (RDF) between the oxygen of the functional groups and the carbon of the single-walled carbon nanotube (SWCNT) with respect to the moisture concentration: (**a**) the oxygen of the hydroxyl groups; and (**b**) the ether oxygen, Figure S3: The RDF between the oxygen of the functional groups and the oxygen of the water molecule for the neat epoxy with respect to the moisture concentration [42]. The RDFs for the water-hydroxyl interactions display higher peaks than those for the water-ether interactions, as demonstrated in the enlarged picture. Meanwhile, the intensity of the peak decreases monotonically with the increasing moisture concentration for both water-functional group interactions, Figure S4: After the equilibration run, the two-dimensional distribution map of the water molecules in the plane perpendicular to the SWCNT axis with respect to the moisture concentration: (**a**) 1.0 wt%; (**b**) 2.0 wt%; (**c**) 3.0 wt%; (**d**) 4.0 wt%. The red dot denotes one atom, and the black dot denotes two atoms at the particular location. For clarity, the SWCNT is labeled by the solid circle and the cutoff distance from the SWCNT surface is labeled by the dashed circle, Figure S5: The distribution of the hydrogen bond (H-bond) probability for a given water cluster in the epoxy/SWCNT nanocomposite with respect to the moisture concentration: (**a**) 1.0 wt%; (**b**) 2.0 wt%; (**c**) 3.0 wt%; (**d**) 4.0 wt%. The single water clusters do not form the H-bond, the Type I water clusters form one water-nanocomposite H-bond, and the Type II water clusters form more than one water-nanocomposite H-bond, Figure S6: The mean squared displacement (MSD) of the water molecules as a function of the simulation time in the epoxy/SWCNT nanocomposite with respect to the moisture concentration: with the increasing concentration, the diffusion of the water molecules inside the structure is restricted, Figure S7: Procedures of the cross-linking process [19]: (**a**) the oxygens in the epoxide groups are treated as the potential reactive oxygens O (labeled in red), and the carbons in the methylene bridge of the epoxide groups are treated as the potential reactive carbons C (labeled in blue); when the distance between the O and C of different epoxide groups (indicated by the dashed line) is less than the reaction radius (indicated by the dashed circle), they are recognized as the reactive pair; (**b**) the two epoxide groups comprising the recognized reactive pair are open by removing the bond between the oxygen and the carbon in the methylene bridge of each epoxide group, respectively; (**c**) the recognized reactive pair O and C of the two open epoxide groups are connected by a newly created bond to form the cross-link between the two monomers; (**d**) the unreacted atoms in the two open epoxide groups are saturated with the hydrogen.

Acknowledgments: This work was supported by the National Natural Science Foundation of China (51608020) and by the Thousand Talents Plan (Young Professionals) in China.

Author Contributions: Lik-ho Tam performed the simulation and collected the data; and Lik-ho Tam and Chao Wu analyzed the data and wrote the manuscript.

Conflicts of Interest: The authors declare no conflict of interest.

References

1. Coleman, J.N.; Khan, U.; Blau, W.J.; Gun'ko, Y.K. Small but strong: A review of the mechanical properties of carbon nanotube-polymer composites. *Carbon* **2006**, *44*, 1624–1652. [CrossRef]
2. Liu, C. Recent developments in polymer mems. *Adv. Mater.* **2007**, *19*, 3783–3790. [CrossRef]
3. Soutis, C. Fibre reinforced composites in aircraft construction. *Prog. Aerosp. Sci.* **2005**, *41*, 143–151. [CrossRef]
4. Gay, D. *Composite Materials: Design and Applications*, 3rd ed.; CRC Press-Taylor & Francis: Boca Raton, FL, USA, 2015.
5. Campo, A.D.; Greiner, C. Su-8: A photoresist for high-aspect-ratio and 3D submicron lithography. *J. Micromech. Microeng.* **2007**, *17*, R81. [CrossRef]
6. Zhou, J.; Lucas, J.P. Hygrothermal effects of epoxy resin. Part I: The nature of water in epoxy. *Polymer* **1999**, *40*, 5505–5512. [CrossRef]
7. Prusty, R.K.; Rathore, D.K.; Ray, B.C. CNT/polymer interface in polymeric composites and its sensitivity study at different environments. *Adv. Colloid Interface Sci.* **2017**, *240*, 77–106. [CrossRef] [PubMed]
8. Kim, J.-K.; Hu, C.; Woo, R.S.; Sham, M.-L. Moisture barrier characteristics of organoclay-epoxy nanocomposites. *Compos. Sci. Technol.* **2005**, *65*, 805–813. [CrossRef]
9. Logakis, E.; Pandis, C.; Peoglos, V.; Pissis, P.; Stergiou, C.; Pionteck, J.; Pötschke, P.; Mičušík, M.; Omastová, M. Structure-property relationships in polyamide 6/multi-walled carbon nanotubes nanocomposites. *J. Polym. Sci. Part B Polym. Phys.* **2009**, *47*, 764–774. [CrossRef]
10. Prolongo, S.G.; Gude, M.R.; Ureña, A. Water uptake of epoxy composites reinforced with carbon nanofillers. *Compos. Part A Appl. Sci. Manuf.* **2012**, *43*, 2169–2175. [CrossRef]
11. Starkova, O.; Buschhorn, S.T.; Mannov, E.; Schulte, K.; Aniskevich, A. Water transport in epoxy/MWCNT composites. *Eur. Polym. J.* **2013**, *49*, 2138–2148. [CrossRef]

12. Gkikas, G.; Douka, D.D.; Barkoula, N.M.; Paipetis, A.S. Nano-enhanced composite materials under thermal shock and environmental degradation: A durability study. *Compos. Part B Eng.* **2015**, *70*, 206–214. [CrossRef]

13. Shin, P.-S.; Kwon, D.-J.; Kim, J.-H.; Lee, S.-I.; DeVries, K.L.; Park, J.-M. Interfacial properties and water resistance of epoxy and CNT-epoxy adhesives on GFRP composites. *Compos. Sci. Technol.* **2017**, *142*, 98–106. [CrossRef]

14. Hwang, J.; Ihm, J.; Lee, K.-R.; Kim, S. Computational evaluation of amorphous carbon coating for durable silicon anodes for lithium-ion batteries. *Nanomaterials* **2015**, *5*, 1654–1666. [CrossRef] [PubMed]

15. Kityk, V.I.; Fedorchuk, O.A.; Ozga, K.; AlZayed, S.N. Band structure simulations of the photoinduced changes in the MgB$_2$:Cr films. *Nanomaterials* **2015**, *5*, 541–553. [CrossRef] [PubMed]

16. Manara, M.R.; Tomasio, S.; Khalid, S. The nucleotide capture region of alpha hemolysin: Insights into nanopore design for DNA sequencing from molecular dynamics simulations. *Nanomaterials* **2015**, *5*, 144–153. [CrossRef] [PubMed]

17. Fan, Y.-C.; Fang, T.-H.; Chen, T.-H. Stress waves and characteristics of zigzag and armchair silicene nanoribbons. *Nanomaterials* **2016**, *6*, 120. [CrossRef] [PubMed]

18. Zhang, L.; Wang, X. DNA sequencing by hexagonal boron nitride nanopore: A computational study. *Nanomaterials* **2016**, *6*, 111. [CrossRef] [PubMed]

19. Tam, L.-h.; Lau, D. A molecular dynamics investigation on the cross-linking and physical properties of epoxy-based materials. *RSC Adv.* **2014**, *4*, 33074–33081. [CrossRef]

20. Tam, L.-h.; Lau, D. Moisture effect on the mechanical and interfacial properties of epoxy-bonded material system: An atomistic and experimental investigation. *Polymer* **2015**, *57*, 132–142. [CrossRef]

21. Yang, L.; Greenfeld, I.; Wagner, H.D. Toughness of carbon nanotubes conforms to classic fracture mechanics. *Sci. Adv.* **2016**, *2*, e1500969. [CrossRef] [PubMed]

22. Zeng, Q.; Yu, A.; Lu, G. Multiscale modeling and simulation of polymer nanocomposites. *Prog. Polym. Sci.* **2008**, *33*, 191–269. [CrossRef]

23. Pal, G.; Kumar, S. Modeling of carbon nanotubes and carbon nanotube-polymer composites. *Prog. Aerosp. Sci.* **2016**, *80*, 33–58. [CrossRef]

24. Chen, Z.; Gu, Q.; Zou, H.; Zhao, T.; Wang, H. Molecular dynamics simulation of water diffusion inside an amorphous polyacrylate latex film. *J. Polym. Sci. Part B Polym. Phys.* **2007**, *45*, 884–891. [CrossRef]

25. Erdtman, E.; Bohlén, M.; Ahlström, P.; Gkourmpis, T.; Berlin, M.; Andersson, T.; Bolton, K. A molecular-level computational study of the diffusion and solubility of water and oxygen in carbonaceous polyethylene nanocomposites. *J. Polym. Sci. Part B Polym. Phys.* **2016**, *54*, 589–602. [CrossRef]

26. Kim, D.-H.; Kim, H.-S. Investigation of hygroscopic and mechanical properties of nanoclay/epoxy system: Molecular dynamics simulations and experiments. *Compos. Sci. Technol.* **2014**, *101*, 110–120. [CrossRef]

27. Mijovic, J.; Zhang, H. Molecular dynamics simulation study of motions and interactions of water in a polymer network. *J. Phys. Chem. B* **2004**, *108*, 2557–2563. [CrossRef]

28. Mani, S.; Khabaz, F.; Godbole, R.V.; Hedden, R.C.; Khare, R. Structure and hydrogen bonding of water in polyacrylate gels: Effects of polymer hydrophilicity and water concentration. *J. Phys. Chem. B* **2015**, *119*, 15381–15393. [CrossRef] [PubMed]

29. Panhuis, M.I.H.; Maiti, A.; Dalton, A.B.; van den Noort, A.; Coleman, J.N.; Mccarthy, B.; Blau, W.J. Selective interaction in a polymer-single-wall carbon nanotube composite. *J. Phys. Chem. B* **2003**, *107*, 478–482. [CrossRef]

30. Chen, X.; Zhang, L.; Zheng, M.; Park, C.; Wang, X.; Ke, C. Quantitative nanomechanical characterization of the van der Waals interfaces between carbon nanotubes and epoxy. *Carbon* **2015**, *82*, 214–228. [CrossRef]

31. Lu, J.P. Elastic properties of single and multilayered nanotubes. *J. Phys. Chem. Solids* **1997**, *58*, 1649–1652. [CrossRef]

32. Nasibulin, A.G.; Pikhitsa, P.V.; Jiang, H.; Kauppinen, E.I. Correlation between catalyst particle and single-walled carbon nanotube diameters. *Carbon* **2005**, *43*, 2251–2257. [CrossRef]

33. Tian, Y.; Jiang, H.; Pfaler, J.V.; Zhu, Z.; Nasibulin, A.G.; Nikitin, T.; Aitchison, B.; Khriachtchev, L.; Brown, D.P.; Kauppinen, E.I. Analysis of the size distribution of single-walled carbon nanotubes using optical absorption spectroscopy. *J. Phys. Chem. Lett.* **2010**, *1*, 1143–1148. [CrossRef]

34. Ionita, M. Multiscale molecular modeling of SWCNTs/epoxy resin composites mechanical behaviour. *Compos. Part B Eng.* **2012**, *43*, 3491–3496. [CrossRef]

35. Jiang, C.; Zhang, J.; Lin, S.; Ju, S.; Jiang, D. Effects of free organic groups in carbon nanotubes on glass transition temperature of epoxy matrix composites. *Compos. Sci. Technol.* **2015**, *118*, 269–275. [CrossRef]

36. Arash, B.; Wang, Q.; Varadan, V.K. Mechanical properties of carbon nanotube/polymer composites. *Sci. Rep.* **2014**, *4*, 6479. [CrossRef] [PubMed]

37. Prolongo, S.G.; Campo, M.; Gude, M.R.; Chaos-Morán, R.; Ureña, A. Thermo-physical characterisation of epoxy resin reinforced by amino-functionalized carbon nanofibers. *Compos. Sci. Technol.* **2009**, *69*, 349–357. [CrossRef]

38. Khare, K.S.; Khare, R. Effect of carbon nanotube dispersion on glass transition in cross-linked epoxy-carbon nanotube nanocomposites: Role of interfacial interactions. *J. Phys. Chem. B* **2013**, *117*, 7444–7454. [CrossRef] [PubMed]

39. Alian, A.R.; Kundalwal, S.I.; Meguid, S.A. Multiscale modeling of carbon nanotube epoxy composites. *Polymer* **2015**, *70*, 149–160. [CrossRef]

40. Connolly, M.L. Solvent-accessible surfaces of proteins and nucleic acids. *Science* **1983**, *221*, 709–713. [CrossRef] [PubMed]

41. Feng, R.; Farris, R.J. The characterization of thermal and elastic constants for an epoxy photoresist SU8 coating. *J. Mater. Sci.* **2002**, *37*, 4793–4799. [CrossRef]

42. Tam, L.-H.; Lau, D.; Wu, C. Understanding the moisture effect on the cross-linked epoxy via molecular dynamics simulations. *J. Mol. Model..* under review.

43. Chakraborty, S.; Roy, S. Structural, dynamical, and thermodynamical properties of carbon nanotube polycarbonate composites: A molecular dynamics study. *J. Phys. Chem. B* **2012**, *116*, 3083–3091. [CrossRef] [PubMed]

44. Larin, S.V.; Falkovich, S.G.; Nazarychev, V.M.; Gurtovenko, A.A.; Lyulin, A.V.; Lyulin, S.V. Molecular-dynamics simulation of polyimide matrix pre-crystallization near the surface of a single-walled carbon nanotube. *RSC Adv.* **2014**, *4*, 830–844. [CrossRef]

45. Wu, C.; Xu, W. Atomistic simulation study of absorbed water influence on structure and properties of crosslinked epoxy resin. *Polymer* **2007**, *48*, 5440–5448. [CrossRef]

46. Yin, Q.; Zhang, L.; Jiang, B.; Yin, Q.; Du, K. Effect of water in amorphous polyvinyl formal: Insights from molecular dynamics simulation. *J. Mol. Model.* **2015**, *21*, 2. [CrossRef] [PubMed]

47. Tonsing, T.; Oldiges, C. Molecular dynamic simulation study on structure of water in crosslinked poly(*n*-isopropylacrylamide) hydrogels. *Phys. Chem. Chem. Phys.* **2001**, *3*, 5542–5549. [CrossRef]

48. Zhao, Z.-J.; Wang, Q.; Zhang, L.; Wu, T. Structured water and water-polymer interactions in hydrogels of molecularly imprinted polymers. *J. Phys. Chem. B* **2008**, *112*, 7515–7521. [CrossRef] [PubMed]

49. Goudeau, S.; Charlot, M.; Vergelati, C.; Müller-Plathe, F. Atomistic simulation of the water influence on the local structure of polyamide 6, 6. *Macromolecules* **2004**, *37*, 8072–8081. [CrossRef]

50. Pandiyan, S.; Krajniak, J.; Samaey, G.; Roose, D.; Nies, E. A molecular dynamics study of water transport inside an epoxy polymer matrix. *Comput. Mater. Sci.* **2015**, *106*, 29–37. [CrossRef]

51. Zhou, J.; Lucas, J.P. Hygrothermal effects of epoxy resin. Part II: Variations of glass transition temperature. *Polymer* **1999**, *40*, 5513–5522. [CrossRef]

52. Qi, D.; Hinkley, J.; He, G. Molecular dynamics simulation of thermal and mechanical properties of polyimide-carbon-nanotube composites. *Modell. Simul. Mater. Sci. Eng.* **2005**, *13*, 493. [CrossRef]

53. Mokashi, V.V.; Qian, D.; Liu, Y. A study on the tensile response and fracture in carbon nanotube-based composites using molecular mechanics. *Compos. Sci. Technol.* **2007**, *67*, 530–540. [CrossRef]

54. Sul, J.-H.; Prusty, B.G.; Kelly, D.W. Application of molecular dynamics to evaluate the design performance of low aspect ratio carbon nanotubes in fibre reinforced polymer resin. *Compos. Part A Appl. Sci. Manuf.* **2014**, *65*, 64–72. [CrossRef]

55. Liu, N.; Zeng, X.; Pidaparti, R.; Wang, X. Tough and strong bioinspired nanocomposites with interfacial cross-links. *Nanoscale* **2016**, *8*, 18531–18540. [CrossRef] [PubMed]

56. Xia, W.; Ruiz, L.; Pugno, N.M.; Keten, S. Critical length scales and strain localization govern the mechanical performance of multi-layer graphene assemblies. *Nanoscale* **2016**, *8*, 6456–6462. [CrossRef] [PubMed]

57. Accelrys Software Inc. *Materials Studio*; Accelrys Software Inc.: San Diego, CA, USA, 2009.

58. Plimpton, S. Fast parallel algorithms for short-range molecular dynamics. *J. Comput. Phys.* **1995**, *117*, 1–19. [CrossRef]

59. Maple, J.R.; Dinur, U.; Hagler, A.T. Derivation of force fields for molecular mechanics and dynamics from ab initio energy surfaces. *Proc. Natl. Acad. Sci. USA* **1988**, *85*, 5350–5354. [CrossRef] [PubMed]

60. Dauber-Osguthorpe, P.; Roberts, V.A.; Osguthorpe, D.J.; Wolff, J.; Genest, M.; Hagler, A.T. Structure and energetics of ligand binding to proteins: *Escherichia coli* dihydrofolate reductase-trimethoprim, a drug-receptor system. *Proteins Struct. Funct. Bioinform.* **1988**, *4*, 31–47. [CrossRef] [PubMed]

61. Varshney, V.; Roy, A.; Michalak, T.; Lee, J.; Farmer, B. Effect of curing and functionalization on the interface thermal conductance in carbon nanotube-epoxy composites. *JOM* **2013**, *65*, 140–146. [CrossRef]

62. Donald, W.B.; Olga, A.S.; Judith, A.H.; Steven, J.S.; Boris, N.; Susan, B.S. A second-generation reactive empirical bond order (REBO) potential energy expression for hydrocarbons. *J. Phys. Condens. Matter* **2002**, *14*, 783.

63. Vu-Bac, N.; Lahmer, T.; Zhang, Y.; Zhuang, X.; Rabczuk, T. Stochastic predictions of interfacial characteristic of polymeric nanocomposites (PNCs). *Compos. Part B Eng.* **2014**, *59*, 80–95. [CrossRef]

64. Zhang, Y.; Zhuang, X.; Muthu, J.; Mabrouki, T.; Fontaine, M.; Gong, Y.; Rabczuk, T. Load transfer of graphene/carbon nanotube/polyethylene hybrid nanocomposite by molecular dynamics simulation. *Compos. Part B Eng.* **2014**, *63*, 27–33. [CrossRef]

65. Jorgensen, W.L.; Chandrasekhar, J.; Madura, J.D.; Impey, R.W.; Klein, M.L. Comparison of simple potential functions for simulating liquid water. *J. Chem. Phys.* **1983**, *79*, 926–935. [CrossRef]

66. Ryckaert, J.-P.; Ciccotti, G.; Berendsen, H.J. Numerical integration of the cartesian equations of motion of a system with constraints: Molecular dynamics of *n*-alkanes. *J. Comput. Phys.* **1977**, *23*, 327–341. [CrossRef]

67. Hockney, R.W.; Eastwood, J.W. *Computer Simulation using Particles*; CRC Press: Boca Raton, FL, USA, 1988.

68. Oie, T.; Maggiora, G.M.; Christoffersen, R.E.; Duchamp, D.J. Development of a flexible intra- and intermolecular empirical potential function for large molecular systems. *Int. J. Quantum Chem.* **1981**, *20*, 1–47. [CrossRef]

69. Hirschl, C.; Biebl-Rydlo, M.; DeBiasio, M.; Mühleisen, W.; Neumaier, L.; Scherf, W.; Oreski, G.; Eder, G.; Chernev, B.; Schwab, W. Determining the degree of crosslinking of ethylene vinyl acetate photovoltaic module encapsulants—A comparative study. *Sol. Energy Mater. Sol. Cells* **2013**, *116*, 203–218. [CrossRef]

70. Chernev, B.S.; Hirschl, C.; Eder, G.C. Non-destructive determination of ethylene vinyl acetate cross-linking in photovoltaic (PV) modules by Raman spectroscopy. *Appl. Spectrosc.* **2013**, *67*, 1296–1301. [CrossRef] [PubMed]

nanomaterials

MDPI

Article

Hierarchical AuNPs-Loaded Fe₃O₄/Polymers Nanocomposites Constructed by Electrospinning with Enhanced and Magnetically Recyclable Catalytic Capacities

Rong Guo [1,2], Tifeng Jiao [1,2,*], Ruirui Xing [2,3], Yan Chen [2], Wanchun Guo [2], Jingxin Zhou [2,*], Lexin Zhang [2] and Qiuming Peng [1]

[1] State Key Laboratory of Metastable Materials Science and Technology, Yanshan University, Qinhuangdao 066004, China; guorong@stumail.ysu.edu.cn (R.G.); pengqiuming@ysu.edu.cn (Q.P.)
[2] Hebei Key Laboratory of Applied Chemistry, School of Environmental and Chemical Engineering, Yanshan University, Qinhuangdao 066004, China; rrxing@ipe.ac.cn (R.X.); chenyan@ysu.edu.cn (Y.C.); wc-g@ysu.edu.cn (W.G.); zhanglexin@ysu.edu.cn (L.Z.)
[3] State Key Laboratory of Biochemical Engineering, Institute of Process Engineering, Chinese Academy of Sciences, Beijing 100190, China
*   Correspondence: tfjiao@ysu.edu.cn (T.J.); zhoujingxin@ysu.edu.cn (J.Z.); Tel.: +86-335-8056854 (T.J.); +86-335-8061569 (J.Z.)

Received: 1 October 2017; Accepted: 10 October 2017; Published: 12 October 2017

Abstract: Gold nanoparticles (AuNPs) have attracted widespread attention for their excellent catalytic activity, as well as their unusual physical and chemical properties. The main challenges come from the agglomeration and time-consuming separation of gold nanoparticles, which have greatly baffled the development and application in liquid phase selective reduction. To solve these problems, we propose the preparation of polyvinyl alcohol(PVA)/poly(acrylic acid)(PAA)/Fe₃O₄ nanocomposites with loaded AuNPs. The obtained PVA/PAA/Fe₃O₄ composite membrane by electrospinning demonstrated high structural stability, a large specific surface area, and more active sites, which is conducive to promoting good dispersion of AuNPs on membrane surfaces. The subsequently prepared PVA/PAA/Fe₃O₄@AuNPs nanocomposites exhibited satisfactory nanostructures, robust thermal stability, and a favorable magnetic response for recycling. In addition, the PVA/PAA/Fe₃O₄@AuNPs nanocomposites showed a remarkable catalytic capacity in the catalytic reduction of p-nitrophenol and 2-nitroaniline solutions. In addition, the regeneration studies toward p-nitrophenol for different consecutive cycles demonstrate that the as-prepared PVA/PAA/Fe₃O₄@AuNPs nanocomposites have outstanding stability and recycling in catalytic reduction.

Keywords: Au nanoparticles; composite materials; catalytic reduction; electrospinning; p-nitrophenol

---

1. Introduction

Au has long been considered to be invaluable precious metals; this did not change until 1973, when Bond et al. revealed the potential application of small-sized Au in hydrogenation reactions [1]. Haruta and Hutchigns et al. in 1987 found that Au nanoparticle catalysts with a size of about 5 nm have good activity in catalyzing oxidation of CO and the reaction of acetylene to vinyl chloride, respectively [2,3]. After that, more and more attention was paid to the nanoscale Au catalysts due to their unusual physical and chemical properties for a variety of catalytic reactions [4–8]. Moreover, in addition to the excellent performances in CO low-temperature oxidation [9–11], the epoxidation of propylene [12–14], and water gas shift reactions [15–17], Au nanoparticles (AuNPs) show

Nanomaterials **2017**, *7*, 317

outstanding catalytic ability in liquid phase selective oxidation [18–20] and selective reduction [21–23]. However, the development of applications in liquid phase selective reduction of AuNPs catalysts have been critically restricted because the massive agglomeration of AuNPs results from their high surface energy and strong van der Waals attraction [24], so the catalytic activity shows a foreseeable sharp decrease in the liquid selective catalytic reduction system. In addition, the significant disadvantages of nanoscale AuNPs are their time-consuming separation [25], which provides an obstruction to facile catalyst recovery and recycling. Once the AuNPs catalyst is applied to industrial practical applications, the separation of AuNPs from the catalytic reaction system requires a faster approach. Considering the above problems, AuNPs immobilization on solid supports is regarded as a conventional and feasible method [26–31]. Chairam et al. synthesized mung bean starch-AuNPs composite, which acted as both the reducing and stabilizing agents [32]. Zhu et al. immobilized AuNPs on a 2D graphene oxide/SiO$_2$ hybrid, showing excellent dispersion and catalytic performance [33]. Kuroda et al. directly deposited AuNPs on poly(methyl methacrylate) beads and the average diameter was 6.9 nm [34]. Zhang et al. obtained Au nanostructures/GO nanocomposites, also exhibiting good catalytic activity by using tannic acid as a reducing and immobilizing agent [35]. Ye et al. synthesized reduced graphene oxide wrapped by polydopamine on which the Pt–Au dendrimer-like nanoparticles were loaded [36]. The nanocomposites exhibit higher catalytic activity, which is substantially affected by Pt-to-Au molar ratios and a superior efficiency for the purification of water containing 4-nitrophenol. Jin et al. coated conducting polymer polyaniline (PANI) on SiO$_2$ templates assembled by Fe$_3$O$_4$ and Au nanoparticles and fabricated Au@Fe$_3$O$_4$@PANI hybrid shells followed by the removal of the SiO$_2$ template [37]. This structure has high stability, recyclability, and largely improves the catalytic activity toward the reduction of 4-nitrophenol.

On the other hand, electrospinning technology can produce continuous fibers with micro/nanoscale diameters, which have drawn wide interest in recent decades by using a suspended droplet of polymer solution or melt at high voltage [38–40]. The electrospun fibers have many outstanding merits, such as good specific surface area [41,42], favorable porosity [43], and great flexibility [44,45], as well as remarkable controllable thickness and diverse architecture [46,47]. Therefore, on the basis of the research of solid supports and many interesting advantages of electrospun fibers, we devote our effort to solving the agglomeration and separation of AuNPs on the premise of guaranteeing small size and high activity. The as-prepared PVA/PAA/Fe$_3$O$_4$ membranes were neatly synthesized by taking advantage of electrospinning, while the in situ Au nanoparticles from the HAuCl$_4$ and NaBH$_4$ solution are firmly immobilized on the surface of the nanofibers with the aid of hydrogen bonds. Better specific surface areas and more active sites in the obtained electrospinning membrane promote better dispersion of AuNPs on the surface of the membranes. Thus, the possibility of agglomeration of AuNPs is enormously declined and the stability of catalysts during the catalytic reduction process is constantly in good condition. In addition, Fe$_3$O$_4$ nanoparticles can contribute to the magnetic recyclability of the nanocomposite membrane in the liquid reaction system, which seems helpful in terms of solving the problems of separation and recovery of the PVA/PAA/Fe$_3$O$_4$@AuNPs catalyst. Moreover, the preparation process of solid supports via electrospinning is highly eco-friendly and easy to operate and regulate, which reflects the dominant position of this nanocomposite in potential large-scale applications of selective catalytic reduction of gold catalysts. Compared to the previous literature summarized in Table 1 [24,32,33,48–51], our PVA/PAA/Fe$_3$O$_4$@AuNPs nanocomposites have the advantages of high activity, high stability, and recyclability, which is crucial to the performance evaluation of catalysts. Moreover, presently prepared nanocomposites also have the characteristics of low cost, easy preparation, and environmental friendliness, demonstrating important and potential applications in catalysis fields.

Table 1. Comparative characteristics and catalytic performance of catalyzers in the reported literature.

No.	Catalyzer	Catalytic Performance $\ln(C_t/C_0)$ min^{-1}	Preparation Method	Characteristics
1	Au@CPF-1 hybrid [24]	0.303	AuNPs synthesized on the activated CPF-1.	Complexed and costly preparation.
2	Starch-supported gold nanoparticles [32]	-	Mix HAuCl$_4$ and MBS in DI water.	Weak reducibility of polysaccharides, weak catalytic activity, simple process, and environmentally friendly.
3	Graphene oxide/SiO$_2$/AuNPs hybrid nanomaterials [33]	1.04	Graphene oxide/SiO$_2$ via a sol–gel process, activated by SnCl$_2$, mixed with HAuCl$_4$.	Remarkable catalytic capacity, accompanying adsorption process, inconvenient preparation process.
4	TiO$_2$/ZnO/AuNF nanofibers [48]	-	Calcined electrospinning nanofibers, SnCl$_2$ activated, adding HAuCl$_4$ solution.	Toxic solvent in preparation, unfriendly to environment.
5	Fe$_3$O$_4$@TiO$_2$@Ag–Au microspheres [49]	0.1148	3-Aminopropyltrimethoxysilane modified Fe$_3$O$_4$@TiO$_2$ microspheres, Ag nanoparticles replacement, Ag–Au bimetallic nanostructures.	Complexed replacement of Au/Ag, weak catalytic activity.
6	Au/Fe$_3$O$_4$@hollow TiO$_2$ nanoreactor [50]	0.46	AuNPs loaded on magnetic SiO$_2$ nanospheres, Fe$_3$O$_4$ modified, covered with TiO$_2$ shell.	Impacted catalytic capacity due to the coverage and isolation of the TiO$_2$ shell.
7	Double-shelled sea urchin-like yolk-shell Fe$_3$O$_4$/TiO$_2$/Au microspheres [51]	1.84	Synthesis of Fe$_3$O$_4$/SiO$_2$/TiO$_2$ core-shell microspheres by sol–gel process, SiO$_2$ shell removed by acid post-treatment, AuNPs loaded.	Remarkable catalytic performance, complexed preparation, negative effect in acid post-treatment.
8	Present work	0.441	AuNPs-loaded, magnetically Fe$_3$O$_4$ support by electrospinning.	Eco-friendly prepared process, high stability, and good catalytic performance.

2. Materials and Methods

2.1. Materials

Polyvinyl alcohol (PVA, 98–99% hydrolyzed, average M.W. 57,000–66,000), poly(acrylic acid) (PAA, M.W. ~2000) and ferric chloride hexahydrate (FeCl$_3$·6H$_2$O, 98%) was purchased from Aladdin Reagent (Shanghai, China). Anhydrous sodium acetate was supplied by Guangzhou Guanghua Chemical Reagent Co. Ltd. (Guangzhou, China). Anhydrous ethanol and ethylene glycol was acquired from the Tianjin Guangfu Fine Chemical Research Institute (Tianjin, China). Chloroauric acid tetrahydrate (HAuCl$_4$·4H$_2$O), sodium borohydride (NaBH$_4$), 2-nitroaniline (2-NA), and 4-nitrophenol (4-NP) were purchased from Alfa Aesar (Beijing, China). Ultra-pure water was obtained through a Milli-Q Millipore filter system (Millipore Co., Bedford, MA, USA) with a resistivity of 18.2 MΩ cm^{-1}. All chemicals were used as received without further purification.

2.2. Preparation of Electrospun Composites

The 5 g of a 10% aqueous PVA solution was stirred for 8 h at 80 °C. Subsequently, the PVA solution was mixed with 2 g of a 30 wt % aqueous PAA solution, and stirred overnight until the solution was as homogeneous as possible. The volume ratio of the aqueous PVA and PAA solution was 5:2, referring to the previous literature [52]. The Fe$_3$O$_4$ nanoparticles were prepared according to the reference report [53]. As shown in Figure S1, the diameter of Fe$_3$O$_4$ nanoparticles range from 200 to 300 nm with a large number of carboxyl groups on the surface. Then, Fe$_3$O$_4$ nanoparticles (50 mg) were added to a homogeneous aqueous PVA and PAA mixture solution (7 g) and stirred to obtain a well-dispersed solution. The electrospinning precursor solution was held in a 10 mL syringe with the stainless steel needle (20G). During electrospinning, the flow rate was delivered at 0.5 mL·h^{-1}, and an aluminum foil was applied as the collector. In addition, the potential difference between the polymer solution and the collector was 20 kV and the distance was 15 cm from the point of needle to collector. After that, the obtained PVA/PAA@Fe$_3$O$_4$ film sample was dried in a vacuum drying oven at 120 °C for 3 h for heat-induced crosslinking reaction between carboxyl acid groups in PAA and hydroxyl groups in PVA molecules. Aqueous HAuCl$_4$ solution (250 μM, 10 mL) and NaBH$_4$ aqueous solution (0.01 M, 12 mL) was mixed in a beaker with simultaneous vigorous stirring. Apparently, the color of the mixed solution turned red, which means that Au nanoparticles were generated with a pH value of 6.28. Furthermore, excess NaBH$_4$ molecules were removed by centrifugation (8000 rpm, 10 min) and washed with ultrapure water three times. PVA/PAA/Fe$_3$O$_4$ electrospun film was immersed in an AuNPs solution (50 mL) for an hour in room temperature. After that, the PVA/PAA/Fe$_3$O$_4$@AuNPs nanocomposites were washed by ultrapure water several times and dried and stored at room temperature for further use.

2.3. Catalytic Performance Test

The evaluation of catalytic performance of PVA/PAA/Fe$_3$O$_4$@AuNPs electrospun membrane was executed by catalytic reduction of aqueous 4-NP and 2-NA solution [54]. NaBH$_4$ was used as a reducing agent for this catalytic reduction reaction, and all the progress was under the monitoring by UV-VIS spectroscopy at room temperature (Figure 1). The PVA/PAA/Fe$_3$O$_4$@AuNPs electrospun membrane (10 mg) was added in an aqueous 4-NP solution (10 mL, 0.005 M), followed by adding fresh aqueous NaBH$_4$ solution (20 mL, 0.1 M) rapidly. The absorbance was monitored every 3 min by UV-VIS spectroscopy until the solution became colorless. After that, the sample was removed by external magnetic field and washed with ethanol and ultra-pure water for several times. The catalysis of aqueous 2-NA solution (10 mL, 0.005 M) was also applied to evaluate the catalytic capacity of PVA/PAA/Fe$_3$O$_4$@AuNPs electrospun membrane. In order to further characterize the recycling capacity, the sample catalyzed new aqueous 4-NP and NaBH$_4$ mixture solutions 10 times.

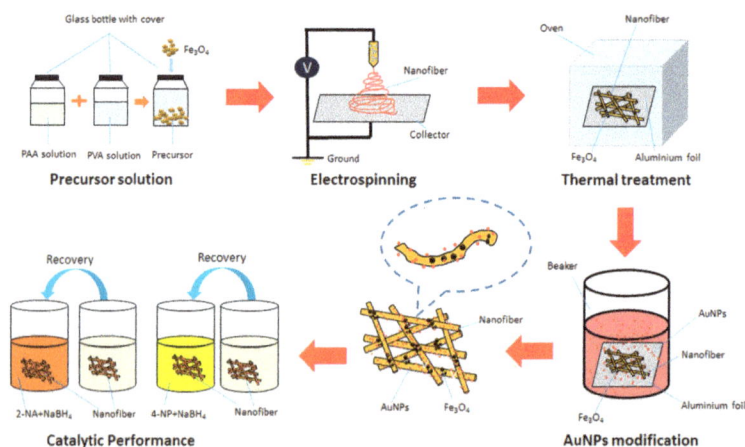

Figure 1. Schematic illustration of the preparation of the PVA/PAA/Fe$_3$O$_4$@AuNPs composite membrane by electrospinning and its catalytic performance.

2.4. Characterization

The microstructure was characterized via scanning electron microscope (SEM) Field Emission Gun FEI QUANTA FEG 250 (FEI Corporate, Hillsboro, OR, USA) with energy dispersive spectroscopy (EDS) for qualitative chemical analysis. All samples have been coated with AuNPs or carbon before SEM measurement. Transmission electron microscopy (TEM, HT7700, High-Technologies Corp., Ibaraki, Japan) was also used to further characterize the obtained samples. High-resolution transmission electron microscopy (HRTEM, Tecnai-G^2 F30 S-TWIN, Philips, Netherlands) were used to observe the morphologies and microstructures of the samples. X-ray diffraction (XRD) analysis was performed on an X-ray diffractometer equipped with a Cu Kα X-ray radiation source and a Bragg diffraction setup (SMART LAB, Rigaku, Akishima, Japan). Thermogravimetry (TG) characterizations were carried out using a NETZSCH STA 409 PC Luxx simultaneous thermal analyzer (Netzsch Instruments Manufacturing Co, Ltd, Seligenstadt, Germany) in an argon gas atmosphere. FT-IR spectra were obtained by Fourier infrared spectroscopy (Thermo Nicolet Corporation, Madison, WI, USA) via the KBr tablet method. X-ray photoelectron spectroscopy (XPS) was measured on an ESCALAB 250Xi XPS (Thermo Fisher Scientific, San Jose, CA, USA) using 200 W monochromated Al Kα radiation. Both survey scans and individual high-resolution scans for characteristic peaks were recorded. The substrate used for XPS testing is a Si plate purchased from Aladdin Reagent (Shanghai, China). The magnetization was measured by a superconducting quantum interference device (SQUID) magnetometer (MPMS-XL, Quantum Design Inc., San Diego, CA, USA) at 300 K.

3. Results and Discussion

3.1. Characterization of Nanocomposites

Firstly, Figure 1 illustrates the scheme for the preparation of PVA/PAA/Fe_3O_4@AuNPs composite membrane. A high-viscosity polymer solution is the key to the success of electrospinning without considering the influence of voltage and other factors. Here, the use of PVA and PAA for electrospinning is proposed based on the following considerations: The selected PVA and PAA reagents with different molecular weights and volume ratios can well form proper spinning solution with suitable viscosity, concentration, and surface tension. In addition, the crosslinking reaction that occurs between PVA and PAA is effective for further application of the obtained electrospinning membranes. According to Figure 1, the PVA and PAA were dissolved in ultra-pure water and magnetically stirred, and Fe_3O_4 nanoparticles were then added. The homogeneous yellow precursor solution was held in a 10 mL syringe with the stainless steel needle (type of 20G) and the PVA/PAA/Fe_3O_4 nanocomposites were obtained by electrospinning and dried in a vacuum oven. Due to all of the weighted Fe_3O_4 nanoparticles added to the PVA/PAA mixed solution to prepare composite films, we speculated that Fe_3O_4 nanoparticles are all in the nanocomposites with complexation efficiency near 100%. After that, PVA/PAA/Fe_3O_4 nanocomposite membranes were immersed in a red Au nanoparticle-containing solution. The synthesized AuNPs in aqueous solution have many hydroxyl groups on the surface of particles. In addition, the environment of the AuNPs aqueous solution is neutral, so hydrogen bonds can be expected to form. In addition, there are large numbers of carboxyl groups in the PAA molecules. The nanofiber membranes also have many excess carboxyl groups on the surface. Thus, AuNPs with many hydroxyl groups on the surface can easily load on the surface of prepared nanofibers mainly due to hydrogen bond interaction. The data of Fourier Transform Infrared Spectoscopy (FT-IR) in Figure S2 can also verify the characteristic chemical groups in the obtained composite membranes. The designed PVA/PAA/Fe_3O_4@AuNPs nanocomposites were thus obtained.

Figure 2 depicts the morphology of the obtained nanocomposites. The size and nanostructure of Fe_3O_4 nanoparticles can be seen in Figure S1. PVA/PAA nanofibers and PVA/PAA/Fe_3O_4 nanofibers have been coated with AuNPs (1–3 nm) before SEM measurement due to organic composites with poor electroconductivity [55–60]. While PVA/PAA/Fe_3O_4@AuNPs nanofibers have been coated with carbon in order to perform Fe/Au elemental mapping and investigate the presence and localization of Fe_3O_4 and AuNPs. The PVA/PAA electrospun fibers present long, straight, and uniform fiber nanostructures with the average diameter of 300 nm according to SEM in Figure 2a. The formed ternary PVA/PAA/Fe_3O_4 nanocomposite membranes also have long and straight nanostructures with substantial nanoparticles on the surface and interior space of the fiber, as is shown in Figure 2b. The carboxyl groups on the surface of Fe_3O_4 nanoparticles can combine with some hydroxyl groups of PVA molecules. After heat treatment, the prepared fibers became insoluble due to a thermal crosslinking reaction. The diameters of each fiber of PVA/PAA/Fe_3O_4 nanocomposites show little differences. In addition, the Fe and Au elemental mapping of PVA/PAA/Fe$_3$O$_4$@AuNPs nanocomposites have been performed and are shown in Figure 2c–e. We can clearly find that a large number of Fe$_3$O$_4$ nanoparticles and AuNPs are well distributed onto the obtained composites fibers. In addition, the images of TEM of all samples have been also measured and are shown in Figure 3. Both PVA/PAA fibers and PVA/PAA/Fe$_3$O$_4$ membranes exhibit long straight fiber nanostructures with Fe$_3$O$_4$ nanoparticles introduced to the nanofiber skeleton, shown in Figure 3a,b. The diameter of the obtained AuNPs ranges from 5 to 10 nm with a mellow shape [61,62], as shown in Figure 3c. The interplanar spacing of Au nanoparticle is 0.2347 nm, which can well match with the (111) crystal surface of Au. Moreover, the Fe/Au elemental mapping of PVA/PAA/Fe$_3$O$_4$@AuNPs nanofibers in Figure 3d further confirm the presence and the good distribution of Fe$_3$O$_4$ and AuNPs in the obtained composite fiber. It can be reasonably speculated that hydrophilic AuNPs successfully load on the surface of PVA/PAA/Fe$_3$O$_4$ fibers via intermolecular hydrogen bonds, which can be expected to exert catalytic activity and good stability in the next recovery and reuse process.

Thermogravimetry (TG) curves of samples were measured under an argon atmosphere, as shown in Figure 4. They were performed to measure the thermal stability of the prepared nanocomposites [63,64]. The weight losses below 150 °C can be regarded as the removal of absorbed water, while from 280 to 500 °C, the sharp loss of weight could be attributed to the thermal decomposition of the carbon skeleton in the PVA and PAA molecules. Above 500 °C, the weight values remain stable. Additionally, it was demonstrated that the PVA/PAA/Fe$_3$O$_4$@AuNPs nanocomposites have better thermal stability. In addition, the weight loss of the PVA/PAA nanofibers was approximately 84.5 wt %, while the PVA/PAA/Fe$_3$O$_4$ and PVA/PAA/Fe$_3$O$_4$@AuNPs nanocomposites lost 81.5 and 79 wt %, respectively. The difference in weight loss can be reasonably explained by the incorporation of Fe$_3$O$_4$ nanoparticles and AuNPs in nanocomposites.

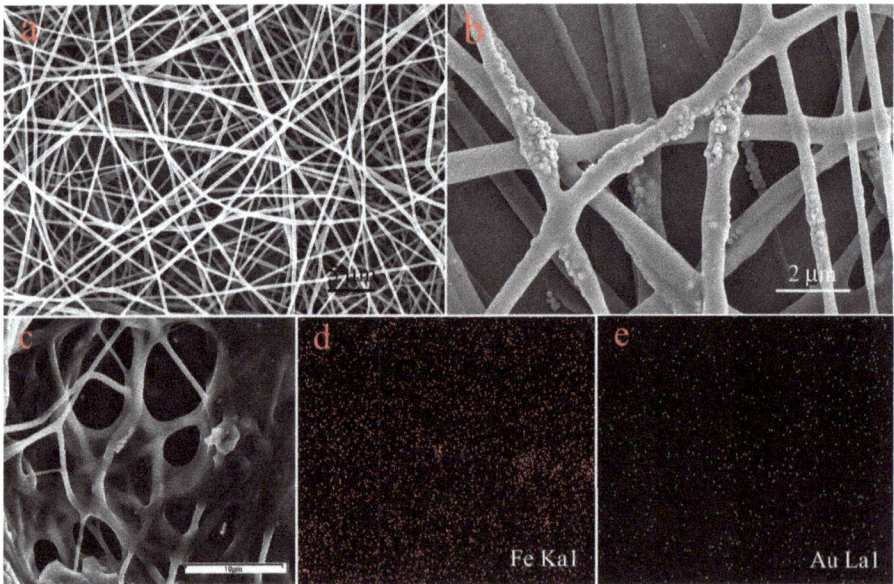

Figure 2. SEM images of the prepared PVA/PAA nanofiber (**a**), PVA/PAA/Fe$_3$O$_4$ nanofiber (**b**), PVA/PAA/Fe$_3$O$_4$@AuNPs nanofibers with coated carbon (**c**), and Fe/Au elemental mapping (**d**,**e**).

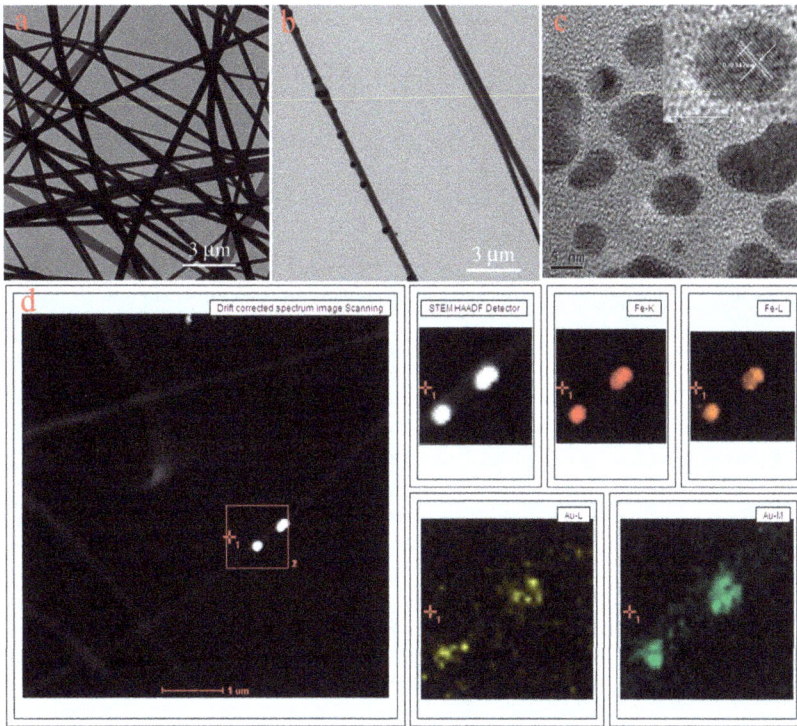

Figure 3. TEM images of the prepared PVA/PAA nanofibers (**a**), PVA/PAA/Fe$_3$O$_4$ nanofiber (**b**), high resolution of AuNPs (**c**), and PVA/PAA/Fe$_3$O$_4$@AuNPs nanofibers with Fe/Au elemental mapping (**d**).

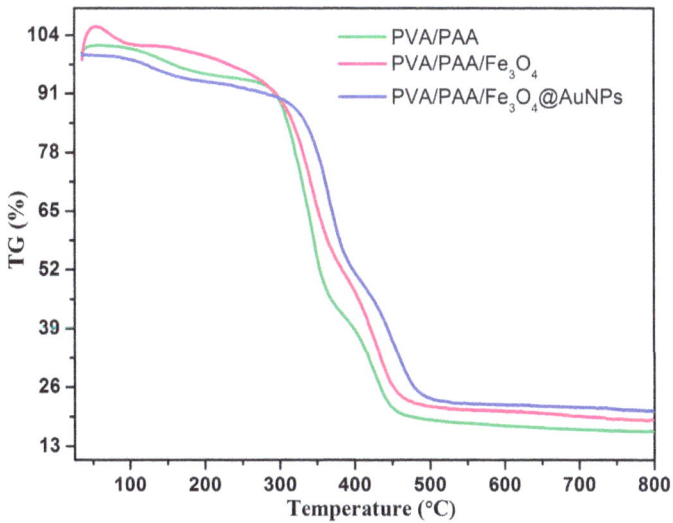

Figure 4. TG curves of PVA/PAA, PVA/PAA/Fe$_3$O$_4$, and PVA/PAA/Fe$_3$O$_4$@AuNPs nanocomposites.

XRD data was also measured to further identify the structure of the membrane, as shown in Figure 5. According to the obtained results, the characteristic absorption peaks of 2θ at 30.0°, 35.3°, 43.0°, 57.0°, and 62.7° can be assigned to the (220), (311), (400), (511), and (400) planes of the face-centered cubic Fe_3O_4 phase. In addition, the $PVA/PAA/Fe_3O_4$ and $PVA/PAA/Fe_3O_4@AuNPs$ nanocomposites both have the same characteristic peaks, which indicates the introduction of Fe_3O_4 nanoparticles in the nanocomposites. The XRD pattern of the $PVA/PAA/Fe_3O_4@AuNPs$ nanocomposite, compared to the XRD patterns of the PVA/PAA nanofibers and $PVA/PAA/Fe_3O_4$ nanocomposites, indicates newly emerging diffraction peaks with 2θ values of 38.9° and 46.1°, which are indexed to the (111) and (200) cubic lattice planes of gold nanoparticles. Similar results about diffraction peaks of AuNPs have been reported in previous reports [54,65]. The signals in the XRD measurements of $PVA/PAA/Fe_3O_4@AuNPs$ nanocomposites are slightly weak mainly due to the thin film state of the nanocomposites containing fewer AuNPs and Fe_3O_4 particles in the measurement process.

In order to verify the XRD spectra and TG results, the composition analysis of the as-prepared $PVA/PAA/Fe_3O_4@AuNPs$ nanocomposite membrane was performed via X-ray photoelectron spectroscopy (XPS), as shown in Figure 6. The survey data demonstrate the characteristic peaks such as C1s, O1s and Au4f in Figure 6a. The Si2p peak came from the Si plate as a substrate [58,66]. In addition, there is a pair of typical spin splitting peaks of Au4f in the spectra with binding energies of 82.5 and 86.3 eV, and the distance between the two characteristic peaks is 3.8 eV, which can be assigned to the $4f_{5/2}$ and $4f_{7/2}$ lines of metallic gold. This is slightly different from the results of a previous study [67] because, in comparison with the main Au^0 species (accounting for 90.4%), there are only 9.6% Au^+ species resulting from residual $HAuCl_4$ molecules that are not completely restored in situ to Au nanoparticles. The peaks located at around 83.5 and 87.0 eV correspond to the spin orbit splitting components of Au4f. Combined with the above characterization, these results represent the Au^0 species that have been successfully incorporated on nanofibers.

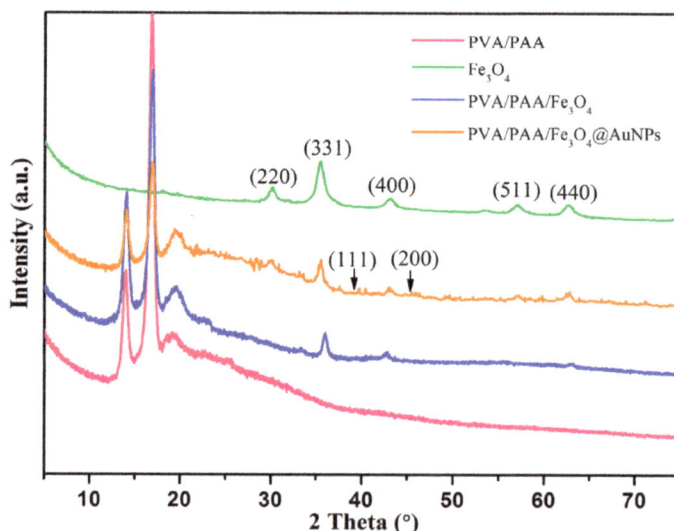

Figure 5. XRD patterns of the obtained PVA/PAA electrospun nanofibers, $PVA/PAA/Fe_3O_4$ nanofibers, $PVA/PAA/Fe_3O_4@AuNPs$ nanocomposites, and Fe_3O_4 nanoparticles.

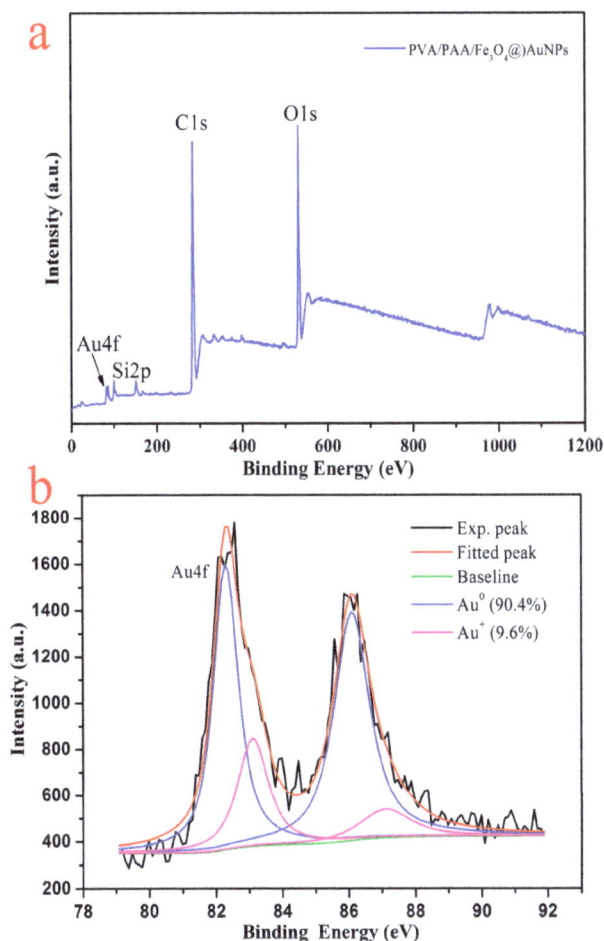

Figure 6. Survey XPS spectra of PVA/PAA/Fe$_3$O$_4$@AuNPs nanocomposites (**a**) and the deconvolution of XPS peaks of the Au4f region (**b**).

Magnetization hysteresis loops, as shown in Figure 7, are further collected to investigate the magnetic performance. The completely reversible field-dependent magnetization curves mean that all of the samples are super-paramagnetic. The saturation magnetization value of Fe$_3$O$_4$ nanoparticles, PVA/PAA/Fe$_3$O$_4$, and PVA/PAA/Fe$_3$O$_4$@AuNPs nanocomposite membranes are 78.5, 40.0 and 32.5 emu/g, respectively. Due to substantial non-magnetic substance of PVA and PAA molecules as fiber skeleton, the saturation magnetization values have significantly reduced. In addition, compared to the PVA/PAA/Fe$_3$O$_4$ nanocomposites, the clear decrease in magnetic response indirectly indicates the incorporation of non-magnetic substance AuNPs into the PVA/PAA/Fe$_3$O$_4$@AuNPs membrane. Although there is obviously a loss of saturation magnetization, this magnetic response can still ensure controllable magnetic recoveries, which shows its great importance in terms of the application of catalytic materials.

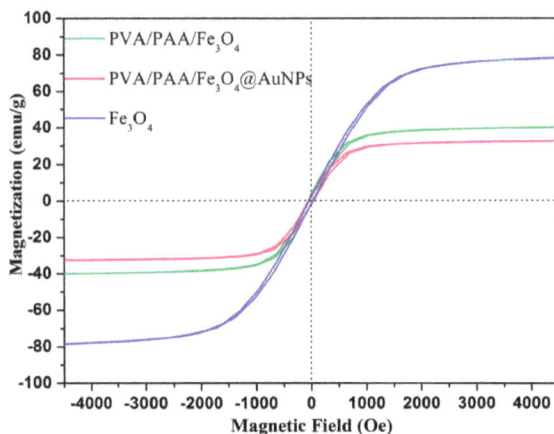

Figure 7. Magnetization hysteresis loops of the obtained PVA/PAA/Fe₃O₄ nanocomposites, PVA/PAA/Fe₃O₄@AuNPs composites, and Fe₃O₄ nanoparticles.

3.2. Catalytic Reduction Performances

The catalytic reduction of 4-NP and 2-NA was carried out to investigate the catalytic activity of the PVA/PAA/Fe₃O₄@AuNPs nanocomposite membrane. The 4-NP solution had a strong characteristic peak at 317 nm, as shown in Figure 8a. After the NaBH₄ solution was added, NaBH₄ molecules provide negative hydrogen ions to attack 4-NP, and the resultant of the reaction was 4-nitorphenolate. The redshift of the characteristic absorption peak at 402 nm can prove the formation of 4-nitrophenolate. The conversion of 4-NP to the 4-nitrophenolate ion takes place within seconds with the help of excess NaBH₄ solution, but further reduction does not progress even over 24 h. After the prepared PVA/PAA/Fe₃O₄@AuNPs nanocomposites were added, the catalytic reaction began and the time was recorded. Then, with the catalytic reaction of composite materials, the nitro group of 4-nitrophenolate was reduced to amino groups with the catalysis of AuNPs, so the adsorption intensity of 4-nitorphenolate decreased. Thus, the visual performance was the descended sharply of characteristic absorption peaks at 402 nm, as shown in Figure 8b. In addition, it is clear that the bright yellow mixed solution became colorless, as shown in Figure 9b. In addition, the catalytic reduction of the 2-NA solution was also applied to further demonstrate the catalytic activity of nanocomposite membrane. No significant changes of 2-NA solution in the color and characteristic absorption peak at 415 nm were observed before or after adding aqueous NaBH₄ solution for 24 h, as shown in Figure 8d. After the PVA/PAA/Fe₃O₄@AuNPs catalyst was added, the absorption band of 2-NA clearly decreased and the system solution became colorless, which demonstrates that this catalyst also exhibits high catalytic activity.

In addition, the PVA/PAA/Fe₃O₄@AuNPs nanocomposites were easily separated by an external magnetic field (Figure 9a), which also validates the previous magnetic measurements. The reaction of the reduction of 4-NP was assumed to be pseudo-first-order kinetics since the concentration of NaBH₄ was significantly higher than that of 4-NP and can be considered constant. As shown in Figure 9c, the linear correlation between $\ln(C_t/C_0)$ and the reaction time (t) confirms the pseudo-first order kinetics of this reaction. C_t and C_0 are the concentrations of 4-NP at time t and the time of the initial concentration, respectively. The pseudo-first-order reaction rate constant (k) was calculated to be 0.441 min^{-1} for the reduction of 4-NP. In order to further study the stability and catalytic activity of the PVA/PAA/Fe₃O₄@AuNPs catalyst, the nanocomposites were allowed to continuously proceed to catalyze a fresh 4-NP and NaBH₄ system eight times to evaluate the recyclable properties, as summarized in Figure 9d. As expected, after eight reductions of 4-NP, the conversion still maintained high catalytic activity and reached a value of 92%. Compared to the first reduction process, the slight decrements of conversion demonstrate excellent stability of PVA/PAA/Fe$_3$O$_4$@AuNPs composite membrane. In addition, the SEM and TEM images with Fe/Au elemental mapping of nanocomposites after the eighth cycle of catalytic reactions are also demonstrated in Figure 10. It can be easily observed that the PVA/PAA/Fe$_3$O$_4$@AuNPs nanocomposites can basically retain the original nanostructure, demonstrating that the prepared composite materials are remarkably stable. After repeated washing and drying in the reuse process, the slightly deformed membrane composite materials still maintained high catalytic performances. Moreover, Fe$_3$O$_4$ nanoparticles still firmly immobilize inside the membrane, which guarantees magnetic performance and recyclability. Thus, the prepared nanocomposites have outstanding stability and demonstrate great potential application in catalysis fields.

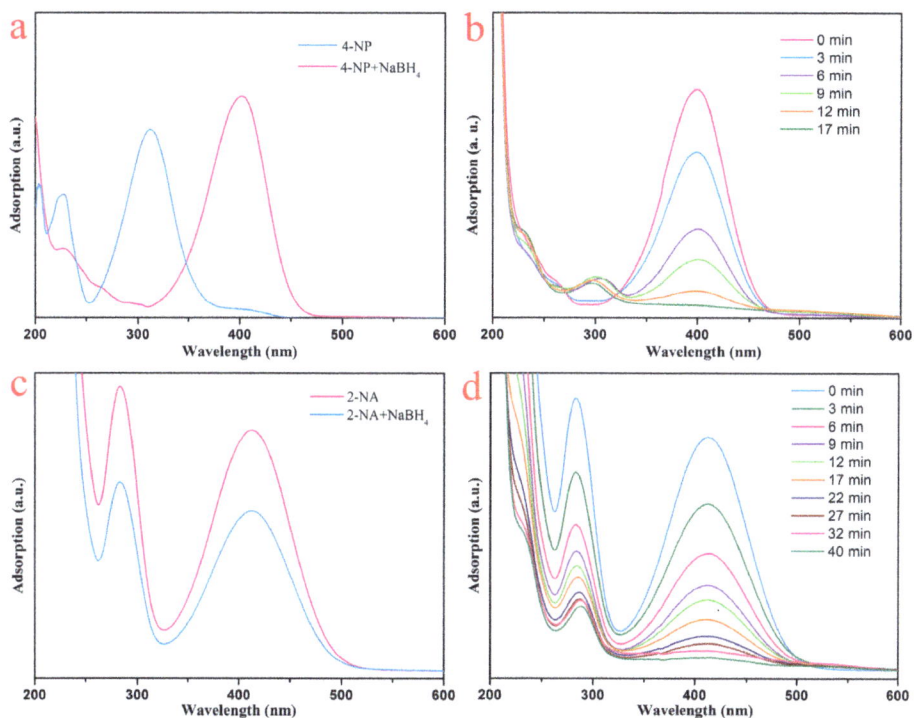

Figure 8. Catalytic reduction of (**a**) 4-NP before and after adding NaBH$_4$ aqueous solution; (**b**) reduction of 4-NP with PVA/PAA/Fe$_3$O$_4$@AuNPs composite; (**c**) 2-NA before and after adding NaBH$_4$ aqueous solution; (**d**) reduction of 2-NA with PVA/PAA/Fe$_3$O$_4$@AuNPs composite.

Such a good catalytic performance of the PVA/PAA/Fe$_3$O$_4$@AuNPs membrane benefits from the use of the electrospun membrane as a support for the gold catalyst. In addition, the loaded AuNPs incorporated on the electrospun PVA/PAA/Fe$_3$O$_4$ composite membrane show a well-dispersed state, which helps to avoid agglomeration and improve catalytic performances. It should be noted that easy aggregations between AuNPs prevent widespread applications. In recent years, various structures and composites with AuNPs have been designed and investigated, as listed in Table 1. In our present system, the PVA/PAA/Fe$_3$O$_4$@AuNPs nanocomposites demonstrate nanostructures with an eco-friendly prepared process and superior catalytic properties, as well as magnetically recyclable capacities, suggesting wide catalytic applications.

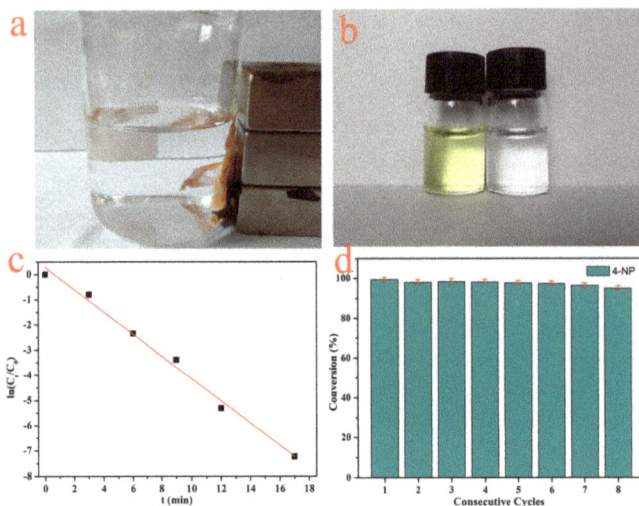

Figure 9. Magnetic recovery of PVA/PAA/Fe$_3$O$_4$@AuNPs nanocomposites with external magnetic field (**a**); comparison of 4-NP solution before and after catalytic reaction (**b**); the relationship between $\ln(C_t/C_0)$ and the reaction time (*t*) of the nanocomposite catalyst (**c**); the reusability test of PVA/PAA/Fe$_3$O$_4$@AuNPs nanocomposites as catalysts for the reduction of 4-NP (**d**).

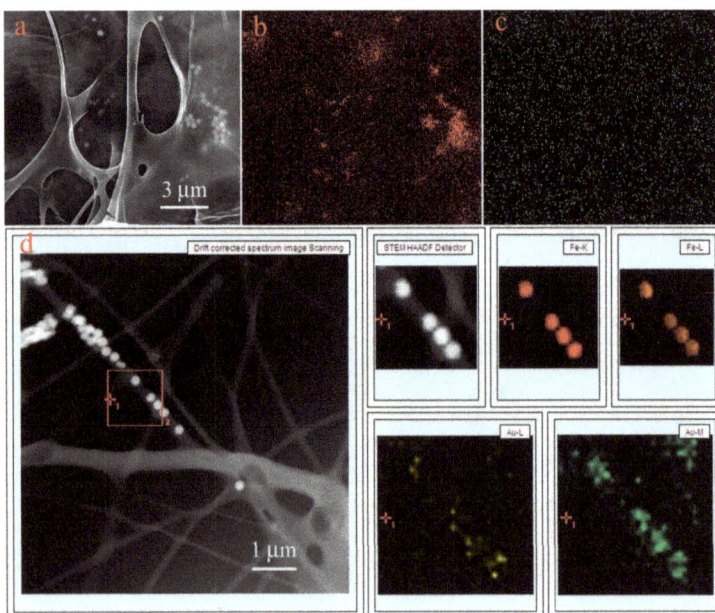

Figure 10. The SEM image (**a**) with Fe/Au elemental mapping (**b,c**) and TEM image with Fe/Au elemental mapping (**d**) of PVA/PAA/Fe$_3$O$_4$@AuNPs nanocomposites after the eighth cycle of catalytic reactions.

Nanomaterials **2017**, *7*, 317

4. Conclusions

The AuNPs-containing PVA/PAA/Fe$_3$O$_4$ nanocomposite materials were successfully prepared via electrospinning and self-assembly. Au nanoparticles were loaded on the surface of a composite membrane via a self-assembly process. The prepared PVA/PAA/Fe$_3$O$_4$ nanocomposites provide good support for AuNPs to be loaded on and effectively avoid agglomeration of AuNPs with improved stability for the next catalytic reduction application. In addition, the introduction of magnetic nanoparticles in the present composite catalysts is advantageous to conveniently separate from the reduction solution and reuse for subsequent recycling. For the catalytic reduction of liquid 4-NP and 2-NA solution, the prepared PVA/PAA/Fe$_3$O$_4$@AuNPs nanocomposite membranes demonstrated significant catalytic activity even after eight cycles for catalytic reduction at room temperature. Thus, the present prepared PVA/PAA/Fe$_3$O$_4$@AuNPs nanocomposites display excellent catalytic activity, good stability, and outstanding magnetic separation. The present research work thus proposes a novel approach to design and prepare new Au nanoparticle-containing composite materials for applications in selective catalytic reduction.

Supplementary Materials: The following are available online at http://www.mdpi.com/2079-4991/7/10/317/s1. Figure S1. SEM (a) with EDX and TEM (b) images of the prepared Fe$_3$O$_4$ nanoparticles. Figure S2. FT-IR of PVA/PAA, PVA/PAA/Fe$_3$O$_4$ and PVA/PAA/Fe$_3$O$_4$@AuNPs nanocomposites.

Acknowledgments: This work was financially supported by the National Natural Science Foundation of China (No. 21473153), the Support Program for the Top Young Talents of Hebei Province, China Postdoctoral Science Foundation (No. 2015M580214), and the Scientific and Technological Research and Development Program of Qinhuangdao City (No. 201701B004).

Author Contributions: Tifeng Jiao and Jingxin Zhou conceived and designed the experiments; Rong Guo performed the experiments; Rong Guo, Ruirui Xing, Tifeng Jiao, and Jingxin Zhou analyzed the data; Yan Chen, Wanchun Guo, Lexin Zhang, and Qiuming Peng contributed reagents/materials/analysis tools; Rong Guo and Tifeng Jiao wrote the paper.

Conflicts of Interest: The authors declare no conflict of interest.

References

1. Bond, G.C.; Sermon, P.A. Gold catalysts for olefin hydrogenation. *Gold Bull.* **1973**, *6*, 102–105. [CrossRef]
2. Haruta, M.; Kobayashi, T.; Sano, H.; Yamada, N. Novel gold catalysts for the oxidation of carbon monoxide at a temperature far below 0 °C. *Chem. Lett.* **1987**, *16*, 405–408. [CrossRef]
3. Hutchings, G.J. Vapor phase hydrochlorination of acetylene: Correlation of catalytic activity of supported metal chloride catalysts. *J. Catal.* **1985**, *96*, 292–295. [CrossRef]
4. Villaverde, G.; Corma, A.; Iglesias, M.; Sanchez, F. Heterogenized gold complexes: Recoverable catalysts for multicomponent reactions of aldehydes, terminal alkynes, and amines. *ACS Catal.* **2016**, *2*, 399–406. [CrossRef]
5. Sanchez, A.; Abbet, S.; Heiz, U.; Schneider, W.D.; Hakkinen, H. When gold is not noble: Nanoscale gold catalysts. *J. Phys. Chem. A* **1999**, *103*, 9573–9678. [CrossRef]
6. Hernández, J.; Sollagullón, J.; Herrero, E.; Aldaz, A.; Feliu, J.M. Electrochemistry of shape-controlled catalysts: Oxygen reduction reaction on cubic gold nanoparticles. *J. Phys. Chem. C* **2015**, *111*, 14078–14083. [CrossRef]
7. Kundu, M.K.; Bhowmik, T.; Barman, S. Gold aerogel supported on graphitic carbon nitride: An efficient electrocatalyst for oxygen reduction reaction and hydrogen evolution reaction. *J. Mater. Chem. A* **2015**, *3*, 23120–23135. [CrossRef]
8. Zheng, G.; Polavarapu, L.; Lizmarzán, L.M.; Pastorizasantos, I.; Perezjuste, J. Gold nanoparticle-loaded filter paper: A recyclable dip-catalyst for real-time reaction monitoring by surface enhanced Raman scattering. *Chem. Commun.* **2015**, *51*, 4572–4575. [CrossRef] [PubMed]
9. Tahir, B.; Tahir, M.; Amin, N.A.S. Gold-indium modified TiO$_2$, nanocatalysts for photocatalytic CO$_2$, reduction with H$_2$, as reductant in a monolith photoreactor. *Appl. Surf. Sci.* **2015**, *338*, 1–14. [CrossRef]
10. Sandoval, A.; Zanella, R.; Klimova, T.E. Titania nanotubes decorated with anatase nanocrystals as support for active and stable gold catalysts for CO oxidation. *Catal. Today* **2017**, *282*, 140–150. [CrossRef]

11. Chen, S.; Luo, L.; Jiang, Z.; Huang, W. Size-dependent reaction pathways of low-temperature CO oxidation on Au/CeO$_2$ catalysts. *ACS Catal.* **2015**, *5*, 75–78. [CrossRef]

12. Sinha, A.K.; Seelan, S.; Tsubota, S.; Haruta, M. A three-dimensional mesoporous titanosilicate support for gold nanoparticles: Vapor-phase epoxidation of propene with high conversion. *Angew. Chem. Int. Ed.* **2004**, *43*, 1546–1548. [CrossRef] [PubMed]

13. Chowdhury, B.; Bravo-Suárez, J.J.; Daté, M.; Tsubota, S.; Haruta, M. Trimethylamine as a gas-phase promoter: Highly efficient epoxidation of propylene over supported gold catalysts. *Angew. Chem. Int. Ed.* **2006**, *45*, 412–415. [CrossRef] [PubMed]

14. Li, Z.S.; Wang, Y.N.; Zhang, J.H.; Wang, D.Y.; Ma, W.H. Better performance for gas-phase epoxidation of propylene using H$_2$ and O$_2$ at lower temperature over Au/TS-1 catalyst. *Catal. Commun.* **2017**, *90*, 87–90. [CrossRef]

15. Chang, M.W.; Sheu, W.S. Water-gas-shift reaction on reduced gold-substituted Ce$_{1-x}$O$_2$ (111) surfaces: The role of Au charge. *Phys. Chem. Chem. Phys.* **2017**, *19*, 2201–2206. [CrossRef] [PubMed]

16. Shi, J.X.; Mahr, C.; Murshed, M.M.; Zielasek, V.; Rosenauer, A.; Guesing, T.; Bäumer, M.; Wittstock, A. A versatile sol-gel coating for mixed oxides on nanoporous gold and their application in the water gas shift reaction. *Catal. Sci. Technol.* **2016**, *6*, 5311–5319. [CrossRef]

17. Yao, S.Y.; Zhang, X.; Zhou, W.; Gao, R.; Xu, W.Q.; Ye, Y.F.; Lin, L.L.; Wen, X.D.; Liu, P.P.; Chen, B.B.; et al. Atomic-layered Au clusters on α-MoC as catalysts for the low-temperature water-gas shift reaction. *Science* **2017**, *357*, 389–393. [CrossRef] [PubMed]

18. Kapkowski, M.; Bartczak, P.; Korzec, M.; Sitko, R.; Szade, J.; Balin, K.; Lelatko, J.; Polanski, J. SiO$_2$-, Cu-, and Ni-supported Au nanoparticles for selective glycerol oxidation in the liquid phase. *J. Catal.* **2014**, *319*, 110–118. [CrossRef]

19. Dong, W.; Reichenberger, S.; Chu, S.; Weide, P.; Ruland, H.; Barcikowski, S.; Wagener, P.; Muhler, M. The effect of the Au loading on the liquid-phase aerobic oxidation of ethanol over Au/TiO$_2$, catalysts prepared by pulsed laser ablation. *J. Catal.* **2015**, *330*, 497–506. [CrossRef]

20. Wang, T.; Yuan, X.; Li, S.; Gong, J. CeO$_2$-modified Au@SBA-15 nanocatalysts for liquid-phase selective oxidation of benzyl alcohol. *Nanoscale* **2015**, *7*, 593–602. [CrossRef] [PubMed]

21. Heeskens, D.; Aghaei, P.; Kaluza, S.; Strunk, J.; Muhler, M. Selective oxidation of ethanol in the liquid phase over Au/TiO$_2$. *Phys. Status Solidi B* **2013**, *250*, 1107–1118. [CrossRef]

22. Evangelista, V.; Acosta, B.; Miridonov, S.; Smolentseva, E.; Fuentes, S.; Simakov, A. Highly active Au-CeO$_2$@ZrO$_2$, yolk-shell nanoreactors for the reduction of 4-nitrophenol to 4-aminophenol. *Appl. Catal. B Environ.* **2015**, *166*, 518–528. [CrossRef]

23. Wang, Y.; Li, H.; Zhang, J.; Yan, X.; Chen, Z. Fe$_3$O$_4$ and Au nanoparticles dispersed on the graphene support as a highly active catalyst toward the reduction of 4-nitrophenol. *Phys. Chem. Chem. Phys.* **2016**, *18*, 615–623. [CrossRef] [PubMed]

24. Ding, Z.D.; Wang, Y.X.; Xi, S.F.; Li, Y.X.; Li, Z.J.; Ren, X.H.; Gu, Z.G. A hexagonal covalent porphyrin framework as an efficient support for gold nanoparticles toward catalytic reduction of 4-Nitrophenol. *Chem. Eur. J.* **2016**, *22*, 17029–17036. [CrossRef] [PubMed]

25. Liu, H.X.; Yang, Y.X.; Ma, M.G.; Wang, X.M.; Zhen, D.X. Self-assembled gold nanoparticles coating for solid-phase microextraction of ultraviolet filters in environmental water. *Chin. J. Anal. Chem.* **2015**, *43*, 207–211. [CrossRef]

26. Zhou, Y.; Zhu, Y.H.; Yang, X.; Huang, J.; Chen, W.; Lv, X.M.; Li, C.Y.; Li, C.Z. Au decorated Fe$_3$O$_4$@TiO$_2$ magnetic composites with visible light-assisted enhanced catalytic reduction of 4-Nitrophenol. *RSC Adv.* **2015**, *5*, 50454–50461. [CrossRef]

27. Lau, M.; Ziefuss, A.; Komossa, T.; Barcikowski, S. Inclusion of supported gold nanoparticles into their semiconductor support. *Phys. Chem. Chem. Phys.* **2015**, *17*, 29311–29318. [CrossRef] [PubMed]

28. Tvauri, I.V.; Gergieva, B.E.; Magkoeva, V.D.; Grigorkina, G.S.; Bliev, A.P.; Ashkhotov, O.G.; Spzaev, V.A.; Fukutani, K.; Magkoev, T.T. Carbon monoxide oxidation on lithium fluoride supported gold nanoparticles: A significance of F-centers. *Solid State Commun.* **2015**, *213–214*, 42–45. [CrossRef]

29. Liu, X.; Zhang, J.; Guo, X.; Wu, S.; Wang, S. Porous alpha-Fe$_2$O$_3$ decorated by Au nanoparticles and their enhanced sensor performance. *Nanotechnology* **2010**, *21*. [CrossRef]

30. Korotcenkov, G.; Brinzari, V.; Cho, B.K. Conductometric gas sensors based on metal oxides modified with gold nanoparticles: A review. *Microchim. Acta* **2016**, *183*, 1033–1054. [CrossRef]

31. Ma, X.; Yang, J.; Cai, W.; Zhu, G.; Liu, J. Preparation of Au nanoparticles decorated polyaniline nanotube and its catalytic oxidation to ascorbic acid. *Chem. Res. Chin. Univ.* **2016**, *32*, 1–7. [CrossRef]

32. Chairam, S.; Konkamdee, W.; Parakhun, R. Starch-supported gold nanoparticles and their use in 4-nitrophenol reduction. *J. Saudi. Chem. Soc.* **2015**. [CrossRef]

33. Zhu, C.; Han, L.; Hu, P.; Dong, S. Loading of well-dispersed gold nanoparticles on two-dimensional graphene oxide/SiO composite nanosheets and their catalytic properties. *Nanoscale* **2012**, *4*, 1641–1646. [CrossRef] [PubMed]

34. Kuroda, K.; Ishida, T.; Haruta, M. Reduction of 4-nitrophenol to 4-Aminophenol over Au nanoparticles deposited on PMMA. *J. Mol. Catal. A Chem.* **2009**, *298*, 7–11. [CrossRef]

35. Zhang, Y.; Liu, S.; Lu, W.; Wang, L.; Tian, J.; Sun, X. In situ green synthesis of Au nanostructures on graphene oxide and their application for catalytic reduction of 4-Nitrophenol. *Catal. Sci. Technol.* **2011**, *1*, 1142–1144. [CrossRef]

36. Ye, W.; Yu, J.; Zhou, Y.; Gao, D.; Wang, D.; Wang, C.M.; Xie, D.S. Green synthesis of Pt-Au dendrimer-like nanoparticles supported on polydopamine-functionalized graphene and their high performance toward 4-nitrophenol reduction. *Appl. Catal. B Environ.* **2016**, *181*, 371–378. [CrossRef]

37. Jin, C.; Han, J.; Chu, F.; Guo, R. Fe₃O₄@PANI hybrid shell as a multifunctional support for Au nanocatalysts with a rmarkably improved catalytic performance. *Langmuir* **2017**, *33*, 4520–4527. [CrossRef] [PubMed]

38. Reneker, D.H.; Chun, I. Nanometre diameter fibres of polymer, produced by electrospinning. *Nanotechnology* **1996**, *7*, 216–223. [CrossRef]

39. Yarin, A.L. Coaxial electrospinning and emulsion electrospinning of core-shell fibers. *Polym. Adv. Technol.* **2015**, *22*, 310–317. [CrossRef]

40. Brown, T.D.; Dalton, P.D.; Hutmacher, D.W. Melt electrospinning today: An opportune time for an emerging polymer process. *Prog. Polym. Sci.* **2016**, *56*, 116–166. [CrossRef]

41. Chen, M.; Patra, P.K.; Lovett, M.L.; Kaplan, D.L.; Bhowmick, S. Role of electrospun fibre diameter and corresponding specific surface area (SSA) on cell attachment. *J. Tissue Eng. Regen. Med.* **2009**, *3*, 269–279. [CrossRef] [PubMed]

42. Yan, J.; Huang, Y.; Miao, Y.E.; Weng, W.T.; Liu, T. Polydopamine-coated electrospun poly(vinyl alcohol)/poly(acrylic acid) membranes as efficient dye adsorbent with good recyclability. *J. Hazard. Mater.* **2015**, *283*, 730–739. [CrossRef] [PubMed]

43. Xing, R.; Wang, W.; Jiao, T.; Ma, K.; Zhang, Q.; Hong, W.; Qiu, H.; Zhou, J.; Zhang, L.; Peng, Q. Bioinspired polydopamine sheathed nanofibers containing carboxylate graphene oxide nanosheet for high-efficient dyes scavenger. *ACS Sustain. Chem. Eng.* **2017**, *5*, 4948–4956. [CrossRef]

44. He, D.; Hu, B.; Yao, Q.F.; Wang, K.; Yu, S.H. Large-scale synthesis of flexible free-standing SERS substrates with high sensitivity: Electrospun PVA nanofibers embedded with controlled alignment of silver nanoparticles. *ACS Nano* **2009**, *3*, 3993–4002. [CrossRef] [PubMed]

45. Ma, Q.; Wang, J.; Dong, X.; Yu, W.; Liu, G. Fabrication of magnetic-fluorescent bifunctional flexible coaxial nanobelts by electrospinning using a modified coaxial spinneret. *Chempluschem* **2014**, *79*, 290–297. [CrossRef]

46. Villarreal-Gómez, L.J.; Cornejo-Bravo, J.M.; Vera-Graziano, R.; Grande, D. Electrospinning as a powerful technique for biomedical applications: A critically selected survey. *J. Biomater. Sci. Polym. Ed.* **2016**, *27*, 157–176. [CrossRef] [PubMed]

47. Hou, C.; Ma, K.; Jiao, T.; Xing, R.; Li, K.; Zhou, J.; Zhang, L. Preparation and dye removal capacities of porous silver nanoparticle-containing composite hydrogels via poly(acrylic acid) and silver ions. *RSC Adv.* **2016**, *6*, 110799–110807. [CrossRef]

48. Zhang, P.; Shao, C.; Li, X.; Zhang, M.Y.; Zhang, X.; Sun, Y.Y.; Liu, X. In situ assembly of well-dispersed Au nanoparticles on TiO₂/ZnO nanofibers: A three-way synergistic heterostructure with enhanced photocatalytic activity. *J. Hazard. Mater.* **2012**, *237*, 331–338. [CrossRef] [PubMed]

49. Shen, J.; Zhou, Y.; Huang, J.; Zhu, Y.H.; Zhu, J.R.; Yang, X.L.; Chen, W.; Yao, Y.F.; Qian, S.H.; Jiang, H.; et al. In-situ SERS monitoring of reaction catalyzed by multifunctional Fe₃O₄@TiO₂@Ag-Au microspheres. *Appl. Catal. B Environ.* **2017**, *205*, 11–18. [CrossRef]

50. Cheng, J.; Zhao, S.; Gao, W.; Jiang, P.B.; Li, R. Au/Fe₃O₄@TiO₂, hollow nanospheres as efficient catalysts for the reduction of 4-nitrophenol and photocatalytic degradation of rhodamine B. *React. Kinet. Mech. Catal.* **2017**, *121*, 797–810. [CrossRef]

51. Li, J.; Tan, L.; Wang, G.; Yang, M. Synthesis of double-shelled sea urchin-like yolk-shell Fe_3O_4/TiO_2/Au microspheres and their catalytic applications. *Nanotechnology* **2015**, *26*, 095601. [CrossRef] [PubMed]

52. Hou, C.L.; Jiao, T.F.; Xing, R.R.; Chen, Y.; Zhou, J.X.; Zhang, L.X. Preparation of TiO_2 nanoparticles modified electrospun nanocomposite membranes toward efficient dye degradation for wastewater treatment. *J. Taiwan Inst. Chem. Eng.* **2017**, *78*, 118–126. [CrossRef]

53. Jiao, J.; Wang, H.X.; Guo, W.C.; Li, R.F.; Tian, K.S.; Xu, Z.P.; Jia, Y.; Wu, Y.H.; Cao, L. In situ confined growth based on a self-templating reduction strategy of highly dispersed Ni nanoparticles in hierarchical yolk-shell Fe@SiO_2 structures as efficient catalysts. *Chem. Asian J.* **2016**, *11*, 3534–4350. [CrossRef] [PubMed]

54. Guo, W.C.; Wang, Q.; Wang, G.; Yang, M.; Dong, W.J.; Yu, J. Facile hydrogen-bond-assisted polymerization and immobilization method to synthesize hierarchical Fe_3O_4@poly(4-vinylpyridine-co-divinylbenzene)@Au nanostructures and their catalytic applications. *Chem. Asian J.* **2013**, *8*, 1160–1167. [CrossRef] [PubMed]

55. Xing, R.R.; Liu, K.; Jiao, T.F.; Zhang, N.; Ma, K.; Zhang, R.Y.; Zou, Q.; Ma, G.; Yan, X. An injectable self-assembling collagen-gold hybrid hydrogel for combinatorial antitumor photothermal/ohotodynamic therapy. *Adv. Mater.* **2016**, *28*, 3669–3676. [CrossRef] [PubMed]

56. Liu, Y.; Ma, K.; Jiao, T.; Xing, R.; Shen, G.; Yan, X. Water-insoluble photosensitizer nanocolloids stabilized by supramolecular interfacial assembly towards photodynamic therapy. *Sci. Rep.* **2017**, *7*, 42978. [CrossRef] [PubMed]

57. Xing, R.; Jiao, T.; Liu, Y.; Ma, K.; Zou, Q.; Ma, G.; Yan, X. Co-assembly of graphene oxide and albumin/photosensitizer nanohybrids towards enhanced photodynamic therapy. *Polymers* **2016**, *8*, 181. [CrossRef]

58. Zhao, X.N.; Ma, K.; Jiao, T.F.; Xing, R.R.; Ma, X.L.; Hu, J.; Huang, H.; Zhang, L.; Yan, X. Fabrication of hierarchical layer-by-layer assembled diamond-based core-shell nanocomposites as highly efficient dye absorbents for wastewater treatment. *Sci. Rep.* **2017**, *7*, 44076. [CrossRef] [PubMed]

59. Guo, H.; Jiao, T.; Zhang, Q.; Guo, W.; Peng, Q.; Yan, X. Preparation of graphene oxide-based hydrogels as efficient dye adsorbents for wastewater treatment. *Nanoscale Res. Lett.* **2015**, *10*, 272. [CrossRef] [PubMed]

60. Zhang, R.; Xing, R.; Jiao, T.; Ma, K.; Chen, C.; Ma, G.; Yan, X. Carrier-free, chemo-photodynamic dual nanodrugs via self-assembly for synergistic antitumor therapy. *ACS Appl. Mater. Interfaces* **2016**, *8*, 13262–13269. [CrossRef] [PubMed]

61. Wang, Y.Z.; Li, X.Y.; Sun, G.; Zhang, G.L.; Liu, H.; Du, J.S.; Yao, S.T.; Bai, J.; Yang, Q.B. Fabrication of Au/PVP nanofiber composites by electrospinning. *J. Appl. Polym. Sci.* **2010**, *105*, 3618–3622. [CrossRef]

62. Kundu, S.; Gill, R.S.; Saraf, R.F. Electrospinning of PAH nanofiber and deposition of Au NPs for nanodevice fabrication. *J. Phys. Chem. C* **2011**, *115*, 15845–15852. [CrossRef]

63. Liu, Y.; Wang, T.; Huan, Y.; Li, Z.; He, G.; Liu, M. Self-assembled supramolecular nanotube yarn. *Adv. Mater.* **2013**, *25*, 5875–5879. [CrossRef] [PubMed]

64. Shen, Z.; Wang, T.; Liu, M. Macroscopic chirality of supramolecular gels formed from achiral tris(ethyl cinnamate) benzene-1,3,5-tricarboxamides. *Angew. Chem. Int. Ed.* **2014**, *53*, 13424–13428. [CrossRef] [PubMed]

65. Deng, Y.H.; Cai, Y.; Sun, Z.K.; Liu, J.; Liu, C.; Wei, J.; Li, W.; Liu, C.; Wang, Y.; Zhao, D.Y. Multifunctional mesoporous composite microspheres with well-designed nanostructure: A highly integrated catalyst system. *J. Am. Chem. Soc.* **2010**, *132*, 8466–8473. [CrossRef] [PubMed]

66. Huo, S.; Duan, P.; Jiao, T.; Peng, Q.; Liu, M. Full- and white color circularly polarized luminescent quantum dots via supramolecular self-assembly. *Angew. Chem. Int. Ed.* **2017**, *129*, 12342–12346. [CrossRef]

67. Garcia-Serrano, J.; Galindo, A.G.; Pal, U. Au-Al_2O_3 nanocomposites: XPS and FTIR spectroscopic studies. *Sol. Energy Mater. Sol. Cell* **2004**, *82*, 291–298. [CrossRef]

nanomaterials

MDPI

Article

Ionic Liquids as Surfactants for Layered Double Hydroxide Fillers: Effect on the Final Properties of Poly(Butylene Adipate-*Co*-Terephthalate)

Sébastien Livi [1,*], Luanda Chaves Lins [1,*], Jakub Peter [2], Hynek Benes [2], Jana Kredatusova [2], Ricardo K. Donato [3] and Sébastien Pruvost [1]

1   Univ Lyon, INSA Lyon, UMR CNRS 5223, IMP Ingénierie des Matériaux Polymères, F-69621 Villeurbanne, France; sebastien.pruvost@insa-lyon.fr

2   Institute of Macromolecular Chemistry AS CR, v.v.i., Heyrovsky Sq. 2, 162 06 Prague 6, Czech Republic; peter@imc.cas.cz (J.P.); benesh@imc.cas.cz (H.B.); kredatusova@imc.cas.cz (J.K.)

3   MackGraphe–Graphene and Nanomaterials Research Center, Mackenzie Presbyterian University, Rua da Consolação 896, São Paulo 01302-907, Brazil; ricardo.donato@mackenzie.br

*   Correspondence: sebastien.livi@insa-lyon.fr (S.L.); luandaqmc@gmail.com (L.C.L.); Tel.: +33-472-438-291 (S.L.)

Received: 16 August 2017; Accepted: 25 September 2017; Published: 28 September 2017

Abstract: In this work, phosphonium ionic liquids (ILs) based on tetra-alkylphosphonium cations combined with carboxylate, phosphate and phosphinate anions, were used for organic modification of layered double hydroxide (LDH). Two different amounts (2 and 5 wt %) of the organically modified LDHs were mixed with poly(butylene adipate-*co*-terephthalate) (PBAT) matrix by melt extrusion. All prepared PBAT/IL-modified-LDH composites exhibited increased mechanical properties (20–50% Young's modulus increase), decreased water vapor permeability (30–50% permeability coefficient reduction), and slight decreased crystallinity (10–30%) compared to the neat PBAT.

Keywords: ionic liquids; poly(butylene adipate-*co*-terephthalate); layered double hydroxide; nanocomposites

---

1. Introduction

In the world of polymer nanocomposites, the continual challenge is to develop high-performance materials at a low cost [1–5]. Thus, polymer nanocomposites based on layered silicates, such as montmorillonite (MMT) or layered double hydroxide (LDH), have long received attention from academic and industrial research. Consequently, many studies have reported the use of organically modified clays in biodegradable, natural or biosourced matrices, such as polylactide (PLA), poly(3-caprolactone) (PCL), poly(alkyl succinates), and poly(butylene adipate-*co*-terephtalate) (PBAT), in order to produce polymer materials having excellent thermal stability, good mechanical properties as well as good water vapor and gas barrier properties for compostable films, food packaging or tissue engineering applications [6–12]. For example, for compostable film applications, different composites based on PBAT and different lignocellulose fibers (pupunha, munguba, kenaf fibers) were studied to increase the performances of this matrix [13–15]. Several surface treatment pathways have been studied: (i) grafting of organosilanes [16,17], (ii) cationic exchange [18–20], and (iii) anionic exchange [21,22]. A vast number of LDH modification methods are available [23], where calcination/rehydration is often a suitable approach. In this case, anionic exchange usually requires two steps composed of an initial calcination step followed by the counter anion intercalation within a solvent medium, where various conventional anions, surfactant adsorbents, and active pharmaceutical ingredients have already been used [24].

Recently, ionic liquids (ILs), which are ionocovalent based organic molecules presenting low melting temperature (≤ 100 °C) and composed of ion pairs, have been used in the field of nanocomposites as a surfactant, interfacial agent or as dispersant of nanoparticles [25–28]. They present numerous advantages such as excellent thermal stability, negligible vapor pressure, and good affinity with organic and/or inorganic materials [25,26]. They present especially good affinity to silicates, allowing dramatic morphology changes when in situ applied to sensible processes such as the Sol-Gel, mainly when applied to its first hydrolytic step [29]. Thus, various authors have studied them as alternatives to the conventional ammonium salts well-known to have low thermal stability (<180 °C), which limits their use in polymer matrices requiring high curing or processing temperatures; e.g. to obtain the best of their mechanical properties epoxy-silica nanocomposites need a strong interphase control, thus an interphase agent is necessary, but they are often post-cured at temperatures above their glass transition, demanding agents that handle long-term exposition to high temperature [30]. The ILs not only resist the curing process without degrading but also allow morphology and mechanical properties tuning, and shape memory effect, which was demonstrated to be strongly influenced by the intimate IL-silica interaction even in such complex systems [27]. These effects were also shown to extend to layered silicates, e.g., Livi et al. demonstrated that the imidazolium and phosphonium ILs play a dual role as surfactant and compatibilizing agent of layered silicates between the polymer matrix and the nanoparticle [19,26]. Thus, they highlighted a good dispersion of the treated-MMT as well as an improvement of the thermal and mechanical properties in high density polyethylene (HDPE) and poly(vinylidene fluoride) (PVDF) matrices [25,26]. More recently, various authors used ILs as surfactant agents of LDHs leading to polymer materials with enhanced properties [9,31]. Bugatti et al. developed PLA films with excellent water barrier properties for food packaging applications [31]. Later, Kredatusova et al. developed a new, fast and environmentally-friendly process based on microwave irradiation leading to an exfoliation of LDH modified with ILs in PCL matrix [9]. Nevertheless, few works have been reported in the literature on the contribution of these fillers to the polymer matrices.

Due especially to their thermal/chemical stability and interphase adhesion properties, ILs are perfectly suitable to substitute traditional dispersion/stabilization agents in polymer nanocomposites, especially in melting processes, helping to migrate the filler properties to the polymer matrix and avoiding losses by evaporation and degradation processes. Thus, in this work, new LDHs modified with phosphonium ILs were prepared and characterized by thermogravimetric analysis (TGA) and X-ray diffraction (XRD) in order to prove the intercalation of the phosphinate, carboxylate and phosphate counter anions into LDH layers. Then, small amounts (2 and 5 wt %) of these treated-LDHs were introduced into a biodegradable matrix (PBAT), forming nanocomposites where their morphologies as well as their thermal, mechanical and barrier properties were investigated.

2. Results and Discussion

2.1. Characterization of Ionic Liquids

Thermal Behavior

After anionic exchange with phosphonium ILs containing hexanoate, phosphinate, and phosphate anions denoted IL351, IL104, and IL349, the thermal stability of the pure ILs were investigated by TGA (Figure 1).

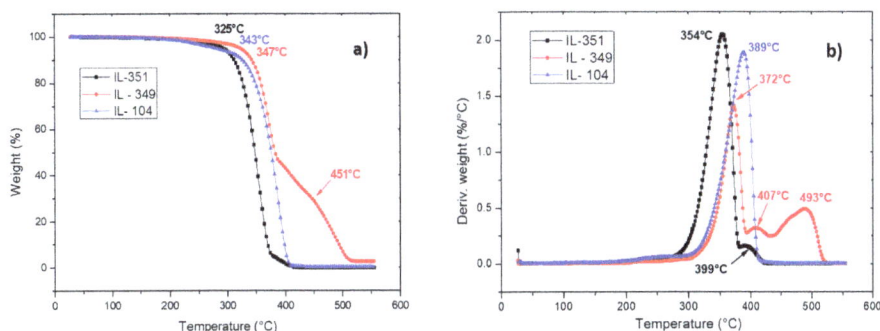

Figure 1. Evolution of weight loss as a function of temperature (thermogravimetric analysis (TGA) (**a**); (**b**) derivative thermo-gravimetric (DTG)) of pure ionic liquids (ILs). (heating rate 20 K·min^{-1}, under nitrogen flow).

All the ILs presented excellent thermal stability (>300 °C), nevertheless, two behaviors are observed in Figure 1. In fact, similar degradation temperatures of the ILs denoted IL 351 and IL 104 of about 340 °C and 350 °C were obtained, respectively. In the case of IL349, three degradation peaks at 372 °C, 407 °C, and 493 °C were observed. This higher degradation temperature can be attributed to the presence of the phosphate anion. According to the literature, the phosphate compounds are commonly used as flame retardants leading to significant inflammability and a better thermal stability [32,33]. Thus, we can assume that the phosphate anion delayed the degradation of IL349 which also explains the presence of residues after 550 °C compared to the other ILs. In summary, these different ILs can be used as more thermal resistant alternatives to the thermally unstable ammonium salts, with stronger potential for polymer nanocomposites processing applications at high temperatures [34,35].

According to the literature, it is well known that pristine LDH have three degradation steps corresponding to: (i) the loss of physisorbed and intercalated water between LDH layers, which takes place between 50 °C and 250 °C [36–38], and (ii) the removal of interlayer carbonate anion, and (iii) dehydroxylation of –OH groups which is between 250 °C and 500 °C [39,40]. In our previous work on the surface treatment of LDH by these three ionic liquids, the presence of hexanoate, phosphinate, and phosphate anions as well as the presence of carbonate anion was also proved where characterization techniques such as TGA, Fourier transform infrared spectroscopy (FTIR) and X-ray photoelectron spectroscopy (XPS) were used. Thus, Kredatusova et al. highlighted that during the surface treatment, the regeneration of the crystalline structure of LDH induced the absorption of carbonate anions [9,38,41–43].

Based on these previous results, the influence of the surface treatment of the LDHs on the final properties of the PBAT matrix is studied hereafter.

2.2. Characterization of PBAT/Modified-LDH Nanocomposites

2.2.1. Morphology

The influence of the counter anions, i.e., hexanoate, phosphinate and phosphate on the morphologies of PBAT nanocomposites containing only 5 wt % of LDH-ILs was investigated. Thus, transmission electron microscopy (TEM) micrographs and the distribution area of PBAT filled with LDH-351, LDH-104, and LDH-349 are shown in Figures 2 and 3.

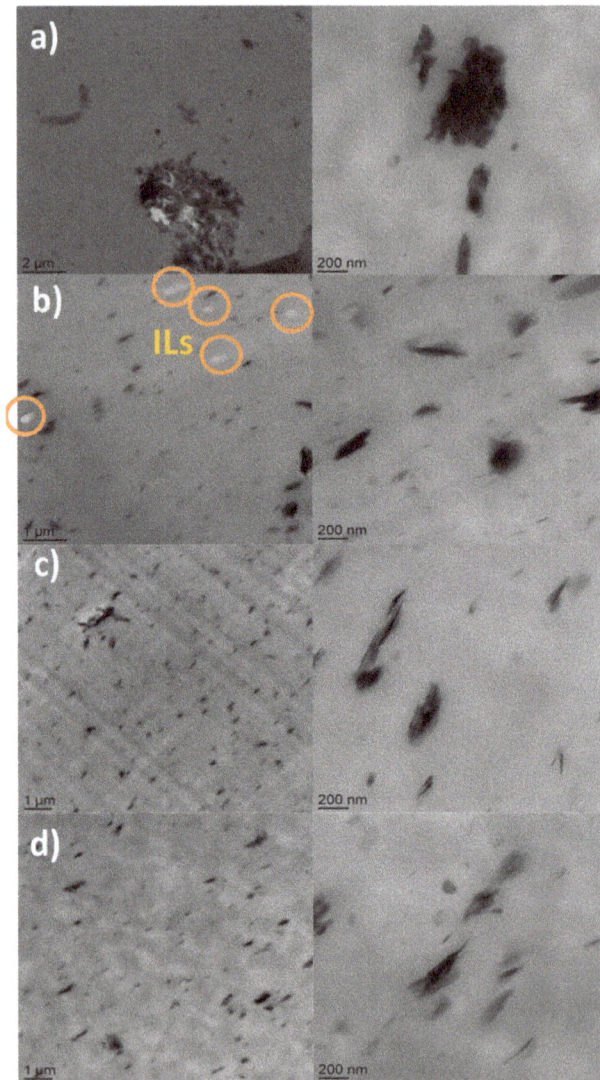

Figure 2. Transmission electron microscopy (TEM) micrographs of poly(butylene adipate-*co*-terephthalate) (PBAT) nanocomposites with 5 wt % of treated-LDHs: (**a**) PBAT/LDH; (**b**) PBAT/LDH-104; (**c**) PBAT/LDH-351; (**d**) PBAT/LDH-349.

In the case of PBAT filled with unmodified LDH, a very poor dispersion of LDH layers is obtained showing the presence of numerous aggregates of several microns. These results have been often reported in the literature concerning polymer nanocomposites containing untreated inorganic fillers, such as layered silicates, especially montmorillonite (MMT) [44,45].

On the other hand, a good distribution of the phosphonium IL-modified LDHs denoted LDH-351, LDH-104, and LDH-349 was observed. In fact, TEM micrographs with one micrometer scale revealed a homogeneous dispersion of LDH-ILs into PBAT matrix with the presence of a few tactoïds having sizes at the most of the order of 1–2 μm. In order to determine more precisely the type of morphology obtained, enlargements were carried out. As can be seen in Figure 2c,d, the addition of 5 wt % of

LDH-351 and LDH-349 led to a mixed morphology, composed of small tactoïds and few well-dispersed clay layers in the PBAT matrix. In the case of LDH-104, an excellent dispersion is highlighted by TEM micrographs corresponding to an intercalated/partially exfoliated morphology characterized by the presence of well-dispersed clay layers combined with very small tactoïds. In addition, the presence of free IL104 defined by white dots (Figure 2b) is highlighted on the TEM images. From a previous paper on PBAT/IL blends, our research group demonstrated that the incorporation of IL104 induced a phase separated morphology due to dipole-dipole interactions between the ion pairs leading to the formation of these ionic clusters [46]. Thus, TEM micrographs confirm the presence of physisorbed IL on the LDH surface which then diffuses into the PBAT matrix during the processing of the nanocomposites by extrusion. However, the chemical nature of the counter anion plays a key role on the miscibility of ILs in polymer matrix which may explain the impossibility to observe the presence of free IL349 and IL351.

Finally, in order to determine the level of dispersion of the untreated and treated LDHs into PBAT matrix, image analysis was carried out. Thus, the area of distribution is presented in the form of a box and whiskers plot (Figure 3).

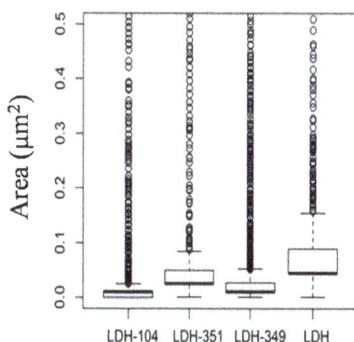

Figure 3. Box and whiskers plot area distribution of untreated and treated layered double hydroxides (LDHs) into PBAT matrix.

Independently of the system studied, image analyses are in agreement with TEM micrographs highlighting the poor area distribution of untreated LDHs (0.091 μm^2) compared to IL treated ones. In addition, smaller distribution areas are obtained for LDH-349 (0.027 μm^2) and LDH-104 (0.011 μm^2) corresponding to a finer and more homogeneous dispersion of these modified LDHs.

In conclusion, the surface treatment of the LDH by using phosphonium ILs as interfacial agents induced a good distribution of LDH-ILs in the PBAT matrix by using an extrusion process.

2.2.2. Thermal Behavior

To investigate the influence of IL-treated LDHs on the PBAT nanocomposite thermal behavior, TGA was performed. Thus, the TGA and DTG curves of PBAT and PBAT containing 2 and 5 wt % of untreated and treated LDHs are presented in Figures 4 and 5.

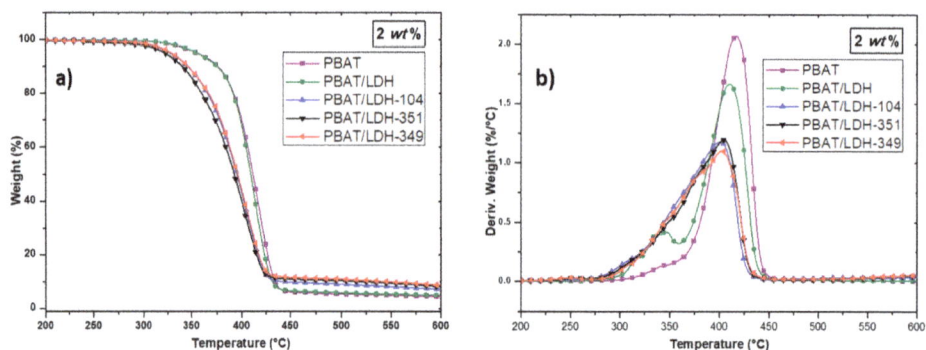

Figure 4. Evolution of weight loss obtained by TGA (**a**) and DTG (**b**) of the neat PBAT and PBAT-LDH (2 wt %) nanocomposites (heating rate 20 K·min^{-1} under nitrogen flow).

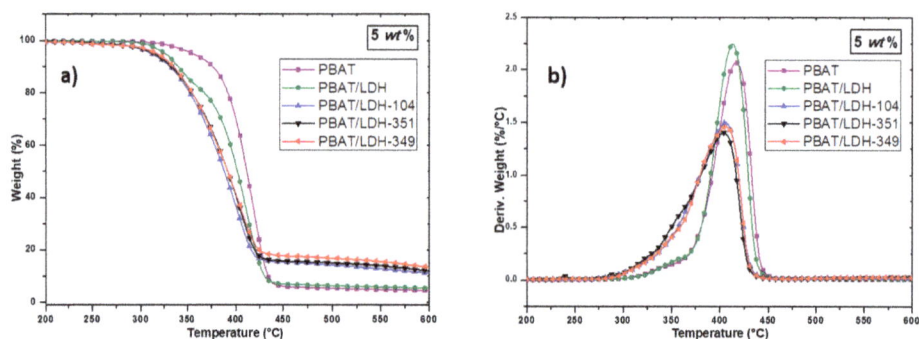

Figure 5. Evolution of weight loss obtained by TGA (**a**) and DTG (**b**) of the neat PBAT and PBAT-LDH (5 wt %) nanocomposites (heating rate 20 K·min^{-1} under nitrogen flow).

In all cases, the incorporation of pristine LDH or IL-modified LDH denoted LDH-349, LDH-351 and LDH-104 led to a decrease in the neat PBAT degradation temperature of about 10–15 °C. According to the literature, this phenomenon is attributed to the presence of water molecules in the LDH layers, leading to an acceleration of the PBAT degradation [41]. In fact, since LDH-ILs were introduced into the PBAT matrix without being previously dried and that the processing of the nanocomposites required a temperature of 160 °C, the presence of water in the LDHs can explain this phenomenon. Our previous work confirms these assumptions in which a weight loss between 5 and 7 wt % corresponding to the loss of physisorbed water between LDH layers is observed for LDH-349, LDH-351, and LDH-104, respectively [9]. In fact, many authors have also observed the same trend in polymer nanocomposites based on PLA and LDH [41,47]. Nevertheless, deeper investigation is still required. In summary, the incorporation of organically modified LDHs has only a slight influence on the thermal stability of the PBAT matrix.

Secondly, the impact of treated LDHs on the thermo-mechanical properties of PBAT was investigated by Dynamic Mechanical Analysis (DMA). The storage moduli G′ and the main relaxation peak evaluated as the maximum of tan δ peak are displayed in Figure 6.

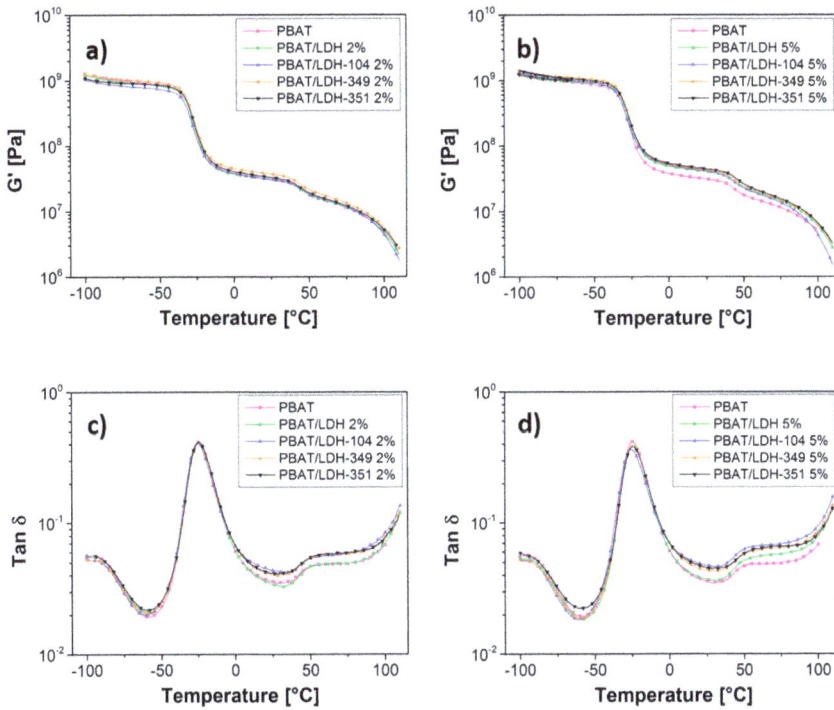

Figure 6. Storage moduli G′ (**a,b**), main relaxation peak of PBAT and the resulting nanocomposites evidenced on tan δ diagrams (**c,d**) recorded at 1 Hz.

In all cases, a main relaxation peak at −24.9 °C and a shoulder peak at 55 °C were highlighted by DMA corresponding well with the literature [48]. Thus, Nayak et al. [48] demonstrated that the first transition corresponds to the motion of the polybutylene adipate unit while the second transition is attributed to the terephthalate unit. Then, the addition of pristine LDH as well as IL-treated LDH in amounts of 2 wt % or 5 wt % did not influence the temperature of these two relaxations.

As the DSC data showed no influence of the LDH-ILs on the glass transition temperature T_g and melting temperature T_m of the PBAT nanocomposites, the differential scanning calorimetry (DSC) curves highlighting the crystallization temperatures T_c of the PBAT and the resulting nanocomposites containing 5 wt % of LDH and LDH-ILs are represented in Figure 7.

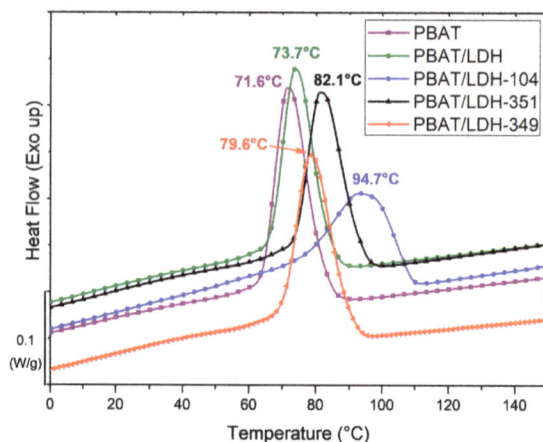

Figure 7. Differential scanning calorimetry (DSC) curves showing the crystallization temperatures T_c of neat PBAT and PBAT containing 5 wt % of untreated LDH and LDH-ILs.

In the case of neat PBAT, a glass transition temperature of $-34.6\,^\circ$C and a melting temperature of $123\,^\circ$C were obtained, which is in agreement with the literature [46,49]. The incorporation of only 2 wt % or 5 wt % of treated-LDHs showed no influence on the T_g or T_m of PBAT. However, differences can be observed for the crystallization temperatures. Whatever the amount of untreated LDH introduced into the PBAT matrix, no influence on the crystallization temperature ($73\,^\circ$C) was observed. These results are similar to those observed by Xie et al. which also showed that the incorporation of LDH containing nitrate ions in a PBAT matrix led to a decrease in the crystallization temperature [50]. In fact, this phenomenon can be attributed to the poor dispersion (Figure 2a) of the pristine LDH in the polymer thus impeding the crystallization of PBAT. Conversely, the use of LDH-ILs induced significant increases of T_c from $73\,^\circ$C to $95\,^\circ$C. Thus, crystallization temperatures of $80\,^\circ$C, $82\,^\circ$C, and $95\,^\circ$C were obtained for LDH-349, LDH-351, and LDH-104, respectively. These results can be explained by the good dispersion of LDH-ILs in the PBAT matrix (Figure 2) inducing a heterogeneous nucleation effect as well as a lamellar ordering effect in the polymer matrix [50,51]. Thus, these DSC data confirm the presence of phosphonium IL at the surface of LDHs leading to a better affinity between LDH and PBAT. In addition, the good distribution of LDH-104 combined with the formation of IL104 ionic clusters (see Figure 2b) may explain the T_c value of $95\,^\circ$C.

In summary, the presence of physisorbed and intercalated phosphonium ILs plays a key role on the compatibilization between the polymer matrix and the hydrotalcites.

2.2.3. Mechanical Properties

In order to reveal the impact of the modified LDHs denoted LDH-104, LDH-349, and LDH-351 on the mechanical performances of the PBAT matrix, uniaxial tensile properties were performed and the fracture properties and moduli are summarized in Table 1.

Table 1. Mechanical performances of neat poly(butylene adipate-co-terephthalate) (PBAT) and the resulting nanocomposites.

Nomenclature	Young Modulus (Mpa)	Stress (Mpa)	Strain at Break (%)
PBAT	47 ± 1	24 ± 1	511 ± 17
PBAT/LDH	56 ± 1	24 ± 1	577 ± 18
PBAT/LDH-349 2%	62 ± 1	22 ± 1	462 ± 30
PBAT/LDH-349 5%	72 ± 2	22 ± 1	400 ± 13
PBAT/LDH-104 2%	52 ± 2	22 ± 1	567 ± 7

Table 1. *Cont.*

Nomenclature	Young Modulus (Mpa)	Stress (Mpa)	Strain at Break (%)
PBAT/LDH-104 5%	50 ± 1	22 ± 1	940 ± 10
PBAT/LDH-351 2%	57 ± 2	21 ± 1	455 ± 30
PBAT/LDH-351 5%	61 ± 1	20 ± 2	440 ± 15

Independently of the amount (2 or 5 wt %) of untreated LDH incorporated into PBAT, no significant influence was observed on the final mechanical performances of the polymer matrix. These results are in agreement with the literature where different authors have demonstrated that the addition of unmodified LDH had no impact on the mechanical properties of the matrix [37,50]. Nevertheless, the use of phosphonium ILs modified LDHs led to different behaviors depending on the IL's chemical nature. Thus, IL351 and IL349 led to increases in the Young's modulus of the order of 20–30% for only 2 wt % of fillers and 30–50% when 5 wt % of modified LDHs are used. In addition, only slight decreases of the fracture behavior are obtained (of the order of 10–15%). These results can be explained by their respective morphologies (Figure 2) where a good dispersion was observed but also due to the good affinity between LDH-ILs and PBAT. The results are promising compared to those reported in the literature [50,52]. In fact, different authors have highlighted that the organic modification of LDH by stearic acid, lauryl alcohol phosphoric acid ester potassium, dodecylsulfate or decanoate counter anions has no effect on the mechanical behavior of polyester nanocomposites based on PLA [50,52]. For example, Xie et al. demonstrated an increase of only 12.5% of the elongation at break with 7.5% of the tensile strength [50]. In the case of IL104, increases in the strain at break without reducing the Young Modulus of the neat PBAT are observed going from 511% to 570% and 940% when 2 wt % and 5 wt % were used, respectively. This phenomenon can be attributed to the well dispersed ionic clusters (IL104, Figure 2), as observed when ionomers or ILs are used [53,54].

In conclusion, the use of phosphonium ILs as interfacial agents of LDH led to a significant improvement of the mechanical performances of PBAT matrix, which offers potential advances in the field of compostable films or food packaging applications.

2.2.4. Influence of Modified-LDHs on the Gas Transport Properties of PBAT Matrix

The LDH surface treatment influence, as well as the previously obtained different morphologies, were investigated by the permeability coefficients of various gases in neat PBAT and PBAT nanocomposites. The permeability coefficients of H_2, O_2, N_2, CO_2, H_2O and the ideal selectivities of selected gas pairs are presented in Table 2. Diffusion and solubility coefficients and their corresponding ideal selectivities are presented in the supplementary material (Tables S1 and S2).

Table 2. Dependence of gas and water vapor permeability coefficients and corresponding ideal selectivities in PBAT polymer materials containing different amount of pristine and modified-LDHs.

Materials	Permeability Coefficient (Barrer) [a]					Ideal Selectivity				
	H_2	O_2	N_2	CO_2	H_2O	H_2/N_2	H_2O/O_2	O_2/N_2	CO_2/H_2	CO_2/N_2
PBAT	4.92	1.22	0.33	12.5	2580	15.1	2110	3.7	2.5	38.3
PBAT/LDH 2%	4.53	1.04	0.31	11.0	1970	14.8	1900	3.4	2.4	36.1
PBAT/LDH 5%	4.38	0.95	0.31	10.0	2210	14.3	2320	3.1	2.3	32.7
PBAT/LDH-349 2%	4.01	0.95	0.29	9.9	1630	13.8	1730	3.3	2.5	33.9
PBAT/LDH-104 2%	3.74	0.93	0.27	9.5	1360	13.7	1470	3.4	2.5	34.8
PBAT/LDH-349 5%	4.97	1.08	0.34	11.7	1400	14.6	1300	3.2	2.4	34.5
PBAT/LDH-351 5%	4.20	1.00	0.28	10.1	1520	15.2	1520	3.6	2.4	36.5
PBAT/LDH-104 5%	4.27	1.05	0.32	10.4	1710	13.2	1630	3.2	2.4	32.1

[a] Barrer = 1×10^{-10} cm^3 (STP) cm/(cm^2 s cm Hg) = 3.3539×10^{-16} mol s^{-1} m^{-1} Pa^{-1}.

In all cases, the permeability coefficients of H_2, O_2, N_2, CO_2, H_2O increases in the following order:

$$N_2 < O_2 < H_2 < CO_2 << H_2O$$

Rigid polymer structures of glassy polymers promote molecules with higher diffusion coefficients and because $D(H_2) >> D(CO_2)$ permeability coefficients are usually larger than those of CO_2 [55]. In this case, flexible blocks in PBAT and relatively high polarity led to strong interaction between PBAT and CO_2 as well as the polymer matrix being above its T_g. In other words CO_2 exhibits significant interactions with the polymer which means high solubility (sorption coefficients) in the polymer and this phenomenon causes CO_2 to permeate significantly faster than H_2 (Tables S1 and S2 in supplementary material). These results are in agreement with the literature where different authors have highlighted this phenomenon which can be explained by the interactions between CO_2 (polar molecule) with the ester groups of PBAT [49,56]. In addition, due to this excellent affinity, other research groups investigated the foaming of PBAT under supercritical CO_2 [57]. Nitrogen, on the other hand, interacts with the polymer very slightly. Thus, we can observe relatively high selectivities for the CO_2/N_2 gas pair (Table 2) which makes these materials potentially suitable for CO_2/N_2 separations. Other gas pairs exhibited only moderate values comparable with most general polymers.

From transport data in Table 2 we can surprisingly see a slight influence of all untreated and treated fillers on both permeabilities and selectivities. In fact a slight decrease in permeabilities can be observed with addition of untreated and treated LDHs. From the morphological point of view most of the gas transport occurs in the neat polymer matrix due to low filler contents, especially in the case of untreated LDHs which creates agglomerates. Agglomerates (seen in Figure 2a) themselves can contain empty spaces where the gas can diffuse freely and thus these spaces promote permeabilities and maintain selectivity, which is determined mostly by the polymer matrix itself. On the other hand treated LDHs tends to disperse well in the polymer matrix, as well as to exfoliate some of its layers which can act as barriers for the gas diffusion. Exfoliation can also make intercalated IL more accessible to interactions with diffusing gases. In our case no such effects played a significant role except for the water vapor (see below). This is in agreement with the theory of transport properties of polymeric materials filled with impermeable or low permeability particles. Filler particles, especially layered ones, partially block pathways for gas molecules penetrating in free volume among the PBAT macromolecules [58,59]. As XRD data have shown the basal spacing of the LDH is not much influenced by the anion type, the same could be concluded for gas permeabilities [9].

Moreover, even stronger interactions between gas and polymer are observed for water vapor. Highly polar water molecules with small molecule dimensions cause several orders of magnitude of higher water vapor permeabilities in PBAT materials compared with N_2 and O_2, which are the main components of air. However, the incorporation of modified LDHs into the PBAT matrix led to significant decreases in water vapor permeability varying with function of the chemical nature of the phosphonium LIs as well as the morphologies previously generated (see Figure 2). Indeed, LDH-104 and LDH-349 led to the most significant decreases in water vapor permeability coefficient. These results can be explained by their respective homogeneous semi-exfoliated morphologies. In the case of PBAT/LDH-104, a decrease of 50% is obtained independently of the amount of modified-LDH used. This is a result of the excellent dispersion of the lamellar fillers as well as of the dispersion of ionic clusters Higher surface of highly polar and well dispersed semi-exfoliated particles create more sorption sites, where the water molecules could sorb resulting in a large number of strong interactions slowing down the water molecules' diffusion through the membrane. Water vapor measurements were performed at water partial pressures close to the saturated one—this means completely non-Fickean diffusion so the diffusion coefficient D and the solubility S parameters in Tables S1 and S2 are only apparent. Concerning PBAT/LDH-351, presenting an intercalated morphology combined with the presence of few aggregates, this decrease is of the order of 30%. On the other hand, the poor dispersion of unmodified LDH into the PBAT matrix resulted in a very slight decrease of only 10%.

In summary, the presence of IL modified-LDHs has no measurable effect on gas transport due to similar diffusion and sorption coefficients of gases in ILs and PBAT matrix, but a significant influence on the water vapor permeation has been highlighted. Thus, these nanocomposites can be good candidates for food packaging applications.

3. Materials and Methods

An Layered Double Hydroxide LDH (aluminum magnesium hydroxy carbonate) denoted PURAL® MG 63 HT) was chosen as pristine anionic clay and was provided by Sasol Performance Chemicals (Hamburg, Germany). PBAT used in this study was supplied by BASF (Ludwigshafen, Germany) under the trade name of Ecoflex. The ILs denoted IL104, IL351, and IL349 based on tributyltetradecylphosphonium cation associated with phosphinate, carboxylate and phosphate counter anions were kindly provided by Cytec Industries Inc. (Thorold, ON, Canada)

Organic modification of LDH: First, the pristine LDH was heated for 24 h at 500 °C to obtain calcined LDH. Then, based on the Anionic Exchange Capacity (AEC = 3.35 meq/g) of the LDH used [9,37,60], LDH and 2 AEC of phosphonium ILs were dispersed in 200 mL of deionized water/tetrahydrofuran mixture (300/100 mL). After, the suspensions were stirred and mixed at 60 °C over 24 h. The resulting precipitate was filtered and washed 5 times with THF. The residual solvent was removed by evaporation under vacuum and finally, the treated LDH was dried overnight at 80°C. The phosphonium ILs used for the anionic exchange and the following abbreviations used to designate the various treated-LDHs are summarized in Table 3.

Table 3. Designation of ionic liquids (ILs) used for the surface treatment of layered double hydroxide (LDH).

Ionic Liquid	Chemical Structure	Designation
Trihexyl(tetradecyl)phosphonium bis(2,4,4-trimethylpentyl)phosphinate		LDH-104
Trihexyl(tetradecyl)phosphonium 2-ethylhexanoate		LDH-351
Trihexyl(tetradecyl)phosphonium bis(2-ethylhexyl)phosphate		LDH-349

Nanocomposites based on PBAT/organically treated LDHs (2 and 5 wt %) i.e., PBAT/LDH, PBAT/LDH-104, PBAT/LDH-351, PBAT/LDH-349 were processed using a 15 g-capacity DSM micro-extruder (DSM Research, Heerlen, The Netherlands) with co-rotating screws (Lenght/Diameter L/D ratio of 18). The mixture was sheared over 3 min with a 100 rpm speed at 160 °C and injected into a 10 cm³ mold at 30 °C to obtain dumbbell-shaped specimens.

Thermogravimetric analysis (TGA) of ILs, untreated and treated-LDH and nanocomposites were performed on a Q500 thermogravimetric analyzer (TA instruments, New Castle, DE, USA). The samples were heated from room temperature to 600 °C at a rate of 20 K·min^{-1} under nitrogen flow.

Differential Scanning Calorimetry measurements *(DSC)* of PBAT and the resulting nanocomposites were performed on a Q20 (TA instruments) from −60 °C to 180 °C. The samples were kept for 1 min at 180 °C to erase the thermal history before being heated or cooled at a rate of 10 K·min^{-1} under nitrogen flow of 50 mL·min^{-1}. The crystallinity was calculated with the heat of fusion for PBAT of 114 J/g [61].

Wide-angle X-ray diffraction spectra (WAXD) were collected on a D8 Advance X-ray diffractometer (Bruker, Billerica, MA, USA) at the Henri Longchambon diffractometry center. A bent quartz monochromator was used to select the Cu K$_{\alpha 1}$ radiation (λ = 0.15406 nm) and run under operating conditions of 45 mA and 33 kV in Bragg-Brentano geometry. The angle range scanned is 1–10° 2θ for the modified clays and for the nanocomposite materials.

Transmission electron microscopy *(TEM)* was carried out at the Center of Microstructures (University of Lyon, France) on a Philips CM 120 field emission scanning electron microscope (Philips, Amsterdam, The Netherlands) with an accelerating voltage of 80 kV. The samples were cut using an ultramicrotome (Leica, Weitzlar, Germany) equipped with a diamond knife, to obtain 60 nm-thick ultrathin sections. Then, the sections were set on copper grids.

Uniaxial Tensile Tests were carried out on a MTS 2/M electromechanical testing system (MTS, Eden Prairie, MN, USA) at 22 °C ± 1 °C and 50 ± 5% relative humidity at crosshead speed of 50 mm·min^{-1}.

Dynamic mechanical analyses were performed on ARES G2 rheometer (TA Instruments). The temperature dependence of the complex shear modulus of rectangular samples (dimension: 20 × 5 × 1.5 mm^3) was measured by oscillatory shear deformation at a frequency of 1 Hz and a heating rate of 3 °C·min^{-1} in a temperature range of −100 to +100 °C. The temperature of a relaxation was evaluated as the maximum of tan δ peak.

Gas transport properties of PBAT with various content of LDH were examined by the time-lag permeation method [62]. Samples in the form of thin films (prepared in a hot-press at 160 °C) were inserted into a membrane cell which was then placed into a permeation apparatus. The sample was then exposed to high vacuum (10–4 mbar) and temperature 45 °C for 12 h. After evacuation the temperature was set on 30 °C. Then 2–3 samples of each membrane were measured. Feed pressure p_i was 1.5 bar. The permeability coefficient P was determined from the increase of the permeate pressure ΔP_p per time interval Δt in a calibrated volume V_p of the product part during the steady state of permeation. For calculation of permeability coefficient, the following formula was used:

$$P = \frac{\Delta p_p}{\Delta t} \times \frac{V_p l}{A p_i} \times \frac{1}{RT} \tag{1}$$

where l is the membrane thickness, p_i feed pressure, A the area, T the temperature, and R the gas constant. Relative standard deviations (SD) of ΔP_p and Δt were lower than 0.3% (given by the 10 mbar MKS Barratron pressure transducer precision). Relative SD of membrane thickness measurement was 1%, relative SD of calibrated volume was lower than 0.5%, and relative SD of feed pressure was 0.3%. Therefore P values had the relative SD 2.4%, very low values of P (below 0.01 Barrers) had relative SD up to 15%, due to lower (nonlinear) precision of the MKS Baratron at very low pressures.

Gas diffusivities were estimated from the time-lag data, using the relation:

$$D = \frac{l^2}{6\theta} \tag{2}$$

where D is the diffusion coefficient, l is the film thickness and θ is the time-lag. Relative standard deviation of diffusion coefficients was 4%. A precision of 0.1 s for the time-lag determination allowed the determination of the diffusion coefficients of hydrogen with relative standard deviation of 8%. Apparent solubility coefficients were calculated using the following equation:

$$S = P/D \tag{3}$$

The overall selectivity of a polymer membrane for a pair of gases *i* and *j* is commonly expressed in terms of an ideal separation factor, *αij*, defined by the following relation:

$$\alpha_{ij} = \frac{P_i}{P_j} = \frac{S_i}{S_j} \times \frac{D_i}{D_j} \tag{4}$$

where P_i and P_j are pure gas permeabilities, D_i/D_j is the diffusion selectivity and S_i/S_j is the solubility selectivity.

4. Conclusions

In this work, phosphonium ILs were used as LDH interfacial agents to prepare thermally stable LDHs by the extrusion process. Firstly, these modified-LDHs were characterized by TGA and X-ray diffraction, highlighting the intercalation of the counter anions among the LDH layers as well as the presence of physisorbed ILs on the LDHs surface. Then, different amounts (2 and 5 wt %) of these organically modified LDHs were introduced into a PBAT matrix by melt extrusion, resulting in partially exfoliated morphologies. Thus, a mechanical performances increase (20% to 50% of the Young's modulus) combined with a water barrier properties increase (30% to 50%) was obtained for all IL-modified-LDH based nanocomposites. These results demonstrated the influence the surfactant's chemical nature, particularly the counter anion, has on the morphologies as well as on the final properties of the PBAT matrix. Altogether, these PBAT based nanocomposites open new possibilities for the field of compostable films and food packaging applications.

Supplementary Materials: The dependence of gas and water vapor diffusion and of gas and water vapor solubility coefficients in PBAT polymer materials containing different amount of pristine and modified-LDHs following are available online at http://www.mdpi.com/2079-4991/7/10/297/s1, Table S1, Dependence of gas and water vapor diffusion coefficients in PBAT polymer materials containing different amount of pristine and modified-LDHs, Table S2: Dependence of gas and water vapor solubility coefficients in PBAT polymer materials containing different amounts of pristine and modified-LDHs.

Acknowledgments: The authors gratefully thank the Grant Agency of the Czech Republic (project 17-08273S) for financial support.

Author Contributions: Sébastien Livi, Luanda C. Lins, Sébastien Pruvost conceived the paper and designed the experiments. Hynek Benes, Jana Kredatusova, Jakub Peter performed the thin films, the mechanical and the gas transport properties. Sébastien Livi, Sébastien Pruvost performed the organic modification of LDHs and their characterization (XRD, TGA) but also the TGA of polymer/LDH nanocomposites. Luanda C. Lins and Ricardo K. Donato performed the extrusion process, DSC and TEM experiments. Sébastien Livi, Luanda C. Lins, Peter Jakub analyzed the data and also wrote the paper.

Conflicts of Interest: The authors declare no conflict of interest.

References

1. Giannelis, E.P. Polymer Layered Silicate Nanocomposites. *Adv. Mater.* **1996**, *8*, 29–35. [CrossRef]
2. Le Baron, P.C.; Zhen, W.; Pinnavaia, J. Polymer-Layered Silicate Nanocomposites: An overview. *Appl. Clay Sci.* **1999**, *15*, 11–29. [CrossRef]
3. Chang, J.H.; An, Y.U.; Cho, D.; Giannelis, E.P. Poly(lactid acid) nanocomposites: Comparison of their properties with montmorillonite and synthetic mica (II). *Polymer* **2003**, *44*, 3715–3720. [CrossRef]
4. Ray, S.S.; Okamoto, M. Polymer/layered silicate nanocomposites: A review from preparation to processing. *Prog. Polym. Sci.* **2003**, *28*, 1539–1641.
5. Alexandre, M.; Dubois, P. Polymer-layered silicate nanocomposites: Preparation, properties and uses of a new class of materials. *Mater. Sci. Eng. R Rep.* **2000**, *28*, 1–63. [CrossRef]
6. Lepoittevin, B.; Devalckenaere, M.; Pantoussier, N.; Alexandre, M.; Kubies, D.; Calberg, C.; Jérôme, R.; Dubois, P. Poly(ε-caprolactone)/clay nanocomposites prepared by melt intercalation: Mechanical, thermal and rheological properties. *Polymer* **2002**, *43*, 4017–4023. [CrossRef]

7. Paul, M.-A.; Alexandre, M.; Degée, P.; Henrist, C.; Rulmont, A.; Dubois, P. New nanocomposite materials based on plasticized poly(L-lactide) and organo-modified montmorillonites: Thermal and morphological study. *Polymer* **2003**, *44*, 443–450. [CrossRef]
8. Yang, Z.; Peng, H.; Wang, W.; Liu, T. Crystallization behavior of poly(ε-caprolactone)/layered double hydroxide nanocomposites. *J. Appl. Polym. Sci.* **2010**, *116*, 2658–2667. [CrossRef]
9. Kredatusová, J.; Beneš, H.; Livi, S.; Pop-Georgievski, O.; Ecorchard, P.; Abbrent, S.; Pavlova, E.; Bogdał, D. Influence of ionic liquid-modified LDH on microwave-assisted polymerization of ε-caprolactone. *Polymer* **2016**, *100*, 86–94. [CrossRef]
10. Romeo, V.; Gorrasi, G.; Vittoria, V.; Chronakis, I.S. Encapsulation and Exfoliation of Inorganic Lamellar Fillers into Polycaprolactone by Electrospinning. *Biomacromolecules* **2007**, *8*, 3147–3152. [CrossRef] [PubMed]
11. Peng, H.; Han, Y.; Liu, T.; Tjiu, W.C.; He, C. Morphology and thermal degradation behavior of highly exfoliated CoAl-layered double hydroxide/polycaprolactone nanocomposites prepared by simple solution intercalation. *Thermochim. Acta* **2010**, *502*, 1–7. [CrossRef]
12. Jager, E.; Donato, R.K.; Perchacz, M.; Jager, A.; Surman, F.; Hocherl, A.; Konefal, R.; Donato, K.Z.; Venturini, C.G.; Bergamo, V.Z.; et al. Biocompatible succinic acid-based polyesters for potential biomedical applications: Fungal biofilm inhibition and mesenchymal stem cell growth. *RSC Adv.* **2015**, *5*, 85756–85766. [CrossRef]
13. Pinheiro, I.F.; Morales, A.R.; Mei, L.H. Polymeric biocomposites of poly (butylene adipate-co-terephthalate) reinforced with natural Munguba fibers. *Cellulose* **2014**, *21*, 4381–4391. [CrossRef]
14. Silva, J.S.P.; Silva, J.M.F.; Soares, B.G.; Livi, S. Fully biodegradable composites based on poly(butylene adipate-co-terephthalate)/peach palm trees fiber. *Composites Part B* **2017**, *129*, 117–123. [CrossRef]
15. Russo, P.; Carfagna, C.; Cimino, F.; Acierno, D.; Persico, P. Biodegradable composites reinforced with kenaf fibers: Thermal, mechanical, and morphological issues. *Adv. Polym. Technol.* **2013**, *32*, E313–E322. [CrossRef]
16. Shen, W.; He, H.; Zhu, J.; Yuan, P.; Frost, R.L. Grafting of montmorillonite with different functional silanes via two different reaction systems. *J. Colloid Interface Sci.* **2007**, *313*, 268–273. [CrossRef] [PubMed]
17. He, H.; Duchet, J.; Galy, J.; Gerard, J.-F. Grafting of swelling clay materials with 3-aminopropyltriethoxysilane. *J. Colloid Interface Sci.* **2005**, *288*, 171–176. [CrossRef] [PubMed]
18. Kornmann, X.; Lindberg, H.; Berglund, L.A. Synthesis of epoxy–clay nanocomposites: Influence of the nature of the clay on structure. *Polymer* **2001**, *42*, 1303–1310. [CrossRef]
19. Livi, S.; Duchet-Rumeau, J.; Pham, T.-N.; Gérard, J.-F. A comparative study on different ionic liquids used as surfactants: Effect on thermal and mechanical properties of high-density polyethylene nanocomposites. *J. Colloid Interface Sci.* **2010**, *349*, 424–433. [CrossRef] [PubMed]
20. Heinz, H.; Vaia, R.A.; Krishnamoorti, R.; Farmer, B.L. Self-Assembly of Alkylammonium Chains on Montmorillonite: Effect of Chain Length, Head Group Structure, and Cation Exchange Capacity. *Chem. Mater.* **2007**, *19*, 59–68. [CrossRef]
21. Okamoto, K.; Sasaki, T.; Fujita, T.; Iyi, N. Preparation of highly oriented organic-LDH hybrid films by combining the decarbonation, anion-exchange, and delamination processes. *J. Mater. Chem.* **2006**, *16*, 1608–1616. [CrossRef]
22. Crepaldi, E.L.; Pavan, P.C.; Valim, J.B. A new method of intercalation by anion exchange in layered double hydroxides. *Chem. Commun.* **1999**, *2*, 155–156. [CrossRef]
23. Wang, Q.; O'Hare, D. Recent Advances in the Synthesis and Application of Layered Double Hydroxide (LDH) Nanosheets. *Chem. Rev.* **2012**, *112*, 4124–4155. [CrossRef] [PubMed]
24. Donato, R.K.; Luza, L.; da Silva, R.F.; Moro, C.C.; Guzatto, R.; Samios, D.; Matějka, L.; Dimzoski, B.; Amico, S.C.; Schrekker, H.S. The role of oleate-functionalized layered double hydroxide in the melt compounding of polypropylene nanocomposites. *Mater. Sci. Eng. C* **2012**, *32*, 2396–2403. [CrossRef]
25. Livi, S.; Duchet-Rumeau, J.; Gérard, J.-F. Tailoring of interfacial properties by ionic liquids in a fluorinated matrix based nanocomposites. *Eur. Polym. J.* **2011**, *47*, 1361–1369. [CrossRef]
26. Livi, S.; Duchet-Rumeau, J.; Pham, T.N.; Gérard, J.-F. Synthesis and physical properties of new surfactants based on ionic liquids: Improvement of thermal stability and mechanical behavior of high density polyethylene nanocomposites. *J. Colloid Interface Sci.* **2011**, *354*, 555–562. [CrossRef] [PubMed]
27. Perchacz, M.; Donato, R.K.; Seixas, L.; Zhigunov, A.; Konefał, R.; Serkis-Rodzeń, M.; Beneš, H. Ionic Liquid-Silica Precursors via Solvent-Free Sol–Gel Process and Their Application in Epoxy-Amine Network: A Theoretical/Experimental Study. *ACS Appl. Mater. Interfaces* **2017**, *9*, 16474–16487. [CrossRef] [PubMed]

28. Donato, R.K.; Matejka, L.; Schrekker, H.S.; Plestil, J.; Jigounov, A.; Brus, J.; Slouf, M. The multifunctional role of ionic liquids in the formation of epoxy-silica nanocomposites. *J. Mater. Chem.* **2011**, *21*, 13801–13810. [CrossRef]

29. Donato, R.K.; Lavorgna, M.; Musto, P.; Donato, K.Z.; Jager, A.; Štěpánek, P.; Schrekker, H.S.; Matějka, L. The role of ether-functionalized ionic liquids in the sol–gel process: Effects on the initial alkoxide hydrolysis steps. *J. Colloid Interface Sci.* **2015**, *447*, 77–84. [CrossRef] [PubMed]

30. Donato, R.K.; Perchacz, M.; Ponyrko, S.; Donato, K.Z.; Schrekker, H.S.; Benes, H.; Matejka, L. Epoxy-silica nanocomposite interphase control using task-specific ionic liquids via hydrolytic and non-hydrolytic sol-gel processes. *RSC Adv.* **2015**, *5*, 91330–91339. [CrossRef]

31. Bugatti, V.; Livi, S.; Hayrapetyan, S.; Wang, Y.; Estevez, L.; Vittoria, V.; Giannelis, E.P. Deposition of LDH on plasma treated polylactic acid to reduce water permeability. *J. Colloid Interface Sci.* **2013**, *396*, 47–52. [CrossRef] [PubMed]

32. Nguyen, T.K.L.; Livi, S.; Soares, B.G.; Benes, H.; Gérard, J.-F.; Duchet-Rumeau, J. Toughening of Epoxy/Ionic Liquid Networks with Thermoplastics Based on Poly(2,6-dimethyl-1,4-phenylene ether) (PPE). *ACS Sustain. Chem. Eng.* **2017**, *5*, 1153–1164. [CrossRef]

33. Sonnier, R.; Dumazert, L.; Livi, S.; Nguyen, T.K.L.; Duchet-Rumeau, J.; Vahabi, H.; Laheurte, P. Flame retardancy of phosphorus-containing ionic liquid based epoxy networks. *Polym. Degrad. Stab.* **2016**, *134*, 186–193. [CrossRef]

34. Awad, W.H.; Gilman, J.W.; Nyden, M.; Harris, R.H.; Sutto, T.E.; Callahan, J.; Trulove, P.C.; DeLong, H.C.; Fox, D.M. Thermal degradation studies of alkyl-imidazolium salts and their application in nanocomposites. *Thermochim. Acta* **2004**, *409*, 3–11. [CrossRef]

35. Ngo, H.L.; LeCompte, K.; Hargens, L.; McEwen, A.B. Thermal properties of imidazolium ionic liquids. *Thermochim. Acta* **2000**, *357*, 97–102. [CrossRef]

36. Zhao, Y.; Li, F.; Zhang, R.; Evans, D.G.; Duan, X. Preparation of Layered Double-Hydroxide Nanomaterials with a Uniform Crystallite Size Using a New Method Involving Separate Nucleation and Aging Steps. *Chem. Mater.* **2002**, *14*, 4286–4291. [CrossRef]

37. Livi, S.; Bugatti, V.; Estevez, L.; Duchet-Rumeau, J.; Giannelis, E.P. Synthesis and physical properties of new layered double hydroxides based on ionic liquids: Application to a polylactide matrix. *J. Colloid Interface Sci.* **2012**, *388*, 123–129. [CrossRef] [PubMed]

38. Oyarzabal, A.; Mugica, A.; Müller, A.J.; Zubitur, M. Hydrolytic degradation of nanocomposites based on poly(L-lactic acid) and layered double hydroxides modified with a model drug. *J. Appl. Polym. Sci.* **2016**, *133*. [CrossRef]

39. Kanezaki, E. Intercalation of naphthalene-2,6-disulfonate between layers of Mg and Al double hydroxide: Preparation, powder X-Ray diffraction, fourier transform infrared spectra and X-Ray photoelectron spectra. *Mater. Res. Bull.* **1999**, *34*, 1435–1440. [CrossRef]

40. Soares, B.G.; Ferreira, S.C.; Livi, S. Modification of anionic and cationic clays by zwitterionic imidazolium ionic liquid and their effect on the epoxy-based nanocomposites. *Appl. Clay Sci.* **2017**, *135*, 347–354. [CrossRef]

41. Ha, J.U.; Xanthos, M. Novel modifiers for layered double hydroxides and their effects on the properties of polylactic acid composites. *Appl. Clay Sci.* **2010**, *47*, 303–310. [CrossRef]

42. Ding, P.; Kang, B.; Zhang, J.; Yang, J.; Song, N.; Tang, S.; Shi, L. Phosphorus-containing flame retardant modified layered double hydroxides and their applications on polylactide film with good transparency. *J. Colloid Interface Sci.* **2015**, *440*, 46–52. [CrossRef] [PubMed]

43. Leroux, F.; Besse, J.-P. Polymer Interleaved Layered Double Hydroxide: A New Emerging Class of Nanocomposites. *Chem. Mater.* **2001**, *13*, 3507–3515. [CrossRef]

44. Aranda, P.; Ruiz-Hitzky, E. Poly(ethylene oxide)-silicate intercalation materials. *Chem. Mater.* **1992**, *4*, 1395–1403. [CrossRef]

45. Greenland, D.J. Adsorption of polyvinyl alcohols by montmorillonite. *J. Colloid Sci.* **1963**, *18*, 647–664. [CrossRef]

46. Livi, S.; Bugatti, V.; Soares, B.G.; Duchet-Rumeau, J. Structuration of ionic liquids in a poly(butylene-adipate-co-terephthalate) matrix: Its influence on the water vapour permeability and mechanical properties. *Green Chem.* **2014**, *16*, 3758–3762. [CrossRef]

47. Park, K.I.; Xanthos, M. A study on the degradation of polylactic acid in the presence of phosphonium ionic liquids. *Polym. Degrad. Stab.* **2009**, *94*, 834–844. [CrossRef]
48. Nayak, S.K. Biodegradable PBAT/starch nanocomposites. *Polym. Plast. Technol. Eng.* **2010**, *49*, 1406–1418. [CrossRef]
49. Livi, S.; Sar, G.; Bugatti, V.; Espuche, E.; Duchet-Rumeau, J. Synthesis and physical properties of new layered silicates based on ionic liquids: Improvement of thermal stability, mechanical behavior and water permeability of PBAT nanocomposites. *RSC Adv.* **2014**, *4*, 26452–26461. [CrossRef]
50. Xie, J.; Zhang, K.; Wu, J.; Ren, G.; Chen, H.; Xu, J. Bio-nanocomposite films reinforced with organo-modified layered double hydroxides: Preparation, morphology and properties. *Appl. Clay Sci.* **2016**, *126*, 72–80. [CrossRef]
51. Du, L.C.; Qu, B.J.; Zhang, M. Thermal properties and combustion characterization of nylon 6/MgAl-LDH nanocomposites via organic modification and melt intercalation. *Polym. Degrad. Stab.* **2007**, *92*, 497–502. [CrossRef]
52. Dagnon, K.L.; Chen, H.H.; Innocentini-Mei, L.H.; Souza, N.A.D. Poly[(3-hydroxybutyrate)-co-(3-hydroxyvalerate)]/layered double hydroxide nanocomposites. *Polym. Int.* **2009**, *58*, 133–141. [CrossRef]
53. Storey, R.F.; Bauch, D.W. Poly(styrene-*b*-isobutylene-*b*-styrene) block copolymers and ionomers therefrom: Morphology as determined by small-angle X-ray scattering and transmission electron microscopy. *Polymer* **2000**, *41*, 3195–3205. [CrossRef]
54. Livi, S.; Gerard, J.F.; Duchet-Rumeau, J. Ionic liquids: Structuration agents in a fluorinated matrix. *Chem. Commun.* **2011**, *47*, 3589–3591.
55. Freeman, T.B.; Yampolskii, Y.; Pinnau, I. *Materials Science of Membranes for Gas and Vapor Separation*; John Wiley & Sons, Ltd.: Chichester, UK, 2006.
56. Gain, O.; Espuche, E.; Pollet, E.; Alexandre, M.; Dubois, P. Gas barrier properties of poly(ε-caprolactone)/clay nanocomposites: Influence of the morphology and polymer/clay interactions. *J. Polym. Sci. Part B Polym. Phys.* **2005**, *43*, 205–214. [CrossRef]
57. Livi, S.; Lins, C.L.; Sar, G.; Gérard, J.-F.; Duchet-Rumeau, J. Supercritical CO$_2$–Ionic Liquids: A Successful Wedding To Prepare Biopolymer Foams. *ACS Sustain. Chem. Eng.* **2016**, *4*, 461–470. [CrossRef]
58. Labruyère, C.; Gorrasi, G.; Monteverde, F.; Alexandre, M.; Dubois, P. Transport properties of organic vapours in silicone/clay nanocomposites. *Polymer* **2009**, *50*, 3626–3637. [CrossRef]
59. Gorrasi, G.; Tortora, M.; Vittoria, V.; Pollet, E.; Lepoittevin, B.; Alexandre, M.; Dubois, P. Vapor barrier properties of polycaprolactone montmorillonite nanocomposites: Effect of clay dispersion. *Polymer* **2003**, *44*, 2271–2279. [CrossRef]
60. Nobuo, I.; Taki, M.; Yoshiro, K.; Kenji, K. A Novel Synthetic Route to Layered Double Hydroxides Using Hexamethylenetetramine. *Chem. Lett.* **2004**, *33*, 1122–1123.
61. Herrera, R.; Franco, L.; Rodríguez-Galán, A.; Puiggalí, J. Characterization and degradation behavior of poly(butylene adipate-co-terephthalate)s. *J. Polym. Sci. Part A Polym. Chem.* **2002**, *40*, 4141–4157. [CrossRef]
62. Rutherford, S.W.; Do, D.D. Review of time lag permeation technique as a method for characterisation of porous media and membranes. *Adsorption* **1997**, *3*, 283–312. [CrossRef]

nanomaterials

MDPI

Article

Transparent Pullulan/Mica Nanocomposite Coatings with Outstanding Oxygen Barrier Properties

Ilke Uysal Unalan [1,2,3,*], Derya Boyacı [1,4], Silvia Trabattoni [5], Silvia Tavazzi [5] and Stefano Farris [1,6,*]

1   DeFENS, Department of Food, Environmental and Nutritional Sciences—Packaging Division, University of Milan, via Celoria, 2, 20133 Milan, Italy; boyaci.derya@gmail.com
2   Department of Food Engineering, Faculty of Engineering, İzmir University of Economics, İzmir 35330, Turkey
3   School of Packaging, Michigan State University, East Lansing, MI 48824, USA
4   Department of Food Engineering, Izmir Institute of Technology, İzmir 35430, Turkey
5   Department of Materials Science, University of Milano Bicocca, via Cozzi 55, 20125 Milan, Italy; silvia.trabattoni@mater.unimib.it (S.T.); silvia.tavazzi@mater.unimib.it (S.T.)
6   INSTM, National Consortium of Materials Science and Technology, Local Unit University of Milan, via Celoria 2, 20133 Milan, Italy
*   Correspondence: iuysalunalan@gmail.com (I.U.U.); stefano.farris@unimi.it (S.F.); Tel.: +39-25-031-6805 (S.F.)

Received: 31 August 2017; Accepted: 15 September 2017; Published: 19 September 2017

Abstract: This study presents a new bionanocomposite coating on poly(ethylene terephthalate) (PET) made of pullulan and synthetic mica. Mica nanolayers have a very high aspect ratio (α), at levels much greater than that of conventional exfoliated clay layers (e.g., montmorillonite). A very small amount of mica (0.02 wt %, which is $\varphi \approx 0.00008$) in pullulan coatings dramatically improved the oxygen barrier performance of the nanocomposite films under dry conditions, however, this performance was partly lost as the environmental relative humidity (RH) increased. This outcome was explained in terms of the perturbation of the spatial ordering of mica sheets within the main pullulan phase, because of RH fluctuations. This was confirmed by modelling of the experimental oxygen transmission rate (OTR) data according to Cussler's model. The presence of the synthetic nanobuilding block (NBB) led to a decrease in both static and kinetic coefficients of friction, compared with neat PET ($\approx$12% and 23%, respectively) and PET coated with unloaded pullulan ($\approx$26% reduction in both coefficients). In spite of the presence of the filler, all of the coating formulations did not significantly impair the overall optical properties of the final material, which exhibited haze values below 3% and transmittance above 85%. The only exception to this was represented by the formulation with the highest loading of mica (1.5 wt %, which is $\varphi \approx 0.01$). These findings revealed, for the first time, the potential of the NBB mica to produce nanocomposite coatings in combination with biopolymers for the generation of new functional features, such as transparent high oxygen barrier materials.

Keywords: coefficient of friction; haze; mica; modelling; optical properties; oxygen barrier; pullulan

---

1. Introduction

Incorporation of two-dimensional nanomaterials as nanobuilding blocks (NBBs) in polymeric matrices paved the way for cutting-edge composites with unprecedented functional properties. Platelet-like nanoparticles in particular, such as layered silicates, have attracted much attention over the last twenty years. Layered silicate minerals include several classes and many groups that, in turn, account for different mineral species that are potentially suitable to produce nanocomposites. However, only few of these minerals (especially montmorillonite) have been widely exploited thus far [1]. Commercially available natural and organically modified montmorillonite show some disadvantages, such as limited aspect ratios (α) < 100 and high surface charge heterogeneity. This quality leads to non-uniform interlayer reactivity that hampers control over the nanoplatelets' stiffness [2]. The bigger advantage of using synthetic clays is represented by standardized physicochemical properties. Unlike natural clays, micas that can expand with very high aspect ratios have been obtained by synthetic pathways, which eventually promote better and higher quality dispersion in the polymer matrix [1,2]. For this reason, the use of synthetic NBBs as attractive nanomaterials is rising at both the academic and industrial level. Until now, only a few studies have dealt with the use of mica in nanocomposites, including thermoset and thermoplastic polymers to improve barrier performance [3,4] and mechanical properties [2,5–7]. There is, however, a gap in the literature that this work aims to fill, specifically dealing with advances regarding mica-based biopolymer nanocomposite coatings.

The use of nanofiller for this generation of bionanocomposites has enormous potential for overcoming the drawbacks that are exhibited by biopolymers, such as poor mechanical and thermal properties, sensitivity to moist environments, and inadequate barrier properties to gas and vapors. However, most examples were concerned with the incorporation of the inorganic phase directly into the bulky biopolymer. The use of fillers within coatings made of biopolymers has been proposed only very recently, to produce bionanocomposite coatings that improve the properties of a plastic substrate without jeopardizing its original attributes, to optimize cost efficiency, and to reduce environmental impact [1,8]. Bionanocomposite coatings were successfully produced by using natural montmorillonite (Na^+-MMT) [9] and graphene oxide [10] to improve the oxygen barrier properties, even at high relative humidity values, or in combination with microfibrillated cellulose (MFC) [11] and borax [12] to produce biocoatings with enhanced mechanical and permeability properties. In this study, we decided to use the NBB mica in combination with the exopolysaccharide pullulan to develop high-performance oxygen barrier materials that concurrently exhibit outstanding optical, surface, and mechanical properties. Among the variety of biopolymers, Pullulan was selected due to its unique properties, such as high flexibility, excellent transparency, good oxygen barrier properties, and non-toxicity—all of which make it a promising candidate for a next-generation of materials that are totally or partially based on renewable resources [13]. Poly(ethylene terephthalate) (PET) has been used as a plastic substrate because of its widespread use in many different applications, ranging from food packaging (e.g., liquids containers, thermoforming applications, layers for flexible packaging solutions) to energy applications (e.g., solar cells), displays, and medical/biomedical uses.

2. Results and Discussion

2.1. Morphological Characterization of Mica

Transmission electron microscopy (TEM) and atomic force microscopy (AFM) images of mica sheets are displayed in Figure 1, panels a–d. TEM allowed for the gathering of information on the degree of the exfoliation of mica sheets. As shown in Figure 1a, mica particles appeared as large overlapping sheets at the highest concentration (0.2 wt %). Dilution by one order of magnitude revealed the full exfoliation of mica to individual platelets by means of ultrasonication (Figure 1b). Quantitative information on the size of mica nanoparticles was obtained by AFM (Figure 1c,d). After collecting several images, it was possible to quantify both the width and thickness of synthetic mica, being equal to 3.6 ± 1.1 μm and 1.1 ± 0.2 nm, respectively. Eventually, an average aspect ratio (α) of 3615 ± 109 was assigned to the mica sheets, which was higher than the aspect ratio reported by [4] ($\alpha = 1064$). This difference can be plausibly attributed to the different type of mica used by those authors (i.e., a synthetic mica organically modified with di-methyl di-hydrogenated tallow ammonium–chloride as intercalant).

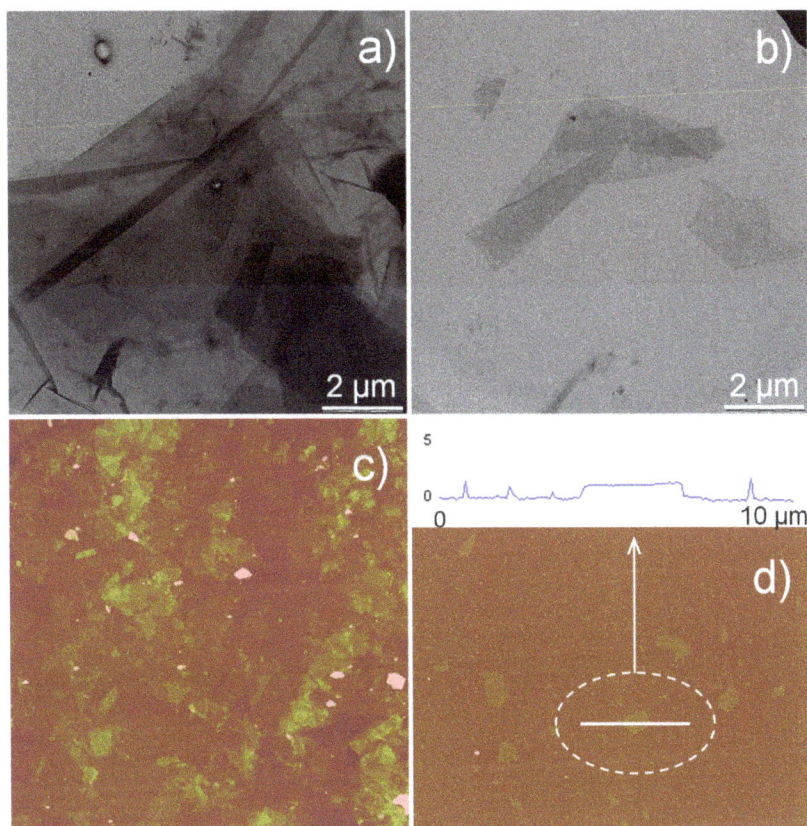

Figure 1. Transmission electron microscopy (TEM) images of mica nanosheets at (**a**) 0.2 wt % and (**b**) 0.02 wt %. Atomic force microscopy (AFM) height images of mica nanosheets: (**c**) at 0.2 wt % and 20 × 20 μm^2; (**d**) at 0.02 wt % and 40 × 40 μm^2.

2.2. Oxygen Barrier Performance

The oxygen barrier performance of bare PET, pullulan-coated PET, and pullulan/mica-coated PET films are reported in Table 1. Pullulan and pullulan nanocomposite coatings had a thickness that ranged from 0.75 μm to 0.80 μm. The deposition of the unloaded pullulan coating decreased the oxygen transmission rate (*OTR*) value of the plastic substrate (from ≈130 mL m^{-2} 24^{-1} to ≈6 mL m^{-2} 24^{-1} at 0% relative humidity (RH)). This has been explained by the tight network formed by pullulan chains due to the extensive hydrogen bonding [10,12]. The addition of mica dramatically increased the oxygen barrier performance of the pullulan coating, even at the lowest concentrations (0.02 and 0.04 wt %), thus yielding an *OTR* decrease of ≈80% in comparison to the bare pullulan coating (RH = 0%). Noticeably, from a concentration of mica of 0.06 wt % (φ = 0.00023) and greater, the *OTR* values of coated PET were below the instrument's detection limit (0.01 mL m^{-2} 24^{-1}). These excellent results suggest the successful exfoliation of mica nanosheets in water, mediated by ultrasonication. This process allowed the achievement of an outstanding performance, even at low loadings, due to both the "tortuosity path" and "organic/inorganic interface" effects [4].

Table 1. Oxygen transmission rate (*OTR*) of uncoated poly(ethylene terephthalate) (PET) and coated PET and bionanocomposite coatings at 0%, 30%, 60%, and 90% relative humidity (RH) for the different filler volume fraction (φ).

Mica Content		l (μm)	OTR (mL m^{-2} 24 h^{-1})			
wt %	φ [†]		0% RH	30% RH	60% RH	90% RH
uncoated PET	-	12.00 ± 0.03 [b]	129.23 ± 2.6 [a]	120.67 ± 0.9 [a]	115.10 ± 2.76 [a]	107.47 ± 0.74 [1]
PET/pullulan	-	12.75 ± 0.07 [a]	5.99 ± 0.02 [b]	26.74 ± 0.3 [b]	45.80 ± 2.65 [bc]	100.73 ± 3.23 [cd]
0.02	0.00008	12.76 ± 0.06 [a]	1.27 ± 0.24 [c]	18.79 ± 2.23 [c]	51.21 ± 4.98 [b]	100.63 ± 1.45 [cd]
0.04	0.00015	12.75 ± 0.07 [a]	1.12 ± 0.29 [c]	14.09 ± 2.74 [d]	50.20 ± 4.28 [b]	100.79 ± 4.27 [cd]
0.06	0.00023	12.78 ± 0.01 [a]	N.D.	13.35 ± 0.51 [d]	50.94 ± 0.1 [b]	99.56 ± 0.63 [cd]
0.08	0.00031	12.80 ± 0.05 [a]	N.D.	11.18 ± 2.24 [de]	49.10 ± 0.31 [b]	97.47 ± 0.42 [d]
0.1	0.00038	12.76 ± 0.05 [a]	N.D.	10.29 ± 0.3 [e]	46.76 ± 1.1 [bc]	96.62 ± 0.58 [d]
0.2	0.00077	12.77 ± 0.05 [a]	N.D.	8.39 ± 0.44 [ef]	41.05 ± 0.65 [cd]	102.40 ± 1.17 [bc]
0.5	0.00256	12.75 ± 0.01 [a]	N.D.	6.05 ± 0.48 [f]	36.01 ± 3.97 [d]	101.11 ± 0.71 [cd]
1	0.00510	12.80 ± 0.06 [a]	N.D.	6.89 ± 0.6 [f]	35.56 ± 5.6 [d]	105.88 ± 3.28 [ab]
1.5	0.01141	12.77 ± 0.06 [a]	N.D.	8.50 ± 0.6 [ef]	24.95 ± 1.16 [e]	103.22 ± 2.04 [abc]

[†] Calculated for a given mica density (ρ) = 2.6 g cm^{-3} and pullulan density (ρ) = 1 g cm^{-3}. [abcdef] Different superscripts within a group (i.e., within each parameter) denote a statistically significant difference ($p < 0.05$). Error around the mean value represents the standard deviation. N.D.: below the instrument detection limit (<0.01 mL m^{-2} 24^{-1}).

Increasing the relative humidity of the environment in contact with the coating led to a different scenario. Starting at 30% RH, the *OTR* values of PET films coated with pullulan increased to a slight extent. The same trend was more evident at 60% and especially 90% RH, insomuch as the benefit arising from the deposition of the pullulan nanocomposite coating disappeared completely at the highest RH value. The detrimental effect of humidity on the barrier properties of pullulan nanocomposite coatings has already been described for other lamellar clays, such as natural montmorillonite [9] and, more recently, graphene oxide [10].

This effect is ascribed to the plasticizing effect of water molecules adsorbed by the polymer surface and bulk, especially in correspondence with the amorphous regions [14,15]. The tight and dense network ensuing from the cooperative adhesion forces at the biopolymer/filler interface is gradually lost, with a concurrent increase in chain mobility and free volume of the nanocomposite network. This occurs until an unconstrained diffusion of the permeant at 90% RH is observed, due to what has been previously defined as the "diluting" effect of water molecules [10]. Eventually, the physical impedance (i.e., the increase of the diffusion path) due to the high aspect ratio of mica sheets has been overcompensated by the increase in free volume and subsequent higher diffusivity of the oxygen molecules through the plasticized network.

Permeability data has been widely modeled to acquire detailed information on the distribution and spatial organization of platy fillers in the main polymer network [16]. In turn, the outcome is a better interpretation of the ultimate O_2-barrier performance of the final material (e.g., coated plastic film), especially when external parameters (e.g., relative humidity) rise as perturbing factors. Figure 2 depicts the experimental *OTR* data of the bionanocomposite coatings at 30% and 60% RH and 23 °C, together with a theoretical prediction based on Cussler's permeation theoretical model. This model describes the permeation phenomenon for impermeable square platelets that are dispersed in a continuous matrix for a semi-dilute regime (i.e., $\alpha\varphi \gg 1$) [17,18]:

$$P_0/P \times (1 - \varphi) = (1 + \alpha\varphi/3)^2 \tag{1}$$

where P_0 is the permeability parameter of the pure biopolymer coating, P is the permeability parameter of the bionanocomposite coatings, α is the aspect ratio of the platelets (the width divided by the thickness), and φ is the volume fraction of the platelets dispersed in the biopolymer matrix. Experimental *OTR* data at 30% RH is compatible with Cussler's prediction for $\alpha = 4000$ (Figure 2a), up to a concentration of the filler that is equal to 0.2 wt %, namely $\varphi = 0.00077$. This clearly suggests that

Cussler's model approached the best fit of experimental data at α = 4000 for mica, which is in line with our experimental values obtained by AFM.

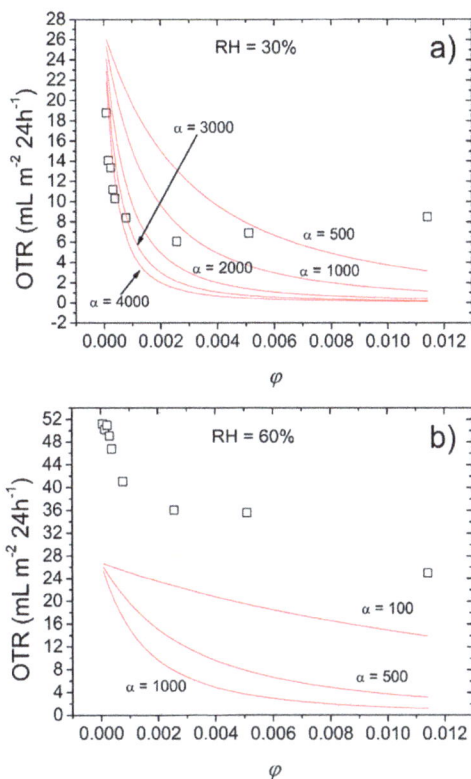

Figure 2. Experimental (symbols) and predicted (solid lines) *OTR* values of bionanocomposite hybrid coatings as a function of filler volume fraction (φ) for different aspect ratio (α) of mica platelets at (**a**) 30% RH and (**b**) 60% RH, according to Cussler's model (Equation (1) in the text).

Above the 0.2 wt % (φ = 0.00077) threshold, a clear deviation was observed between experimental and predicted *OTR* values, which can be reasonably due to the aggregation and stacking of mica sheets at high loading, according to the well-known "self-similar clay aggregation mechanism" [19]. This aspect is further emphasized as the relative humidity increased. At 90% RH, the applied model underestimated the aspect ratio of the filler to be lower than 100, which is in clear contrast with the TEM and AFM observations. This unrealistic prediction can be explained once again in terms of aggregation of the filler, which is exaggerated due to the aforementioned diluting effect of water molecules, insomuch as the tactoid configuration of mica can be thought to be restored.

The model simulation at lower RHs, together with the AFM and TEM results, strongly supports the higher aspect ratio of mica, compared to the widely used inorganic clays (e.g., montmorillonite), which have been reported to have aspect ratios between 10 and 100 for a similar pullulan-based system [9,20]. This aspect turns out to be of great importance during the design of high performance barrier materials (e.g., films and coatings), since using mica would be, in principle, more effective than using clays due to the wider surface (for the same volume fraction) opposed to the permeation.

2.3. Mechanical Properties

2.3.1. Friction Behavior

Control of friction influences a broad range of material applications, such as the packaging industry. Here, low coefficients of friction are desirable in order to avoid blocking the plastic webs that run on packaging lines, which would eventually affect the throughput of the process. For this reason, the presence of nanocoatings that are deposited on the plastic surface may have a relevant impact on the friction behavior of the plastic materials. A few studies have assessed the inherent friction behavior of mica layers [21,22]. However, no studies that investigated the improvement of the frictional performance of mica-based polymeric materials have been found.

Both static (μ_s) and dynamic (μ_k) friction coefficients of PET, pullulan, and pullulan-mica coatings are reported in Table 2. The addition of mica significantly decreased both μ_s and μ_k at a concentration of 0.2 wt %. In particular, a decrease of approximately 12% and 23% for the two coefficients, respectively, was obtained compared to the neat PET. Comparatively, a reduction of ≈26% for both coefficients was achieved compared to the pullulan-coated PET. The explanation for the observed improvement lies in the surface morphology of the nanocomposite coatings. More specifically, the addition of mica roughened the surface of the coatings in comparison to the unloaded pullulan coating, which exhibited a surface roughness (expressed as root-mean-square roughness, or RMS) between 1.2 nm and 3.0 nm [9,23]. In this work, RMS values of 4.0 nm and 7.0 nm were obtained by AFM for the pullulan coatings loaded with mica at concentrations of 0.2 wt % and 1.5 wt %, respectively. Noticeably, increasing the concentration of mica loaded in the main pullulan matrix to be above 0.2 wt % did not result in any appreciable improvement in the friction performance; that is, the loading of 0.2 wt % ($\varphi = 0.00077$) was the minimum and sufficient amount to achieve the reduction in both friction coefficients. Reversely, the fact that concentrations of the filler below 0.2 wt % had no effect on the friction performance is plausibly explained by considering that the tribological behavior of the coatings surface is influenced by the surface morphology and topography at a macro-scale level (i.e., when filler aggregates form on the coating surface). As already reported, differences at the nano-scale level do not seem to have any significant effect [24,25].

2.3.2. Tensile Properties

Mechanical properties of uncoated PET, pullulan-coated PET, and bionanocomposite films with different amounts of mica are summarized in Table 2. In general, the use of clay nanoplatelets as a reinforcing agent in polymeric matrices leads to an increase in stiffness and a concomitant decrease in elongation [26]. However, few studies can be found on the effect of nanocomposite coatings on the tensile properties of the plastic substrate beneath. In this work, the nanocomposite coating deposited on the PET substrate had a significant effect from a mica concentration of 0.2 wt % and greater. The elastic modulus (Emod) increased linearly, with a mica concentration up to a 7 wt % increase at a mica loading of 1.5 wt %, compared to the neat PET and pullulan-coated PET.

Despite the high stiffness of mica platelets (≈50 GPa) [27], the final reinforcing effect of mica was indeed moderate. The reason is that mica was added only to the coating biopolymer matrix (i.e., pullulan), while the ultimate elastic modulus value accounts for the final material, which includes the thin coating (less than 1 μm) and the thicker plastic substrate (≈12 μm).

That is, the greatest contribution on the final performance comes from the substrate, rather than the coating. For the same reason, the elongation at break (%) was not significantly influenced by the addition of synthetic mica to the pullulan coating. As for the tensile strength (TS), a significant improvement was recorded only at a concentration of 0.2 wt %, whereas at higher mica loadings TS decreased again, thus approaching the original values of bare PET and pullulan-coated PET. This behavior is most likely related to stress concentration points at the sharp tactoid edges, resulting in flaws (i.e., mechanical failures) at the polymer/filler interface, as already observed in poly(methyl methacrylate) (PMMA)/clay [27], poly(ethylene terephthalate)/mica [4], and epoxy/clay [7] nanocomposites.

Table 2. Coefficient of friction (COF, static and dynamic), elastic modulus (Emod), elongation at break (ε), and tensile strength (TS) of uncoated PET, pullulan-coated PET, and bionanocomposite coatings for different mica concentrations (wt %).

Sample	COF (Coating/Metal)		Emod (GPa)	ε (%)	TS (MPa)
	μ_s	μ_k			
uncoated PET	0.35 ± 0.01 [a]	0.26 ± 0.02 [a]	3.65 ± 0.20 [a]	15.80 ± 3.23 [a]	104.99 ± 9.65 [ab]
PET/Pullulan	0.42 ± 0.02 [b]	0.27 ± 0.01 [a]	3.65 ± 0.15 [a]	15.52 ± 3.50 [a]	105.33 ± 8.98 [ab]
Mica 0.02%	0.35 ± 0.01 [a]	0.23 ± 0.01 [b]	3.63 ± 0.15 [a]	16.37 ± 1.67 [a]	109.43 ± 2.33 [ab]
Mica 0.2%	0.31 ± 0.02 [c]	0.20 ± 0.02 [c]	3.71 ± 0.12 [ab]	18.82 ± 5.16 [a]	113.45 ± 7.04 [a]
Mica 0.5%	0.31 ± 0.02 [c]	0.20 ± 0.01 [c]	3.79 ± 0.17 [ab]	18.61 ± 2.99 [a]	105.23 ± 7.88 [ab]
Mica 1.0%	0.31 ± 0.02 [c]	0.21 ± 0.02 [bc]	3.81 ± 0.06 [ab]	18.97 ± 2.68 [a]	105.51 ± 10.23 [ab]
Mica 1.5%	0.31 ± 0.01 [c]	0.20 ± 0.01 [c]	3.90 ± 0.20 [b]	16.79 ± 3.15 [a]	102.21 ± 9.81 [b]

[abc] Different superscripts within a group (i.e., within each parameter) denote a statistically significant difference ($p < 0.05$). Error around the mean value represents the standard deviation.

2.4. Optical Properties

The optical properties of materials are of great importance for a large number of applications, such as construction (e.g., reflective glasses), agriculture (e.g., greenhouse windows), energy (e.g., solar cells), display/screen, and food packaging, just to provide a few examples. In many applications, transparent materials are sought in order to have a clear vision of the objects. From a practical point of view, the "see-through" property, which is obtained by reducing the contrast between objects viewed through the material, is actually one of the most important requirements as it can influence the final choice made by consumers [28].

For all of the coating formulations, the haze value of the coated PET is within 3.0 wt %, with the lowest value recorded for the pullulan-coated PET (Table 3). The 3.0 wt % threshold is deemed necessary to warrant a suitable display of the item behind the plastic film [9]. The only exception is represented by the formulation that includes the highest amount of mica (1.5 wt %), for which a final haze of 3.23% was measured. The increase in haze observed for films and coatings can be explained in terms of surface roughness [29] and the presence of scattering centers that come from the reaggregation of the nanoparticles [9]. As discussed previously (see Section 2.3.1), the addition of mica led to an increase in the roughness of the coating surface, in particular for the highest mica concentration (RMS = 7.0 nm). This, along with the reaggregation phenomenon postulated for the high mica concentrations (see Section 2.2), would justify the slight increase in haze arising from the addition of the filler. This observation is consistent with previous works on nanocomposite polymers, including platy [9,10,21] and rod-shape [11] nanoparticles.

Table 3. Haze (H), transmittance (T), and reflectance (R) of uncoated PET, pullulan-coated PET, and bionanocomposite coatings for different mica concentrations (wt %).

Sample	H (%)	T (%) 550 nm	T (%) 400–700 nm [†]	T_{simul} [†]	R (%) 400–700 nm [†]
uncoated PET	2.72 ± 0.08 [bc]	82.88 ± 0.77 [a]	83.4 ± 1.3	83.3 ± 0.1	10.4 ± 0.1
PET-pullulan	2.63 ± 0.22 [bc]	86.30 ± 0.94 [b]	86.9 ± 1.0	86.4 ± 1.2	6.4 ± 1.6
Mica 0.04%	2.81 ± 0.21 [b]	85.04 ± 0.41 [b]	N.A.	N.A.	N.A.
Mica 0.02%	2.69 ± 0.11 [bc]	85.72 ± 1.02 [b]	85.3 ± 1.2	84.9 ± 0.3	8.4 ± 0.4
Mica 1.5%	3.23 ± 0.17 [a]	83.28 ± 0.32 [a]	N.D.	N.D.	N.D.

[†] By ellipsometry. [abc] Different superscripts within a group (i.e., within each parameter) denote a statistically significant difference ($p < 0.05$). Error around the mean value represents the standard deviation. N.D.: not determined.

Transmittance values for uncoated PET and coated PET are reported in Table 3. All of the coating formulations had transmittance values that were higher than the bare PET, meaning that the coating deposition improved the overall performance of the plastic substrate. In particular, the best performance was again recorded for the pullulan-coated PET, thus suggesting the "clarity-enhancer" attribute of this biopolymer, presumably due to anti-reflective properties of this polysaccharide in the form of thin layers. To confirm this, we have decided to carry out some ellipsometry experiments.

In an initial step, the plastic substrate alone has been characterized to gather a proper model enabling the simulation of the PET optical properties.

The ellipsometry results were fitted by assuming a semi-infinite bulk of a transparent material with a refractive index described by the Cauchy model:

$$n(\lambda) = n_A + n_B/\lambda^2 \tag{2}$$

where n_A and n_B are free parameters for the fitting. Experimental evidence was found of interference fringes attributed to a possible thin layer on top of the PET substrate. However, this layer was neglected in the model since its effects were not critical and a reliable fitting of the ellipsometry data were obtained with the simpler model of a semi-infinite bulk material. The fitting procedure provided n_A = 1.8564 and n_B = 0.0019896, corresponding to a mean refractive index of the PET substrate in the visible range (320–700 nm) equal to 1.864 $\pm$ 0.002. The correlation factor between the n_A and n_B obtained by the fitting was relatively large (94.3%). Nevertheless, the model can be considered reliable because: (i) the correlation between n_A and n_B did not strongly affect the mean refractive index in the visible range; and (ii) the obtained refractive index is in reasonable agreement with data reported in the literature [30], notwithstanding the adopted approximation of a simplified model.

The model obtained for the PET semi-infinite substrate was then used to investigate the optical properties of the pullulan coating laid on the PET surface. The stack for the simulation was built by adding a transparent layer on top of the PET semi-infinite substrate. The optical properties of the PET were assumed to be known, as deduced from the previous ellipsometry measurements. The optical properties of the added layer were described by the Cauchy model, with new n_A and n_B parameters. The fitting of the ellipsometry data was performed with four free parameters, namely n_A and n_B, the thickness t of the added layer, and its thickness non-uniformity t_{n-u}. The results of the fitting are reported in the first line of Table 4, together with the corresponding mean-squared-error *(MSE)* value of the fitting and the refractive index of the top layer. The mean refractive index is in reasonable agreement with data reported in the literature for pullulan [31]. The correlation factors between the free parameters of the fitting are reported in Table S1. The relatively large correlation (67.9%) between the parameters n_A and n_B of the Cauchy model does not invalidate the result, for the same reasons as discussed for the PET substrate. The other correlation factors are relatively low. The same procedure and model were also adopted for the pullulan coating loaded with the lowest amount of mica (0.02 wt %). It is noticed that the lowest and highest refractive index is recorded for the uncoated PET and the pullulan-coated PET, respectively. Similarly, the non-uniform thickness of the top layer increased with the addition of mica. Transmittance measurements in the 320–800 nm spectral range were also performed at normal incidence on the PET, PET-pullulan, and PET-pullulan samples loaded with 0.02 wt % and 0.04 wt % of mica (Figure 3). The PET substrate showed a relatively low transmittance, while a marked increase was detected for the PET film coated with pullulan. This further demonstrates the anti-reflection behavior of the pullulan layer. Indeed, the refractive index of pullulan is intermediate between air and PET. With the addition of mica, the refractive index of the top layer increased again, approaching the PET refractive index value. Therefore, the addition of the nanofiller reduced the original anti-reflection behavior of the pullulan coating. This is confirmed by the fact that the transmittance spectra of the PET films coated with the nanocomposite formulations are half-way placed, with the spectrum obtained for the highest concentration of mica (0.02 wt %) closer to the transmittance spectrum of the bare PET.

Table 4. Parameters n_A, n_B, t, and t_{n-u} of the top layer obtained by fitting the ellipsometry $\Psi(\lambda)$ and $\Delta(\lambda)$ experimental data, *MSE* of the fitting, and deduced refractive index of the top layer both at 589 nm (n_{589}) and averaged in the visible range from 400 nm to 700 nm ($n_{mean} \pm$ std dev).

Samples	n_A	n_B (μm^2)	t (nm)	t_{n-u} (%)	*MSE*	n_{589}	$n_{mean} \pm$ std dev
uncoated PET	1.8564	0.0019896	-	-	-	-	1.864 ± 0.002
PET-pullulan	1.5593	0.0020923	470.50	20.6	4.71	1.559	1.569 ± 0.002
Mica 0.04 wt %	1.6552	0.021063	465.53	41.6	1.58	1.716	1.731 ± 0.026

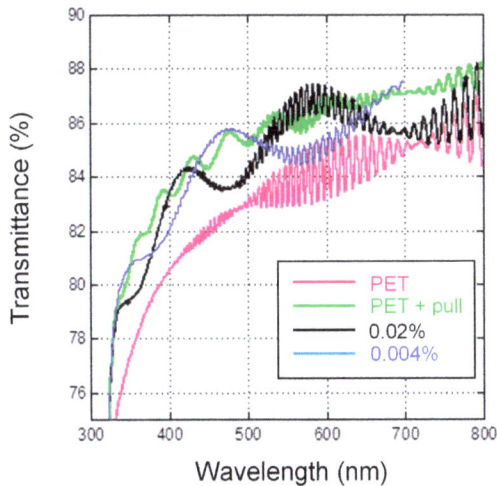

Figure 3. Transmittance spectra of bare PET, pullulan-coated PET, and PET coated with the nanocomposite coatings at different concentrations of mica (0.02 wt % and 0.04 wt %).

The mean value of the transmittance measured in the visible region is reported in Table 3. Transmittance was also simulated by taking into consideration the model and the parameters obtained by ellipsometry for each sample (T_{simul} of Table 3).

A good agreement was found between the measured and simulated values. The lowest transmittance was found for the PET substrate, while the largest one was again recorded for the pullulan-coated PET sample. The presence of mica in the top layer determined a decrease in the transmittance. Even if the reflectance was not experimentally measured, the spectra were calculated for near-normal incidence (20°, polarization s), based on the ellipsometry models (Table 3, last column). As expected, the minimum reflectance was calculated for the pullulan-coated PET sample, due to the anti-reflection behavior of the pullulan layer. Comparatively, an increase was observed when increasing the concentration of mica, which is consistent with previous measurements. These results confirm the anti-reflective properties of pullulan. The same behavior was described in previous works on pullulan-coated bi-oriented polypropylene (BOPP) [24] and low-density polyethylene (LDPE) [28]. The reason for this unique property of pullulan (compared to other biopolymers) lies in its fully amorphous organization, which in turn must be ascribed to its inherent structural flexibility centered on the α-(1→6)-linkage between maltotriose units [13].

3. Materials and Methods

3.1. Raw Materials and Reagents

Pullulan (PI-20 grade, $M_w \approx 200$ kDa) was purchased from Hayashibara Biochemical Laboratories Inc. (Okayama, Japan), which currently belongs to the Nagase Group. The structural characteristics of pullulan (determined by high-performance size-exclusion chromatography equipped with multi-laser scattering and refractive index detectors—HPSEC-MALLS-RI) are weight average molar mass $(M_w) = 2.094 \times 10^5 \pm 0.002$; polydispersity index $(M_w/M_n) = 1.321 \pm 0.02$; and radius of gyration $(R_g) = 24.7 \pm 0.002$ nm [32]. Synthetic swelling type mica NTS-5 (Na-tetrasilic mica, $NaMg_{2.5}Si_4O_{10}F_2$) was purchased as a water dispersion (6 wt %) from Topy Industries Ltd. (Toyohashi, Japan). As a plastic substrate, AryaPET–A410 (JBF RAK LLC, Ras Al Khaimah, United Arab Emirates), kindly provided by Metalvuoto Spa (Roncello, Italy), was used. It is a one-side corona-treated polyester film 12.0 ± 0.5 μm thick, suitable for metallizing, printing, and lamination, with good wettability and excellent machinability. Milli-Q water (18.3 MΩ cm) was used as the only solvent throughout the experiments.

3.2. Preparation of the Bionanocomposite Coatings

A fixed amount of pullulan (10 wt %, wet basis) was dissolved in distilled water at 25 °C for 1 h under gentle stirring (500 rpm). Afterward, 50 mL of the mica dispersion (0.2 wt %, wet basis) was ultrasonicated by means of an UP400S (maximum power = 400 W; frequency = 24 kHz) ultrasonic device (Hielscher, Teltow, Germany), equipped with a cylindrical titanium sonotrode (mod. H14, tip Ø 14 mm, amplitude$_{max}$ = 125 μm; surface intensity = 105 W cm^{-2}) under the following conditions: 0.5 cycle and 50% amplitude for 2 min. In parallel, the resulting mica dispersions were diluted in distilled water (18.3 MΩ cm) under vigorous stirring (500 rpm) for 15 min. The pullulan solution and the inorganic dispersion were then mixed together under gentle stirring (300 rpm) for an additional 60 min. More specifically, the quantity of mica in the pullulan-water solutions was 0.002, 0.004, 0.006, 0.008, 0.01 and 0.02 wt % (wet basis). After drying, the concentrations of mica corresponded to 0.02, 0.04, 0.06, 0.08, 0.1 and 0.2 wt % on dry basis. PET films were treated with a high frequency corona treatment (Arcotec, Ülm, Germany). An aliquot of each bionanocomposite water dispersion was then placed on the corona-treated side of rectangular (24×18 cm^2) PET samples. The deposition of the coating was carried out by using an automatic film applicator (ref 1137, Sheen Instruments, Kingston, UK) at a constant speed of 2.5 mm s^{-1}, according to ASTM D823-07—Practice C. The deposition was performed by using a horizontal steel rod with an engraved pattern, which yielded final coatings of comparable nominal thickness of 1 μm after water evaporation. Water evaporation was performed using a constant and perpendicular flux of mild air (25.0 ± 0.3 °C for 2 min) at a distance of 40 cm from the applicator. The coated films were then stored under controlled conditions (23.0 ± 0.5 °C in a desiccator) for 48 h before measurements.

3.3. Analyses

The thickness of the pullulan/mica nanocomposite coating was obtained by a gravimetric method. A 10×10 cm^2 sample (coated PET) was cut and weighed (M_1, grams). The coating was then mechanically removed by immersion in hot water (80 °C), and the resulting bare PET film was weighed (M_2, grams). The apparent thickness (μm) of the coating was obtained according to the following equation [33]:

$$l = [(M_1 - M_2)/\rho] \times 100 \tag{3}$$

where ρ (g cm^{-3}) is the density of the aqueous dispersion. Three replicates were analyzed for each biopolymer composition.

Mica nanosheets were characterized by both TEM and AFM. The TEM images were acquired by using a LEO 912 AB energy-filtering transmission electron microscope (EFTEM) (Carl Zeiss, Oberkochen, Germany) operating at 80 kV. Digital images were recorded with a ProScan 1K Slow-Scan CCD camera (Proscan, Scheuring, Germany). Samples for TEM analyses were prepared by drop-casting a few millilitres of dispersion onto Formvar-coated Cu grids (400-mesh). The samples rested for 24 h at room temperature to allow water to evaporate. The AFM experiments were carried out to quantify the size features of mica nanosheets (e.g., width and thickness). The analyses were performed with a

Nanoscope V Multimode (Bruker, Karlsruhe, Germany) in intermittent-contact mode after dropping 10 µL of diluted mica water dispersion (0.2 mg mL^{-1} and 0.02 mg mL^{-1}) onto a mica substrate. The images were collected with a resolution of 512 × 512 pixels with silicon tips (force constant 40 N m^{-1}, resonance frequency 300 kHz). Dimensional calculations on the acquired images were conducted with nanoscope software (version 7.30, Bruker, Karlsruhe, Germany). The mean values reported for mica sheet dimensions were calculated over several images.

The oxygen barrier properties of the films were assessed on a 50 cm^2 surface sample using a Multiperm permeability analyzer (Extrasolution Srl, Capannori, Italy) equipped with an electrochemical sensor. The *OTR* data were determined according to the standard method of ASTM F2622-08, with a carrier flow (N$_2$) of 10 ml min^{-1} at 23 °C at 0%, 30%, 60%, and 90% relative humidity (RH) and at 1 atm pressure difference on the two sides of the specimen. During the analyses, the coated side of each sample faced the upper semi-chamber into which the humid test gas (oxygen) was flushed. Each *OTR* value was from three replicates.

Static (μ_s) and kinetic (μ_k) friction coefficients were measured using a dynamometer (model Z005, Zwick Roell, Ulm, Germany), in accordance with the standard method ASTM D 1894-87. The software TestXpert V10.11 (Zwick Roell, Ulm, Germany) was used for data analysis. The friction opposing the onset on relative motion (impending motion) is represented by μ_s, whereas μ_k can be considered as the friction opposing the continuance of the relative motion once that motion has started. In the case of solid-on-solid friction (with or without lubricants), these two types of friction coefficients are conventionally defined as follows:

$$\mu_s = F_s \cdot P \tag{4a}$$

$$\mu_k = F_k \cdot P \tag{4b}$$

where F_s is the force just sufficient to prevent the relative motion between two bodies, F_k is the force needed to maintain the relative motion between the two bodies and P is the force normal to the interface between the sliding bodies. In this study, the motion of each type of film (coated and uncoated) on a metallic rigid surface (a polished stainless steel 150 × 450 × 3 mm^3) was considered. This surface, in addition to acting as a supporting base to guarantee a firm position between the moving crosshead and the force-measuring device, served the purpose of simulating the friction between the plastic web and the metallic parts of the equipment used during the manufacturing processes and operations.

Tensile properties of films were measured according to the ASTM D882-02 by means of a dynamometer (mod. Z005, Zwick Roell, Ulm, Germany) fitted with a 5 kN load cell and connected with two clamps placed at 125 mm apart. Elastic (Young's) modulus (Emod), tensile strength (TS), and elongation (ε) were gathered from the stress-strain curves. For each parameter, the final results are the mean of at least five replicates.

Transparency and haze were determined by using a UV-Vis, high-performance spectrophotometer (Lambda 650, PerkinElmer, Waltham, MA, USA). Transparency was assessed in terms of specular transmittance (i.e., the transmittance value obtained when the transmitted radiant flux includes only the light transmitted in the same direction as that of the incident flux at a 550 nm wavelength) in accordance with the ASTM D1746-88.

Haze was measured within the wavelength range of 780–380 nm, in accordance with ASTM D1003-00, by using a 150-mm integrating sphere coupled with the main spectrophotometer, in order to trap the diffuse transmitted light. Haze is defined as the percentage of transmitted light deviating by more than an angle of 2.5° from the direction of the incident beam. Three replicates were made for each uncoated and coated film sample.

Variable-angle spectroscopic ellipsometry (VASE) measurements were performed in the spectral range from 320 nm to 800 nm (with steps of 2 nm), at different angles of incidence (from 40° to 70°), using the ellipsometer J.A. Woollam Co. Inc. (Lincoln, NE, USA). The ratio between the elements of the Jones matrix (r_{pp}/r_{ss}) = tan (Ψ) e$^{i\Delta}$ were acquired through Ψ and Δ, which depend on the wavelength λ of the incident polarized light. The measured $\Psi(\lambda)$ and $\Delta(\lambda)$ data were analyzed with the software

Nanomaterials **2017**, *7*, 281

WVASE 32 by describing the sample using the following model. The fitting of the experimental data was performed by minimizing the mean-squared-error (*MSE*) defined as:

$$MSE = \sqrt{\frac{\sum_{i=1}^{N}\left[\left(\frac{\psi_i^{mod}-\psi_i^{exp}}{\sigma_{\psi,i}^{exp}}\right)+\left(\frac{\Delta_i^{mod}-\Delta_i^{exp}}{\sigma_{\Delta,i}^{exp}}\right)\right]}{2N-M}} \tag{5}$$

where N is the number of measured Ψ and Δ pairs and M is the total number of model fit parameters.

The statistical significance of differences was determined by one-way analysis of variance (ANOVA), using JMP 5.0.1 software (SAS, Cary, NC, USA). Where appropriate, the mean values were compared by using a least significant difference (LSD) test, with a significance level $p < 0.05$.

4. Conclusions

The fabrication of water-based nanocomposite coatings incorporating mica NBBs into pullulan has been proposed in this study as a feasible and environmentally friendly process. Besides the exceptional oxygen barrier performance (especially at low and middle RHs), the PET films with the nanocomposite coatings exhibited outstanding optical properties, even at high loadings of mica. Interestingly, we have demonstrated that the addition of the nanocomposite coating moderately improved the elastic modulus and friction properties of the final material. The findings that arise from this work suggest the use of mica as a valid alternative to more common (e.g., montmorillonite) or more appealing (e.g., graphene and its derivatives) nanosheets for the design of nanocomposite coatings of potential utility in different fields.

Supplementary Materials: The following are available online at http://www.mdpi.com/2079-4991/7/9/281/s1, Table S1: Correlation factors between the free parameters of the fitting of the ellipsometry data for the different samples.

Acknowledgments: The authors acknowledge the support of the University of Milan-Action "Research Support Plan 2015–2017", Line 2, grant # 15-6-3024000-402.

Author Contributions: Ilke Uysal Unalan and Stefano Farris conceived and designed the experiments; Ilke Uysal Unalan and Derya Boyaci performed the experiments; Ilke Uysal Unalan, Derya Boyaci, and Stefano Farris analyzed the data; Silvia Trabattoni and Silvia Tavazzi contributed analysis tools; Ilke Uysal Unalan and Stefano Farris wrote the paper. Stefano Farris coordinated the overall work.

Conflicts of Interest: The authors declare no conflict of interest.

References

1. Unalan, I.U.; Cerri, G.; Marcuzzo, E.; Cozzolino, C.A.; Farris, S. Nanocomposite films and coatings using inorganic nanobuilding blocks (NBB): Current applications and future opportunities in the food packaging sector. *RSC Adv.* **2014**, *4*, 29393–29428. [CrossRef]
2. Ziadeh, M.; Weiss, S.; Fischer, B.; Förster, S.; Altstädt, V.; Müller, A.H.; Breu, J. Towards completely miscible PMMA nanocomposites reinforced by shear-stiff, nano-mica. *J. Colloid Interface Sci.* **2014**, *425*, 143–151. [CrossRef] [PubMed]
3. Alves, V.D.; Costa, N.; Coelhoso, I.M. Barrier properties of biodegradable composite films based on kappa-carrageenan/pectin blends and mica flakes. *Carbohydr. Polym.* **2010**, *79*, 269–276. [CrossRef]
4. Soon, K.; Harkin-Jones, E.; Rajeev, R.S.; Menary, G.; Martin, P.J.; Armstrong, C.G. Morphology, barrier, and mechanical properties of biaxially deformed poly(ethylene terephthalate)-mica nanocomposites. *Polym. Eng. Sci.* **2012**, *52*, 532–548. [CrossRef]
5. Krzesińska, M.; Celzard, A.; Grzyb, B.; Mareche, J.F. Elastic properties and electrical conductivity of mica/expanded graphite nanocomposites. *Mater. Chem. Phys.* **2006**, *97*, 173–181. [CrossRef]
6. Chang, J.H.; Mun, M.K. Nanocomposite fibers of poly(ethylene terephthalate) with montmorillonite and mica: Thermomechanical properties and morphology. *Polym. Int.* **2007**, *56*, 57–66. [CrossRef]

7. Kothmann, M.H.; Ziadeh, M.; Bakis, G.; de Anda, A.R.; Breu, J.; Altstädt, V. Analyzing the influence of particle size and stiffness state of the nanofiller on the mechanical properties of epoxy/clay nanocomposites using a novel shear-stiff nano-mica. *J. Mater. Sci.* **2015**, *50*, 4845–4859. [CrossRef]

8. Farris, S.; Introzzi, L.; Piergiovanni, L. Evaluation of a bio-coating as a solution to improve barrier, friction and optical properties of plastic films. *Packag. Technol. Sci.* **2009**, *22*, 69–83. [CrossRef]

9. Introzzi, L.; Blomfeldt, T.O.; Trabattoni, S.; Tavazzi, S.; Santo, N.; Schiraldi, A.; Piergiovanni, L.; Farris, S. Ultrasound-assisted pullulan/montmorillonite bionanocomposite coating with high oxygen barrier properties. *Langmuir* **2012**, *28*, 11206–11214. [CrossRef] [PubMed]

10. Unalan, I.U.; Boyacı, D.; Ghaani, M.; Trabattoni, S.; Farris, S. Graphene oxide bionanocomposite coatings with high oxygen barrier properties. *Nanomaterials* **2016**, *6*, 244. [CrossRef] [PubMed]

11. Cozzolino, C.A.; Cerri, G.; Brundu, A.; Farris, S. Microfibrillated cellulose (MFC)-pullulan bionanocomposite films. *Cellulose* **2014**, *21*, 4323–4335. [CrossRef]

12. Cozzolino, C.A.; Campanella, G.; Türe, H.; Olsson, R.T.; Farris, S. Microfibrillated cellulose and borax as mechanical, O$_2$-barrier, and surface-modulating agents of pullulan biocomposite coatings on BOPP. *Carbohydr. Polym.* **2016**, *143*, 179–187. [CrossRef] [PubMed]

13. Farris, S.; Uysal Unalan, I.; Introzzi, L.; Fuentes-Alventosa, J.M.; Cozzolino, C.A. Pullulan-based films and coatings for food packaging: Present applications, emerging opportunities, and future challenges. *J. Appl. Polym. Sci.* **2014**, *131*, 40539–40551. [CrossRef]

14. Aulin, C.; Gällstedt, M.; Lindström, T. Oxygen and oil barrier properties of microfibrillated cellulose films and coatings. *Cellulose* **2010**, *17*, 559–574. [CrossRef]

15. Kurek, M.; Guinault, A.; Voilley, A.; Galić, K.; Debeaufort, F. Effect of relative humidity on carvacrol release and permeation properties of chitosan based films and coatings. *Food Chem.* **2014**, *144*, 9–17. [CrossRef] [PubMed]

16. Sanchez-Garcia, M.D.; Hilliou, L.; Lagaron, J.M. Nanobiocomposites of carrageenan, zein, and mica of interest in food packaging and coating applications. *J. Agric. Food Chem.* **2010**, *58*, 6884–6894. [CrossRef] [PubMed]

17. Lape, N.K.; Nuxoll, E.E.; Cussler, E.L. Polydisperse flakes in barrier films. *J. Membr. Sci.* **2004**, *236*, 29–37. [CrossRef]

18. DeRocher, J.P.; Gettelfinger, B.T.; Wang, J.; Nuxoll, E.E.; Cussler, E.L. Barrier membranes with different sizes of aligned flakes. *J. Membr. Sci.* **2005**, *254*, 21–30. [CrossRef]

19. Alexandre, M.; Dubois, P. Polymer-layered silicate nanocomposites: Preparation, properties and uses of a new class of materials. *Mater. Sci. Eng.* **2000**, *28*, 1–63. [CrossRef]

20. Fuentes-Alventosa, J.M.; Introzzi, L.; Santo, N.; Cerri, G.; Brundu, A.; Farris, S. Self-assembled nanostructured biohybrid coatings by an integrated 'sol–gel/intercalation' approach. *RSC Adv.* **2013**, *3*, 25086–25096. [CrossRef]

21. Jung, J.C.; Chen, T. Measurement of friction force between two mica surfaces with multiple beam interferometry. In Proceedings of the EPJ Web of Conferences, Paris, France, 24–28 May 2010; EDP Sciences: Les Ulis, France, 2010; Volume 6, p. 06002.

22. Sakuma, H. Adhesion energy between mica surfaces: Implications for the frictional coefficient under dry and wet conditions. *J. Geophys. Res. Solid Earth* **2013**, *118*, 6066–6075. [CrossRef]

23. Farris, S.; Introzzi, L.; Biagioni, P.; Holz, T.; Schiraldi, A.; Piergiovanni, L. Wetting of biopolymer coatings: Contact angle kinetics and image analysis investigation. *Langmuir* **2011**, *27*, 7563–7574. [CrossRef] [PubMed]

24. Cozzolino, C.A.; Castelli, G.; Trabattoni, S.; Farris, S. Influence of colloidal silica nanoparticles on pullulan-coated BOPP film. *Food Packag. Shelf Life* **2016**, *8*, 50–55. [CrossRef]

25. Gualtieri, E.; Pugno, N.; Rota, A.; Spagni, A.; Lepore, E.; Valeri, S. Role of roughness on the tribology of randomly nano-textured silicon surface. *J. Nanosci. Nanotechnol.* **2011**, *11*, 9244–9250. [CrossRef] [PubMed]

26. Shah, D.; Maiti, P.; Jiang, D.D.; Batt, C.A.; Giannelis, E.P. Effect of nanoparticle mobility on toughness of polymer nanocomposites. *Adv. Mater.* **2005**, *17*, 525–528. [CrossRef]

27. Fischer, B.; Ziadeh, M.; Pfaff, A.; Breu, J.; Altstädt, V. Impact of large aspect ratio, shear-stiff, mica-like clay on mechanical behaviour of PMMA/clay nanocomposites. *Polymer* **2012**, *53*, 3230–3237. [CrossRef]

28. Introzzi, L.; Fuentes-Alventosa, J.M.; Cozzolino, C.A.; Trabattoni, S.; Tavazzi, S.; Bianchi, C.L.; Schiraldi, A.; Piergiovanni, L.; Farris, S. 'Wetting enhancer' pullulan coating for anti-fogpackaging applications. *ACS Appl. Mater. Interfaces* **2012**, *4*, 3692–3700. [CrossRef] [PubMed]

29. Tilley, R.J.D. *Colour and the Optical Properties of Materials*, 2nd ed.; Wiley: Hoboken, NJ, USA, 2008; p. 42.

30. Laskarakis, A.; Logothetidis, S. Study of the electronic and vibrational properties of poly(ethylene terephthalate) and poly(ethylene naphthalate) films. *J. Appl. Phys.* **2007**, *101*. [CrossRef]

31. Gradwell, S.E. Self-Assembly of Pullulan Abietate on Cellulose Surfaces. Ph.D. Thesis, Virginia Polytechnic Institute and State University, Blacksburg, VA, USA, 2004.

32. Xiao, Q.; Tong, Q.; Zhou, Y.; Deng, F. Rheological properties of pullulan–sodium alginate based solutions during film formation. *Carbohydr. Polym.* **2015**, *130*, 49–56. [CrossRef] [PubMed]

33. Brown, W.E. *Plastics in Food Packaging: Properties: Design and Fabrication*; Dekker: New York, NY, USA, 1992; pp. 200–202.

nanomaterials

MDPI

Article

Water Diffusion through a Titanium Dioxide/Poly(Carbonate Urethane) Nanocomposite for Protecting Cultural Heritage: Interactions and Viscoelastic Behavior

Mario Abbate and Loredana D'Orazio *

Istituto per i Polimeri, Compositi e Biomateriali, Via Campi Flegrei, 34, Fabbricato 70, 80078 Pozzuoli (Naples), Italy; mario.abbate@ipcb.cnr.it
* Correspondence: loredana.dorazio@ipcb.cnr.it; Tel.: +39-081-867-5064

Received: 3 July 2017; Accepted: 7 September 2017; Published: 13 September 2017

Abstract: Water diffusion through a TiO_2/poly (carbonate urethane) nanocomposite designed for the eco-sustainable protection of outdoor cultural heritage stonework was investigated. Water is recognized as a threat to heritage, hence the aim was to gather information on the amount of water uptake, as well as of species of water molecules absorbed within the polymer matrix. Gravimetric and vibrational spectroscopy measurements demonstrated that diffusion behavior of the nanocomposite/water system is Fickian, i.e., diffusivity is independent of concentration. The addition of only 1% of TiO_2 nanoparticles strongly betters PU barrier properties and water-repellency requirement is imparted. Defensive action against penetration of water free from, and bonded through, H-bonding association arises from balance among TiO_2 hydrophilicity, tortuosity effects and quality of nanoparticle dispersion and interfacial interactions. Further beneficial to antisoiling/antigraffiti action is that water-free fraction was found to be desorbed at a constant rate. In environmental conditions, under which weathering processes are most likely to occur, nanocomposite Tg values remain suitable for heritage treatments.

Keywords: Polymer/TiO_2 nanocomposites; thermoplastic polyurethanes; diffusion barrier; sorption; cultural heritage

1. Introduction

Cultural heritage assets are exposed to weather and submitted to influence of environmental parameters in a world where the climate is changing. Physical, chemical, and biological factors interact with constitutive materials inducing changes both in their compositional and structural characteristics [1–3]. The great importance of water as a threat to heritage is acknowledged: in natural conditions atmospheric water is the main agent associated with stone degradation, acting mainly through capillary rising. Rainwater penetrating by absorption is a vehicle of airborne acidic pollutants interacting with stone through chemical reactions of dissolved CO_2, NO_x, and SO_2. Moreover, water changes cohesion properties of the stone crystalline structure through physical/mechanical decay due to thermal excursions in wet conditions (freeze-thaw cycles) [4]. Hence, the need to improve effectiveness and eco-sustainability of preventive conservation and maintenance solutions are grown hugely.

Different classes of polymers have been so far employed as protective coatings of stone heritage without adequate knowledge of the properties of both plain polymer and polymer/substrate system [5–7]. As a result, insufficient efficacy and/or poor weatherability was usually observed. Such polymeric materials in most cases only provide short-term water repellency of the treated surfaces and are intrinsically unstable in photo-oxidative conditions typical of outdoor exposure. Notwithstanding that even polymers with partially fluorinated, side chains were ad hoc synthesized and tested to increase water repellency effectiveness and coating Ultraviolet (UV) radiation stability [8], presently the scientific community is still far from the achievement of materials fulfilling all the fundamental requirements of protective coatings [9].

While the last decade has seen several advancements in the field of polymer nanocomposites for a wide range of mechanical, electronic, magnetic, biological, and optical properties, fewer efforts have been focused on designing such a nanomaterial with optimal macroscale properties for protecting cultural heritage. A nanocomposite's properties depend ultimately upon a myriad of variables that include the quality of dispersion, interfacial adhesion, extent of region between nanoparticles fillers and bulk polymer matrix, processing methods, loading of the particles, modification of the surfaces of nanoparticles, aspect ratio of particles, compatibility of particles and host polymer, size of particles, radius of gyration of the host polymer and the properties of the constituents. Even though in literature structure-property relationships are lacking, it is evident that the properties of polymer nanocomposites are highly sensitive to both the quality of dispersion and region between nanoparticles fillers and bulk polymer matrix and that small changes in processing conditions, particle size, or chemistry dramatically affects these two key factors [10].

Recently results were achieved by matching a polymer with proper end properties, including eco-sustainable usage and non-toxicity, to create an inorganic photocatalytic nanocompound that was efficient in de-soiling and had biocide activities. A polymeric coating for protecting cultural heritage based on a water-dispersed TiO_2/poly (carbonate urethane) nanocomposite was prepared by a low impact procedure, i.e., cold mixing of the single components via sonication [11]. By means of the polymeric nanocomposites technology, highly innovative and outstanding performances were also achieved in terms of stability and durability as compared with other treatments based on acrylic and vinylic polymers widely used in conservation and restoration [6,7].

The next step of our investigation is concerned with applications of nanocomposite water dispersions on a porous degradable stone to demonstrate treatments' aesthetical compatibility, and ability in reducing soiling and biocide properties [12]. For a given nanocomposite concentration (w/v %), water Absorption Coefficients (ACs) of untreated and treated stone samples were also evaluated according to NORmalizzazione MAteriali Lapidei (NORMAL) 11/85 [13] as a function of the application procedure; i.e., air-brush until the stone surface was saturated, following a widespread practice in conservation, and full immersion in nanocomposite dispersions at room temperature for 1 h. The AC values achieved, pertaining to stone characterization, indicated that the treatments performed slowed the rate of water absorption of the stone. Hence, the nanocomposite homogeneous, transparent, colorless film formed by water casting at room temperature was proved to protect stone against water penetration.

The present work is focused, conversely, on water diffusion characteristics through TiO_2/poly (carbonate urethane) nanocomposite film samples, the novelty consisting of an in-depth analysis, on one hand, of nanocomposite diffusivity, and of the other hand, of effects of water uptake amount and nanoparticles/matrix interactions on glass transition temperature (Tg) of Polyurethane (PU) soft and hard domains. As a matter of fact, Tg has a deep influence on transport properties and, for applications of polymer-based materials in the field of cultural heritage, Tg is, as well, a relevant requirement. Coatings with a Tg value considerably higher than room temperature cannot be able to react to dimensional changes of treated items, whereas coatings with Tg values conspicuously lower than room temperature are much too soft for working and moreover are inclined to pick up dirt. Nanocomposite water diffusion coefficients were determined by means of gravimetric techniques combined with on time-resolved Fourier Transform (FT)-Near Infrared (NIR) measurements and compared to that exhibited by pristine PU matrix. Moreover, vibrational spectroscopy was selected as one of the best-suited techniques for probing hydrogen-bonded molecular structures [14] with the aim of gathering information on amount of water uptake, as well as, of species of water molecules absorbed within polymer matrix in presence of TiO_2 nanoparticles. In particular, significant effects of the addition of 1% of TiO_2 nanoparticles on amount of water free from, and strongly bonded through, H-bonding association absorbed/desorbed within the PU matrix, at environmental conditions under which weathering processes are most likely to occur, were highlighted. Correlations between adsorbed water amount and nanocomposite viscoelastic behavior were also established through Dynamic Mechanical Thermal Analysis (DMTA).

2. Results

2.1. Gravimetric Measurements

Because water absorption of a polymer depends on its nature and formulation there are many different behaviors, and hence many different models have been proposed [15,16]. Nevertheless, the most frequent approach to modeling diffusion of small molecules, such as water molecules, through a polymer bulk is to consider Fick's second law applied to simple single-free-phase diffusion [17–19]. Under unsteady state circumstance, Fick's second law describes the diffusion process as given by Equation (1).

$$\frac{\partial C}{\partial t} = \frac{\partial}{\partial x}\left[D\frac{\partial C}{\partial x}\right] \tag{1}$$

where C is the penetrant concentration, D a diffusion coefficient and x the distance of diffusion.

Equation (1) stands for concentration change of penetrant at certain element of the system with respect to the time (t) for one-dimensional model of linear flow of mass in a solid bonded by two parallel planes.

Assuming D constant in the direction of diffusion Equation (1) can be re-written as:

$$\frac{\partial C}{\partial t} = D\frac{\partial^2 C}{\partial x^2} \tag{2}$$

It has been demonstrated by Comyn [16] that for a polymer film of thickness $2l$ immersed into the infinite bath of penetrant, then concentrations, C_t, at any spot within the film at time t is given by Equation (3).

$$\frac{C_t}{C_\infty} = 1 - \frac{4}{\pi}\sum_{n=0}^{\infty}\frac{(-1)^n}{2n+1}\exp\left[\frac{-D(2n+1)^2\pi^2 t}{4l^2}\right]\cos\frac{(2n+1)\pi x}{2l} \tag{3}$$

where C_∞ is the amount of accumulated penetrant at equilibrium, i.e., the saturation equilibrium concentration within the system. $L = 2l$ is the distance between two boundaries layers, x_0 and x_1. Simple schematic representation of the concentration profile of the penetrant during the diffusion process between two boundaries is shown in Figure 1.

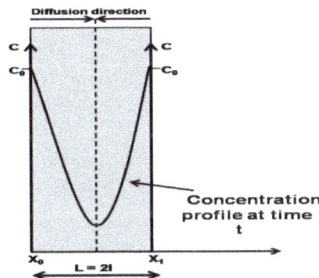

Figure 1. Schematic representation of the concentration profile of penetrant during its diffusion process between two boundaries.

Integrating Equation (3) over the entire thickness yields Equation (4) giving the mass of sorbed penetrant by the film as a function of time t, M_t, and compared with the equilibrium mass, M_∞.

$$\frac{M_t}{M_\infty} = 1 - \sum_{n=0}^{\infty} \frac{8}{(2n+1)^2 \pi^2} \exp\left[\frac{-D(2n+1)^2 \pi^2 t}{4l^2}\right] \tag{4}$$

For M_t/M_∞ ratio ≤ 0.5, Equation (4) can be written as follow:

$$\frac{M_t}{M_\infty} = 1 - \frac{8}{\pi^2} \sum_{n=0}^{\infty} \frac{1}{(2n+1)^2} \exp\left[\frac{-D(2n+1)^2 \pi^2 t}{4l^2}\right] \tag{5}$$

This estimation shows negligible error on the order of 0.1% [20].

Equation (5) was simplified by Shen and Springer [19] showing that the initial absorption is given by:

$$\frac{M_t}{M_\infty} = \frac{4}{L}\left(\frac{Dt}{\pi}\right)^{\frac{1}{2}} \tag{6}$$

where L is the film thickness. By plotting the M_t/M_∞ ratio as a function of time square root/L, the diffusion constant (D) can be calculated according to the following equation:

$$D = 0.0625 \, \pi\theta^2 \tag{7}$$

where θ is the initial slope of the curve in Fick's plot [21,22].

The isothermal sorption curves at 20 °C achieved by gravimetric measurements, shown by both plain poly (carbonate urethane) and nanocomposite wet samples, are typical Fickian diffusion diagrams; i.e., displaying a pronounced linear region in the early stages of the process, afterwards approaching the plateau with a downward concavity (see Figure 2). In Figure 3 the loss in weight due to water desorption as a function of time for the plain poly (carbonate urethane) and nanocomposite wet₁ samples, simulating a second type of environment such materials could be exploited in, is reported. Such samples were immersed in a deionized water bath at 20 °C until they absorbed a water content constant in the time.

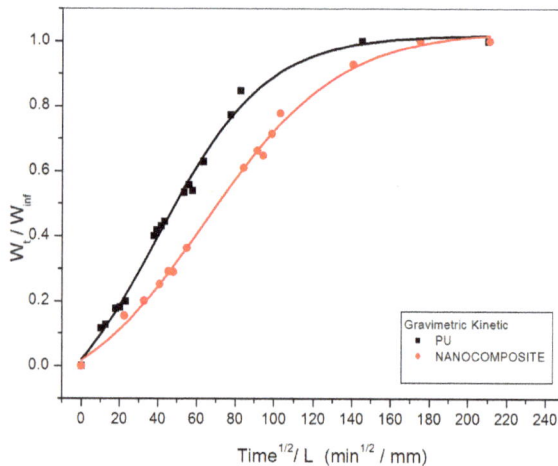

Figure 2. Curves of weight gain versus time for plain poly (carbonate urethane) and TiO₂/poly (carbonate urethane) nanocomposite systems.

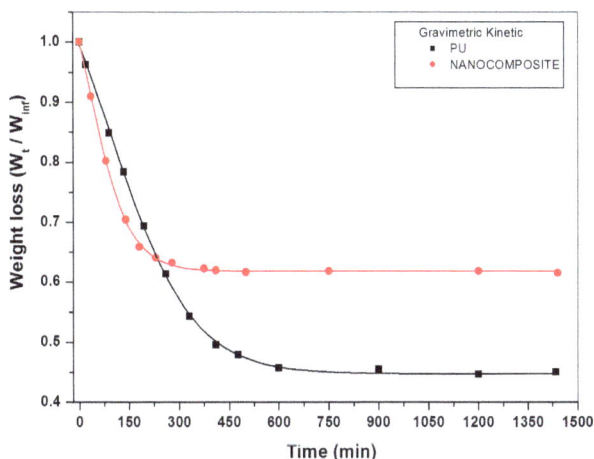

Figure 3. Curves of weight loss versus time for plain poly (carbonate urethane) and TiO$_2$/poly (carbonate urethane) nanocomposite systems.

As clearly shown in Figures 2 and 3 a very different behavior is exhibited by the two systems under investigation. The Fickian diffusion coefficients at 20 °C, calculated from the initial slopes of the gravimetric kinetic curves constructed for plain poly (carbonate urethane) and its nanocomposite, are reported in Table 1. In such a table the percentages (wt. %) of water respectively adsorbed at equilibrium and saturation by the two systems are also compared. It should be underlined that irrespective of the environmental conditions set up, the diffusion coefficients calculated for the plain poly (carbonate urethane) are approximately twice as that calculated for the nanocomposite material. As a matter of fact, the overall amount of absorbed water by the nanocomposite is considerably lower than that absorbed by the plain poly (carbonate urethane); the extent of such a lowering increasing strongly at saturation. Notwithstanding that mass transport in a nanocomposite system is heterogeneous, D value representing an average rate over a macro-volume, such results prove that the addition of 1% (wt. %) of TiO$_2$ nanoparticles imparts water repellency properties to the PU matrix; i.e., coatings consisting of TiO$_2$/poly (carbonate urethane) nanocomposite protect substrates against exposure/ penetration of water and degradation agents conveyed by water.

Table 1. Water diffusion coefficients (D) calculated for the systems under investigation at equilibrium and saturation.

Sample	D (mm^2/min)	Thickness (mm)	Absorbed Water (wt. %)
PU (equilibrium)	Gravimetric: 1.93×10^{-5} Spectroscopic: 2.09×10^{-5}	0.652	15
Nanocomposite (equilibrium)	Gravimetric: 1.04×10^{-5} Spectroscopic: 9.22×10^{-6}	1.037	9.95
PU (saturation)	Gravimetric: 2.00×10^{-4}	1.023	85
Nanocomposite (saturation)	Gravimetric: 1.16×10^{-4}	0.975	52

2.2. FT-NIR Measurements

In Figure 4 the absorbance FT-NIR spectra shown by dry and wet film samples of plain poly (carbonate urethane) and its nanocomposite are respectively reported. Frequencies and assignments of the main absorption bands of the poly (carbonate urethane) phase were reported in a previous work [11], the characteristic absorptions peaks of the plain PU remaining unchanged in presence of the TiO$_2$ nanoparticles.

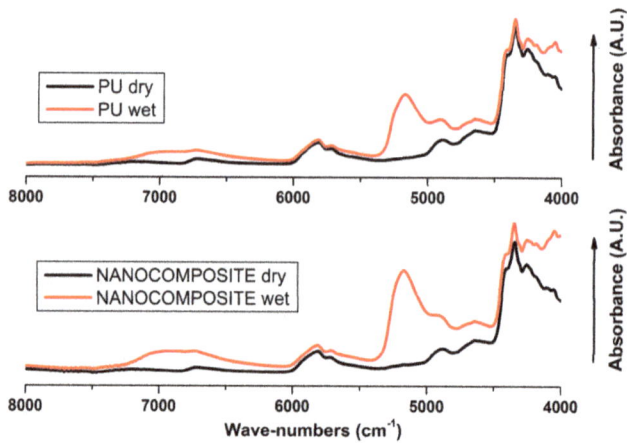

Figure 4. FT-NIR transmission spectra in the wave-number range 8000–4000 cm^{-1} for dry and wet samples of plain poly (carbonate urethane) and TiO$_2$/poly (carbonate urethane) nanocomposite.

A comparison of spectra between dry and wet samples reveal a characteristic peak for absorbed water at 5171 cm^{-1}, which is to be assigned to the combination of asymmetric stretching (ν_{as}) and in-plane deformation (δ) of water that occurred at 3755 cm^{-1} and 1595 cm^{-1} in the vapor phase spectrum [23]. The 5171 cm^{-1} peak, reasonably resolved, was found appropriate for kinetic studies being free from interference by PU phase and showing a change in intensity strong enough to assess quantitatively the water content in each sample. The absorbed water spectra obtained by difference spectroscopy method [24] representing $\nu + \delta$ combination peaks, for plain poly (carbonate urethane) and its nanocomposite are shown in Figure 5. As shown, for both the systems under investigation, the similar profile indicating the presence of different water species is observed. It was found that the normalized absorbance of the water band is considerably higher in plain PU than in nanocomposite suggesting that there is a higher amount of equilibrium water uptake in the PU system, in agreement with the results shown by the gravimetric analysis. Moreover, another multicomponent band for water occurs around 6900 cm^{-1} resulting from the combination of ν_{as} and ν_s fundamentals. This profile is superimposed onto a much stronger absorption due to the first O–H overtone of the hydroxyl group within the PU matrix producing only a slight increase in the intensity and breath of the band in the 7500–6100 cm^{-1} range.

Spectroscopic monitoring of the absorbance of the (ν_{as}) + (δ) peak representing the overall water diffusion process for the poly (carbonate urethane), without and with the TiO$_2$ nanoparticles, was carried out. Time-resolved Fourier Transform Infrared Spectroscopy (FTIR) measurements were performed at different time and the spectrum of water adsorbed was compared with that shown by the dried sample.

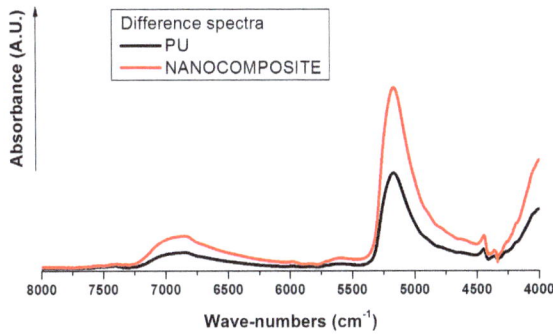

Figure 5. FT-NIR absorbed water spectra for plain poly (carbonate urethane) and TiO$_2$/ poly (carbonate urethane) nanocomposite.

Suppressing the interference of swelling of the samples during the process of diffusion it is possible to calculate the absolute parameters of diffusion using the equation of Fick as follows [18–22]:

$$\frac{A_t - A_o}{A_\infty - A_o} = \frac{C_t - C_o}{C_\infty - C_o} = \frac{M_t}{M_\infty} \qquad (8)$$

where C_0, C_t, C_∞ represent the concentration of water into sample at time 0, t, ∞ at equilibrium. Therefore $C_0 - C_t = M_t$ and $C_0 - C_\infty = M_\infty$ represent the mass of water absorbed from the sample at time t and at equilibrium respectively. The Fick's plot obtained from the spectral data is shown in Figure 6. A calibration plot of the recorded absorbance areas normalized for the sample thickness (reduced absorbance) against the content of adsorbed water in milligrams was constructed [24–26]. The values of the water diffusion coefficients spectroscopically achieved are reported in Table 1. As expected, at equilibrium, in presence of TiO$_2$ nanoparticles the material is confirmed to be comparatively characterized by a lower diffusion coefficient. The finding that D values spectroscopically achieved approach closely D values gravimetrically evaluated (see Table 1) demonstrate, for the systems under investigation, the reliability of FT-NIR way in following the process of water diffusion. It is to be reasonably expected that nanocomposite improved barrier property, observed as a reduction in water uptake, is strongly affected by physico-chemical properties of TiO$_2$/PU film such as higher availability of hydrophilic active sites for hydrogen bonding, TiO$_2$ mode and state of dispersion, particle size, and morphology, etc. Scanning Electron Microscopy (SEM) analysis of cryogenic fracture surfaces of nanocomposite film shows that the TiO$_2$ nanoparticles are homogeneously dispersed and uniformly distributed without significant particle-particle aggregation (see Figure 7); the presence of nanoparticles small clusters resulting in preferential penetrant pathways for water transport [27]. Tortuosity effects of the transport path along with effects of the nanoparticles on PU free volume properties are to be taken also into account.

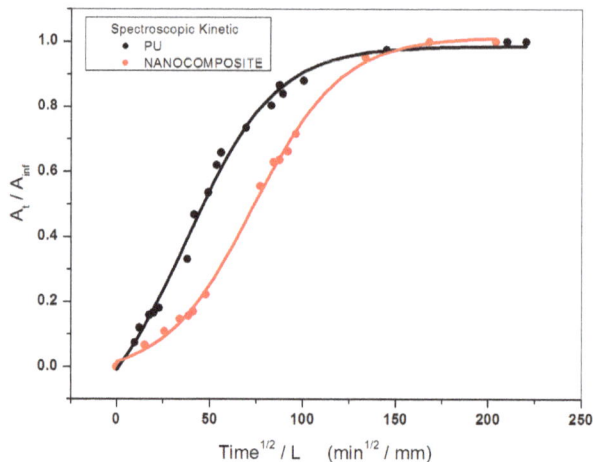

Figure 6. Fick's curves plotted by spectral data for plain poly (carbonate urethane) and TiO_2/poly (carbonate urethane) nanocomposite systems.

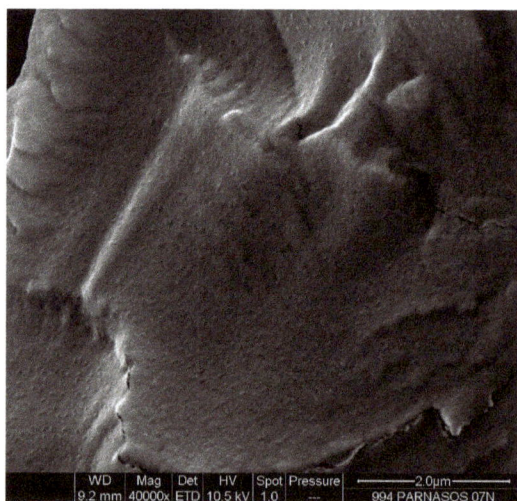

Figure 7. FESEM micrograph of cryogenical fracture surface of TiO_2/poly (carbonate urethane) nanocomposite film sample.

2.3. FT-NIR Curve-Fitting Analysis in the 5400–4600 cm^{-1} Wave-Number Range

Water molecules are well known to dissociate and/or molecularly adsorbed on TiO_2 surfaces [28–34]; water behavior being affected by Titanium dioxide surface chemistry and geometry [35–37]. Hence, to enhance information on the diffusion process of water through the plain PU and its nanocomposite a curve-fitting analysis in the 5400–4600 cm^{-1} range was performed through PerkinElmer IR Data Manager (IRDM) software (Perkin-Elmer, Beaconsfield, UK). The related deconvolution data are reported in Table 2; the χ^2 values, representing the goodness of curve-fitting analysis performed, were 0.049 and 0.045 for plain PU and its nanocomposite respectively.

Table 2. Results of the Curve-Fitting Analysis of the Spectra of Water absorbed by plain PU and its nanocomposite.

Peak	Center (cm^{-1})	Height (a.u.)	Left (cm^{-1})	Right (cm^{-1})	Fwhh [a] (cm^{-1})	Area (a.u.)
			PU			
Peak 1	5212	0.346	5500	4900	123	53.6
Peak 2	5113	0.336	5600	4600	171	72.0
Peak 3	4931	0.099	5500	4500	239	29.8
			Nanocomposite			
Peak 1	5211	0.643	5500	4900	122	98.6
Peak 2	5214	0.630	5600	4600	170	134.5
Peak 3	4931	0.198	5500	4500	246	61.3

[a] Full width at half-height.

As reported in Table 2, for both the systems, a three-water component spectrum is found. Such a finding can be interpreted in terms of a simplified association model, whereby three different water species can be spectroscopically distinguished, on the basis of the strength and the number of H-bonding interactions formed by water with proton accepting groups. In particular, the peak at the higher frequency (5212 cm^{-1}) corresponds to those water molecules in which the hydrogens do not form any interaction of the H-bonding type with the systems under investigation. This is not to say that these water species are to be regarded as completely detached from the surrounding polymer chain. Weaker polymer-penetrant interactions undetectable by vibrational spectroscopy, such as dipole-dipole and charge transfer, may still exist. Such kind of water is mobile being localized into excess free volume elements (microvoids and other morphological defects). The component at 5113 cm^{-1} arises from water molecules forming a single H-bonding interaction, whereas the broad component centered at 4931 cm^{-1} originates from water species having both the hydrogens involved in H-bonding with proton acceptor groups. This species may correspond both to single penetrant molecules bridged to two adjacent proton acceptors and to self-associated water in molecular clusters. In the plain PU matrix role of proton acceptor can be most probably played by free carbonyl groups, whether they are in hard or soft segments (i.e., both urethane and carbonate), according to the extent of soft and hard phase mixing considering that –NH groups in urethane linkage are able to form hydrogen bonds with urethane carbonyl and carbonate carbonyl [38]. In the nanocomposite, additional strong proton acceptors are the oxygen atoms in TiO_2 molecules; the O–H bond being much stronger and more covalent than the O–Ti bond. At the sorption equilibrium, the ratio between the relative fractions of not bonded and bonded water, as calculated by the areas of the absorbance peaks reported in Table 2 for plain poly (carbonate urethane) and nanocomposite, was estimated 0.53 and 0.50 respectively. This finding indicates that the presence of the TiO_2 nanoparticles reduces the overall amount of absorbed water affecting the fraction of absorbed water molecularly bound.

The ratio of the area of the individual component peaks to the total absorbance area for the water spectra collected, representing the relative contributions at sorption equilibrium of not bonded, weakly and strongly interacting water, evaluated by curve fitting analysis, are plotted against the time in Figures 8–10. Hence, for a given water species, the barrier property exhibited by the nanocomposite system is compared to that shown by the pristine PU system. It is interesting to point out that the addition of TiO_2 nanoparticles specifically modifies diffusivity, through the PU matrix, of not bonded and strongly bonded water. As a matter of fact, the fraction of not bonded water expected readily desorbed, for the nanocomposite system decreases following a linear trend (see Figure 8).

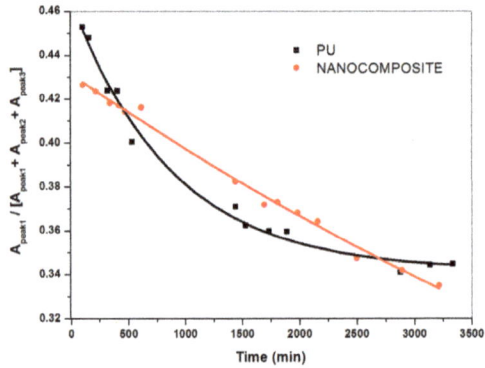

Figure 8. Relative fraction of not bonded water for plain (carbonate urethane) and TiO_2/poly (carbonate urethane) nanocomposite against the time.

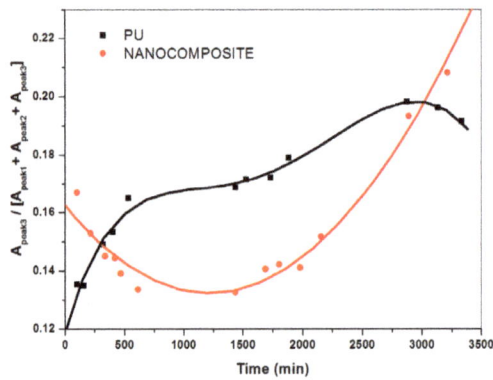

Figure 9. Relative fraction of strongly interacting water for plain (carbonate urethane) and TiO_2/poly (carbonate urethane) nanocomposite against the time.

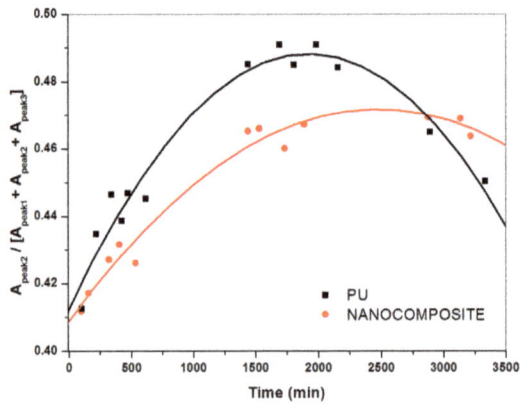

Figure 10. Relative fraction of weakly interacting water for plain (carbonate urethane) and TiO_2/poly (carbonate urethane) nanocomposite against the time.

In presence of TiO$_2$ nanoparticles, for the content investigated at least, the transport of such a kind of water occurs with a constant rate. This finding indicates water tendency to spread perfectly across nanocomposite film surface (high wettability) and/or comparable mean free path of water molecules to pass through the polymer matrix. Regardless, such an effect is beneficial in making surfaces easily washable with a plus of oil absorption resistance; i.e., antigraffiti, antisoiling coatings, etc. What is more, in presence of the TiO$_2$ nanoparticles the relative contribution of strongly bonded water as a function of time results in downward concavity points. Conversely, upward concavities points are found for the poly (carbonate urethane)/water system (see Figure 9) thus revealing that the addition of TiO$_2$ nanoparticles dramatically affects diffusivity of such a kind of water through the polymer matrix. For the pristine matrix system, the upward concavities points are presumably due to the occurrence of water clustering, contributing to an increase in water solubility according to the free volume theory [39–41], and to water molecules forming double hydrogen bonds with two already hydrogen-bonded C=O groups. Puffr and Sebenda [42,43] showed that such water molecules are more firmly bounded than that bridging the gaps between the hydrogen-bonded N–H and C=O groups. By contrast, for the nanocomposite system, water absorption immobilized on specific sites, free volume reduction and tortuosity of diffusion path, which the presence of the inorganic nanoparticles causes, could give an account of downwards concavity points.

As far as the contribution of weakly bonded water, the systems under investigation show similar behavior as a function of time (see Figure 10) suggesting that such species of water molecules could jump from one site to another site, irrespective of the TiO$_2$ nanoparticles presence.

It should be pointed out that the analysis used here provides only a limited insight into the water transport in heterogeneous systems such as a polymer-based nanocomposite. On account of the nanocomposites structural and interactional peculiarities, their diffusion kinetics are rather complicated.

2.4. DMTA Analysis

The dynamic-mechanical spectra in terms of loss factor (tan δ) at 1 Hz for dry, wet and wet$_l$ film samples of plain poly(carbonate urethane) and its nanocomposite are shown in Figure 11. In agreement with previous results [11], for both the materials the tan δ plots (Figure 11a,c) reveal the occurrence of two distinct relaxation processes with increasing temperature. Such relaxations are α transition processes corresponding to the glass transition (*Tg*) of poly (carbonate urethane) soft and hard segments respectively. In order to accomplish more accurate data, *Tg* values were defined through the peaks obtained by loss modulus (E″) plot also shown in Figure 11b,d. The *Tg* values for dry, wet and wet$_l$ samples of the plain poly (carbonate urethane) and its nanocomposite so achieved are reported in Table 3.

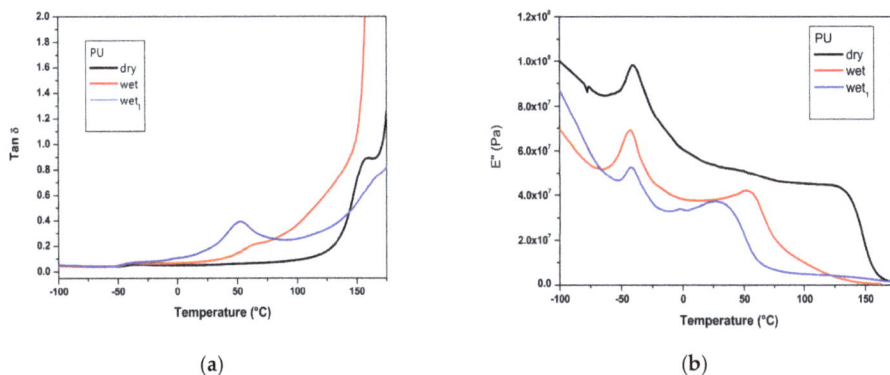

(a)　　　　　　　　　　　　　　(b)

Figure 11. *Cont.*

(c)

(d)

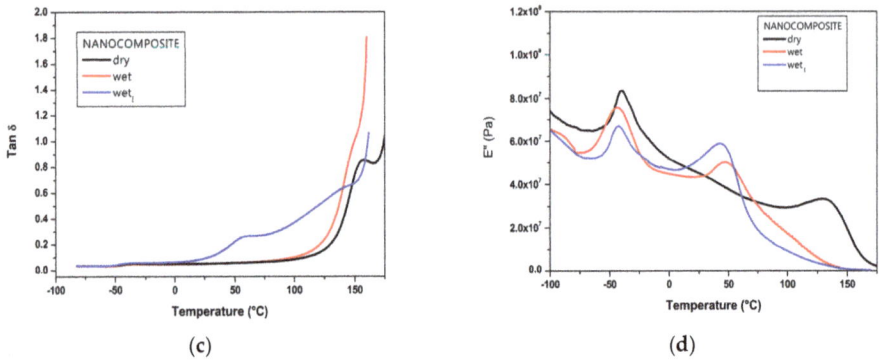

Figure 11. Dynamic-mechanical spectra in terms of loss factor (tan δ) (**a,c**) and loss modulus (E'') (**b,d**) at 1 Hz for dry, wet and wet$_I$ samples of plain poly(carbonate urethane) and TiO_2/poly (carbonate urethane) nanocomposite.

Table 3. *Tg* values for dry, wet and wet$_I$ samples of plain PU and its nanocomposite.

Sample	*Tg* (°C)	*Tg* (°C)
PU$_{dry}$	130	−40
PU$_{wet}$	49	−43
PU$_{wetI}$	30	−42
Nanocomposite$_{dry}$	132	−40
Nanocomposite$_{wet}$	53	−43
Nanocomposite$_{wetI}$	44	−42

It should be noted that in water absence and in presence of TiO_2 nanophase, a slight *Tg* increase of hard domains is found. Such a result is in agreement with that achieved through a DMTA multi-frequency analysis revealing that in nanocomposite film samples molecular motions of poly (carbonate urethane) hard segments are restricted by the presence of TiO_2 nanophase, as higher energy is required for their relaxation [11]. The enhanced *Tg* value could suggest positive PU hard phase-nanoparticles interfacial interactions that reduce cooperative segmental mobility.

Polyurethanes are especially prone to moisture-induced plasticization because water molecules can occupy intermolecular hydrogen bonding sites between chains, which would otherwise act as physical crosslinks and restrict chain mobility. Possible water effects on hydrogen bonding are shown by the schematic models reported in Figure 12.

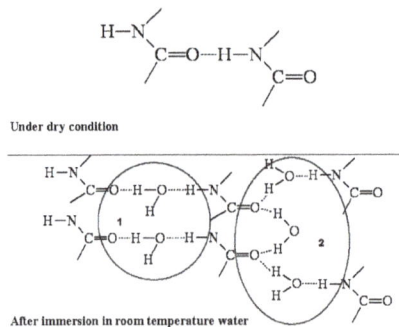

Figure 12. Effects of water on the hydrogen bonding in PUs: (1) weakly bonded water; (2) firmly bonded water.

Absorbed water molecules, bridging the gaps between the hydrogen-bonded N–H and C=O groups, weaken hydrogen bonding between N–H and C=O groups. Decrease in hydrogen bonding forces causes decrease in *Tg* together with the function of water as a plasticizer [44–46]. Splitting water absorption on the basis of the strength and the number of H-bonding interactions formed by water with PU proton accepting groups as schematically shown in Figure 12, it seems reasonably feasible that free water has negligible effect on the glass transition, while bound water reduces it strongly by weakening the hydrogen bonding between N–H and C=O groups.

As shown in Table 3, the presence of water decreases the *Tg* values to be ascribed to poly (carbonate urethane) hard phase strongly; this change being thermally reversible upon heating. In contrast, the *Tg* value of poly (carbonate urethane) soft phase is affected scarcely. It has been reported by various researchers that the hydrogen bonding between polymer and inorganic interface can reduce chain mobility and increase *Tg*; i.e., *Tg* confinement effect due to polymer chains confined between nanofillers interfaces [47,48]. Assuming that hydrophilic TiO$_2$ nanoparticles take up free volume within PU matrix creating a tortuous path for water molecules and reducing swelling by the water of domains of soft phase, the significant *Tg* reduction observed for the poly (carbonate urethane) hard phase is a combination of reduced hydrogen bonding, water plasticization effect, and polymer-TiO$_2$ interactions. The previous two effects could reduce *Tg*, while the last would increase *Tg*.

3. Discussion

In order to counteract external degradation of monuments and buildings caused by the atmospheric pollution and meet the demands of cultural heritage with ecological, economic and social aspects, aqueous dispersions of different nanoparticles with photocatalytic capacity were used. Among them, nano-TiO$_2$ is one of the most common owing to its versatility and green production, eco-compatibility and low-level impact on the chemical composition of materials. Notwithstanding this, relevant issues are still pending regarding the effectiveness and long-term stability of the coatings "in situ" and the impact of nanoparticles on human health and environment. As an alternative, and with outstanding advantages, the present paper shows that inorganic nanoparticles can be dispersed by means of low impact procedures into polymer matrices suitably selected and that modulation of relevant physical chemical properties such as water-repellency of a protective can be obtained.

The mechanism through which water diffuses into polymeric materials can be summarized as either infiltration into the free space or specific molecular interactions. The former is controlled by the free space available such as commonly occurring micro-voids and other morphological defects; an increase in the free space should result in an increase of both the water uptake and diffusivity. The diffusion of water by molecular interaction is, on the other hand, controlled by the available hydrogen bond at hydrophilic sites.

For the diffusion of water at room temperature through film samples of TiO$_2$/poly (carbonate urethane) nanocomposite gravimetric sorption/desorption tests and FTIR spectroscopic analysis demonstrated that, for the composition investigated at least, the diffusion behavior is Fickian, and substantially linear, in so far as the diffusivity is independent of concentration. The mechanism expected when the diffusion rates are much slower than those of polymer relaxations (Fickian diffusion) can be summarized as follows. At temperatures below *Tg*, the polymer backbone is considered to be in a frozen state, segmental chain motions are drastically reduced, the number of free volume holes is fixed and no hole redistribution is likely. Mass transport is, therefore, assumed to take place via fixed (pre-existing) holes. A penetrant molecule must find its way from hole to hole along pathways involving only minor segmental rearrangements. This means that the diffusivity depends largely on the number of the holes with an appropriate size able to accommodate the diffusing molecule. In the rubbery state above *Tg*, the polymer chains are mobile and the free volume holes show a dynamic variation about size, shape, and position. The penetrant molecules diffuse within the fluctuating interstitial free volume with much greater mobility than in the glassy state.

Moreover, it was shown that the addition of only 1% (wt. %) of hydrophilic TiO_2 nanoparticles to a poly (carbonate urethane) matrix strongly betters its barrier property. Water absorption in polymer nanocomposites containing impermeable anisotropic domains has been described in several publications. The most common nanocomposites investigated consist of a variety of polymers, both thermoplastic and thermoset, and nanoclay. Transport properties of PUs with soft segments consisting of polycaprolactone/organically modified montmorillonite nanocomposites have been investigated by Tortora et al. [49]. Diffusivity of heterogeneous systems such as polymer nanocomposites is a complex phenomenon. Impermeable domains affect permeability not only by reducing the volume of material available for flow, but also by creating more sinuous pathways according to a tortuous model. Essentially impermeable nanoparticles act as obstacles forcing penetrant molecules to follow longer and complicated routes to diffuse through the material. At the same time, the incorporation of inorganic nano-fillers into the polymer matrix inevitably changes its morphological features and, consequently, its free volume properties. Effects of nanoparticles on polymer free volume to be expected are interfacial regions, interstitial cavities in the filler agglomerates, chain segmental motion immobilization, insufficient chain packaging, changes of the free volume hole size distribution, changes of the crystallinity of the matrix and change of the cross-linking density of the matrix. Which of them become dominant depends primarily on the degree of interaction between the components, the volume fraction of the filler and the geometrical features of the particles. Several studies carried out on reinforced epoxy nanocomposites showed that the maximum water absorption of a polymer system decreased due the presence of nano-filler [50]. Such a phenomenon was generally ascribed to nano-fillers barrier properties together with a tortuous pathway for water molecules to diffuse. The achieved results indicate that the TiO_2/poly (carbonate urethane) nanocomposite defensive action against penetration of water free from, and bonded through, H-bonding association arises from a balance among TiO_2 hydrophilicity, tortuosity effects and quality of nanoparticles dispersion and positive inter-facial interactions. Hence, the barrier property of such nanocomposite film is governed by a combination of physico-chemical properties including mode and state of dispersion of the minor component, the interaction between TiO_2 nanophase and PU matrix, particle size and structure of TiO_2 nanoparticles, PU morphology and structure, etc. Different analytical techniques, such as Thermo-Gravimetric Analysis–Differential Scanning Calorimetry (TGA-DSC), Field Emission Scanning Electron Microscopy (FESEM), Wide Angle X-ray Scattering (WAXS), DMTA and Attenuated Total Reflectance (ATR)-FTIR were, therefore, applied on both nanocomposite and pristine PU film samples to achieve a thorough characterization [11]. The TiO_2/poly (carbonate urethane) nanocomposite is a multiphase system in which an inorganic phase with an average size of 31.08 nm was dispersed through sonication. Nanocomposite WAXS intensity profile shows a broad diffraction halo to be ascribed to the amorphous polyurethane phase [6]; no Bragg reflection can be seen corresponding to both the TiO_2 crystallographic forms Anatase and Rutile [38]. Such a nanophase gives rise to superficial dissociation and/or adsorption and to specific interactions with the water molecules together with interactions with poly (carbonate urethane) hard segments. In turn, the poly (carbonate urethane) phase itself is to be considered as a two-phase amorphous-amorphous system, in which both hard and soft segments are permeable to the water molecules. The morphology of the hard and soft segments of the plain poly (carbonate urethane) was investigated through a careful examination of –NH and carbonyl peaks of ATR-FTIR spectra. It was found that the most of the amide groups are involved in hydrogen bonding [38]. Work is in progress to investigate effects of the addition of TiO_2 nanoparticles on PU phase separation by means of ATR-FTIR spectroscopy.

It is to be underlined that the amorphous structure of the TiO_2/poly (carbonate urethane) nanocomposite confers material a certain degree of rubber elasticity essential for its applications on items with cultural value. In perspective of our final goal, i.e., showing that treatments based on water dispersions of TiO_2/poly (carbonate urethane) nanocomposite successfully protect outdoor cultural assets stonework, it is to be pointed out that all the effects achieved by the addition of 1% (wt. %) of TiO_2 nanoparticles are beneficial to combat both exposure/penetration of water and degradation agents conveyed by water and soiling and graffiti. Moreover, it is worthy to note that the nanocomposite Tg values, irrespective of water uptake amount, fulfill requirements for protective coatings. At environmental conditions under which weathering processes are most likely to occur, the PU soft phase remains above its Tg in an amorphous rubbery state, balanced by the PU hard phase in a glassy amorphous state below its Tg.

4. Materials and Methods

The raw materials used in this work are reported as follows: a linear aliphatic poly(carbonate urethane) (trade name Idrocap 994) was prepared by ICAP-SIRA (Parabiaco, Milano, Italy) in water dispersion with neutral pH to allow applications on substrates pH sensitive and organic solvents. The prepolymer mixing process followed was reported in a previous work [11]. The Mw values of the poly(carbonate urethane) so achieved are in the range between 30,000 and 50,000 in Gel Permeation Chromatography (GPC) with standard Polystyrene (PS). Titanium dioxide (TiO_2) nanoparticles were synthesized and kindly supplied in water dispersion by the research center CE.RI.Col of Colorobbia Italia (Sovigliana, Vinci, Florence, Italy). [11]. TiO_2 nanoparticles have an average size equal to 31.08 nm by Dynamic Light Scattering (DLS) technique with a polydispersity index of 0.241. All the reactants and solvents were used as received.

Plain poly (carbonate urethane) and nanocomposite film samples 0.60–1.00 mm thick were safely achieved using water-casting at room temperature. The preparation of a TiO_2/poly (carbonate urethane) nanocomposite containing 1% (wt. %) of TiO_2 nanoparticles was performed by cold mixing the single components via sonication following the low impact method elsewhere reported [11]. Also, the plain poly (carbonate urethane) was undergone identical sonication process.

Gravimetric sorption measurements were carried out by the so-called pat-and-weight technique. Film samples 0.60–1.00 mm thick were dried for 3 h at 100 °C under vacuum to achieve complete removal of absorbed water. The total absence of absorbed water was confirmed by means of FT-NIR spectroscopy. A Perkin-Elmer Spectrum 100 spectrophotometer (Perkin-Elmer, Beaconsfield, UK) was used. The instrumental parameters adopted for the FT-NIR monitored tests were as follows: resolution 4 cm^{-1}, spectral range 8000–4000 cm^{-1}.

FT-NIR spectra exhibited by the dried nanocomposite and plain poly (carbonate urethane) materials were also taken as a reference for spectral subtraction analysis.

To deeply investigate the effects related to the presence of TiO_2 nanoparticles on water absorption and desorption kinetics of the poly (carbonate urethane) matrix the following procedures were carried out. Dried film specimens were introduced in an environmental climatic chamber SU250 Angelantoni Industries S.p.a (Cimacolle, Perugia, Italy) at the temperature of 20 °C and 50% of Relative Humidity (RH) simulating weathering. The samples, hereafter wet samples, were removed from the chamber at certain time intervals, weighted in a high precision analytical balance and FT-NIR transmission spectra were collected simultaneously. The amount of absorbed water was calculated by the weight difference. When the content of water remained invariable in the specimens then the kinetics were stopped.

Dried film samples were also immersed in a deionized water bath thermostatically controlled at 20 °C ± 1 °C until they adsorbed a water content constant in the time. The wet samples so achieved, hereafter wet$_I$ samples, were introduced in the chamber SU250 Angelantoni Industries (Angelantoni, Naples, Italy) setting the same conditions of temperature and relative humidity used for weathering simulation. Periodically, the samples were removed, blotted and reweighted, the desorption of water was so monitored. In such a procedure we could not apply FT-NIR technique as the high amounts of water absorbed.

Effects of water diffusion on the visco-elastic behavior of both nanocomposite and plain poly (carbonate urethane) were investigated through dynamic mechanical thermal analysis (DMTA) using a Perkin-Elmer Pyris Diamond DMA apparatus (Perkin-Elmer Italia S.p.A, Monza, Italy). Tests were performed in bending mode, applying a strain of 1%. Single-frequency measurements at 1 Hz were performed at a constant heating rate of 3 °C/min, in the temperature range from −100 °C up to 200 °C.

Mode and state of dispersion of the TiO$_2$ nanoparticles into the poly (carbonate urethane) matrix were analyzed by means of a Fei Quanta 200 field emission Environmental Scanning Electron Microscope (ESEM, FEI, Hillsboro, OR, USA) operating in high vacuum mode.

Acknowledgments: Funds for covering the costs to publish in open access were supported by Istituto per i Polimeri, Compositi e Biomateriali (IPCB)—National Research Council (CNR).

Author Contributions: M.A. and L.D. conceived and designed the experiments, performed the experiments, analyzed the data and wrote the paper.

Conflicts of Interest: The authors declare no conflict of interest.

References

1. Brimblecombe, P. *Urban Pollution and Changes to Materials and Building Surfaces*; Imperial College Press: London, UK, 2016.
2. Dornieden, T.H.; Gorbushina, A.A.; Krumbein, W.E. Biodecay of cultural heritage as a space/time-related ecological situation—An evaluation of a series of studies. *Int. Biodeterior. Biodegrad.* **2000**, *46*, 261–270. [CrossRef]
3. Sabbioni, C.; Brimblecombe, P.; Cassar, M. *The Atlas of Climate Change Impact on European Culture Heritage*; Anthem Press: London, UK, 2010.
4. Lazzarini, L.; Tabasso, M.L. *Il Restauro Della Pietra*; CEDAM: Obernai, France, 1986.
5. UNESCO. Synthetic materials used in the conservation of cultural material. In *The Conservation of the Cultural Property. Museum and Monuments 11*; UNESCO: Paris, France, 1968; pp. 303–331.
6. D'Orazio, L.; Gentile, G.; Mancarella, C.; Martuscelli, E.; Massa, V. Water-dispersed polymers for the conservation and restoration of cultural heritage: A molecular, thermal, structural and mechanical characterisation. *Polym. Test.* **2001**, *20*, 227–240. [CrossRef]
7. Cocca, M.; D'Arienzo, L.; D'Orazio, L.; Gentile, G.; Martuscelli, E. Polyacrylates for conservation: Chemico-physical properties and durability of different commercial products. *Polym. Test.* **2004**, *23*, 333–342. [CrossRef]
8. Alessandrini, G.; Aglietto, M.; Castelvetro, V.; Ciardelli, F.; Peruzzi, R.; Toniolo, L. Comparative evaluation of fluorinated and unfluorinated acrylic copolymers as water-repellent coating materials for stone. *J. Appl. Polym. Sci.* **2000**, *76*, 962–977. [CrossRef]
9. ICR-CNR. *Normal Protocol 20/85. Conservation Works: Planning, Execution and Preventive Evaluation*; ICR-CNR: Rome, Italy, 1985.
10. Paul, D.R.; Robeson, L.M. Polymer nanotechnology: Nanocomposites. *Polymer* **2008**, *49*, 3187–3204. [CrossRef]
11. D'Orazio, L.; Grippo, A. A water dispersed Titanium dioxide/poly (carbonate urethane) nanocomposite for protecting cultural heritage: Preparation and properties. *Prog. Org. Coat.* **2015**, *79*, 1–7. [CrossRef]
12. D'Orazio, L.; Grippo, A. A water dispersed Titanium dioxide/poly (carbonate urethane) nanocomposite for protecting cultural heritage: Eco-sustainable treatments on Neapolitan Yellow Tuff. *Prog. Org. Coat.* **2016**, *99*, 412–419. [CrossRef]
13. National Researh Council (CNR)—Istituto Centrale del Restauro (ICR) Recommendations. *AA. VV.: NORMAL 11/85 Water Absorption by Capillarity*; National Researh Council (CNR)—Istituto Centrale del Restauro (ICR) Recommendations: Rome, Italy, 1985.
14. Pimentel, G.C.; McClellan, A.L. *The Hydrogen Bond*; Freeman: San Francisco, CA, USA, 1960.
15. Crank, J.; Park, G.S. *Diffusion in Polymers*; Academic Press: London, UK; New York, NY, USA, 1968.
16. Comyn, J. *Polymer Permeability*; Elsevier Applied Science Publishers: London, UK, 1985.
17. Crank, J. *The Mathematics of Diffusion*, 2nd ed.; Oxford University Press: Oxford, UK, 1975.
18. Liu, W.; Hoe, S.V.; Pugh, M. Water uptake of epoxy clay nanocomposites: Model development. *Compos. Sci. Technol.* **2008**, *68*, 156–163. [CrossRef]
19. Shen, C.H.; Springer, G.S. Moisture Absorption and Desorption of Composite Materials. *J. Comp. Mater.* **1976**, *10*, 2. [CrossRef]
20. Vergnaud, J.M. *Liquid Transport Processes in Polymeric Materials: Modeling and Industrial Applications*; Prentice Hall: London, UK, 1991.

21. Jacobs, P.M.; Jones, F.R. Diffusion of moisture into two-phase polymers. *J. Mater. Sci.* **1989**, *24*, 2331–2336. [CrossRef]

22. Maggana, C.; Pissis, P. Water sorption and diffusion studies in an epoxy resin system. *J. Polym. Sci. Part B* **1999**, *37*, 1165–1182. [CrossRef]

23. Segtan, V.H.; Šašic, Š.; Isaksson, T.; Ozaki, Y. Studies on the structure of water using two-dimensional Near-Infrared correlation spectroscopy and principal component analysis. *Anal. Chem.* **2001**, *73*, 3153–3161. [CrossRef]

24. Musto, P.; Mensitieri, G.; Lavorgna, M.; Scarinzi, G.; Scherillo, G. Combining gravimetric and vibrational spectroscopy measurements to quantify first and second-shell hydration layers of polyimides with different molecular architectures. *J. Phys. Chem. B* **2012**, *116*, 1209–1220. [CrossRef] [PubMed]

25. Cotugno, S.; Larobina, D.; Mensitieri, G.; Musto, P.; Ragosta, G. A novel spectroscopic approach to investigate transport processes in polymers: The case water-epoxy system. *Polymer* **2001**, *42*, 6431–6438. [CrossRef]

26. Musto, P.; Abbate, M.; Lavorgna, M.; Ragosta, G.; Scarinzi, G. Microstructural features, diffusion and molecular relaxations in polyimide/silica hybrids. *Polymer* **2006**, *47*, 6172–6186. [CrossRef]

27. Diez-Pascual, A.M.; Diez-Vicente, A.L. Poly (3-hydroxybutyrate)/ZnO bionanocomposites with improved mechanical, barrier and antibacterial properties. *Int. J. Mol. Sci.* **2014**, *15*, 10950–10973. [CrossRef] [PubMed]

28. Tien, Y.I.; Wie, K.H. Hydrogen bonding and mechanical properties of segmented montmorillonite/ polyurethane nanocomposite with different hard segment ratios. *Polymer* **2001**, *42*, 3213–3221. [CrossRef]

29. Fujishima, A.; Rao, T.N.; Tryk, D.A. Titanium dioxide photocatalysis. *J. Photochem. Photobiol. C* **2000**, *1*, 1–21. [CrossRef]

30. Fahmi, A.; Minot, C. A theoretical investigation of water adsorption on Titanium dioxide surfaces. *Surf. Sci.* **1994**, *304*, 343–359. [CrossRef]

31. Boehm, H.P.; Herrmann, Z. Über die chemie der oberfläche des Titandioxids. I. Bestimmung des aktiven wasserstoffs, thermische entwässerung und rehydroxy lierung. *Anorg. Allg. Chem.* **1967**, *352*, 156–167. (In German) [CrossRef]

32. Lo, J.W.; Chung, Y.W.; Somorjai, G.A. Electron spectroscopy studies of the chemisorption of O_2, H_2 and H_2O on the TiO_2 (100) surfaces with varied stoichiometry: Evidence for the photogeneration of Ti^{+3} and for its importance in chemisorption. *Surf. Sci.* **1978**, *71*, 199–219.

33. Suda, Y.; Morimoto, T. Molecularly adsorbed water on the bare surfaces of TiO_2. *Langmuir* **1987**, *3*, 786–788. [CrossRef]

34. Sun, C.H.; Liu, L.M.; Selloni, A.; Lu, G.Q.; Smith, S.C. Titania-water interactions: A review of theoretical studies. *J. Mater.Chem.* **2010**, *20*, 9559–9612. [CrossRef]

35. Raju, M.; Kim, S.Y.; van Duin, A.C.T.; Fichthorn, K.A. ReaxFF Reactive Force Field study of the dissociation of water on Titania surfaces. *J. Phys. Chem. C* **2013**, *117*, 10558–10572. [CrossRef]

36. Zhang, H.; Banfield, J.F. Structural characteristics and mechanical and thermodynamic properties of nanocrystalline TiO_2. *Chem. Rev.* **2014**, *114*, 9613–9644. [CrossRef] [PubMed]

37. Henderson, M.A. The interaction of water with solid surfaces: Fundamental aspects revisited. *Surf. Sci. Rep.* **2002**, *46*, 1–308. [CrossRef]

38. Cocca, M.; D'Orazio, L. Novel silver/polyurethane nanocomposite by in situ reduction: Effects of the silver nanoparticles on phase and viscoelastic behavior. *J. Polym. Sci. Part B* **2008**, *46*, 344–350. [CrossRef]

39. Thompson, T.L.; Yates, J.T. Surface science studies on the photoactivation of TiO_2: New photochemical processes. *Chem. Rev.* **2006**, *106*, 4428–4453. [CrossRef] [PubMed]

40. Watanabe, T.; Nakajima, A.; Wang, R.; Minabe, M.; Koizumi, S.; Fujishima, A.; Hashimoto, K. Photocatalytic activity and photoinduced hydrophilicity of titanium dioxide coated glass. *Thin Solid Films* **1999**, *351*, 260–263. [CrossRef]

41. Xu, G.; Gryte, C.C.; Nowick, A.S.; Li, S.Z.; Pak, Y.S.; Greenbaum, S.G. Dielectric relaxation and deuteron NMR of water in polyimide films. *J. Appl. Phys.* **1989**, *66*, 5290–5296. [CrossRef]

42. Lim, L.T.; Britt, U.; Tung, M.A. Sorption and transport of water vapor in nylon 6,6 film. *J. Appl. Polym. Sci.* **1999**, *71*, 197–206. [CrossRef]

43. Puffr, R.; Sebenda, J. On the structure and properties of polyamides. The mechanism of water sorption in polyamides. *J. Polym. Sci. C* **1967**, *16*, 79–93. [CrossRef]

44. Berens, A.R. Solubility of vinyl chloride in poly (vinyl chloride). *Die Angew. Makromol. Chem.* **1975**, *47*, 97–110. [CrossRef]

45. Stapf, S.; Kimmich, R.; Seitter, R.O. Proton and deuteron field cycling NMR relaxometry of liquids in porous glasses: Evidence for Lévy-Walk Statistics. *Phys. Rev. Lett.* **1995**, *75*, 2855–2858. [CrossRef] [PubMed]

46. Sammon, C.; Mura, C.; Yarwood, J.; Everall, N.; Swart, R.; Hodge, D. FTIR-ATR studies on the structure and dynamics of water molecules in polymeric matrixes. A comparison of PET and PVC. *J. Phys. Chem. B* **1998**, *102*, 3402–3411. [CrossRef]

47. Rittigstein, P.; Priestley, R.D.; Broadbelt, L.J.; Torkelson, J.M. Model polymer nanocomposites provide an understanding of confinement effects in real nanocomposites. *Nat. Mater.* **2007**, *6*, 278–282. [CrossRef] [PubMed]

48. Lan, T.; Torkelson, J.M. Methacrylate-based polymer films useful in lithographic applications exhibit different glass transition temperature-confinement effects at high and low molecular weight. *Polymer* **2014**, *55*, 1249–1258. [CrossRef]

49. Tortora, M.; Gorrasi, G.; Vittoria, V.; Galli, G.; Ritrovati, S.; Chiellini, E. Structural characterization and transport properties of organically modified montmorillonite/polyurethane nanocomposites. *Polymer* **2002**, *43*, 6147–6157. [CrossRef]

50. Mohan, T.P.; Kanny, K. Water barrier properties of nanoclay filled sisal fibre reinforced epoxy composites. *Composites Part A* **2011**, *42*, 385–393. [CrossRef]

nanomaterials

MDPI

Article

Nanocomposites Based on PCL and Halloysite Nanotubes Filled with Lysozyme: Effect of Draw Ratio on the Physical Properties and Release Analysis

Valeria Bugatti, Gianluca Viscusi, Carlo Naddeo and Giuliana Gorrasi *

Department of Industrial Engineering, University of Salerno, Via Giovanni Paolo II, 132, 84084
Fisciano (Salerno), Italy; vbugatti@unisa.it (V.B.); g.viscusi@studenti.unisa.it (G.V.); cnaddeo@unisa.it (C.N.)
* Correspondence: ggorrasi@unisa.it; Tel.: +39-089-964146

Received: 4 June 2017; Accepted: 1 August 2017; Published: 4 August 2017

Abstract: Halloysite nanotubes (HNTs) were loaded with lsozyme, as antimicrobial molecule, at a HNTs/lysozyme ratio of 1:1. Such a nano-hybrid was incorporated into a poly (ε-caprolactone) (PCL) matrix at 10 wt % and films were obtained. The nano-composites were submitted to a cold drawn process at three different draw ratios, λ = 3, 4, and 5, where λ is $l_{(\text{final length})}/l_{0(\text{initial length})}$. Morphology, physical, and barrier properties of the starting nanocomposite and drawn samples were studied, and correlated to the release of the lysozyme molecule. It was demonstrated that with a simple mechanical treatment it is possible to obtain controlled release systems for specific active packaging requirements.

Keywords: halloysite nanotubes; PCL; lysozyme; controlled release; active packaging

1. Introduction

The expanding consumer demands of minimally-processed fresh, tasty, and convenient food products on an industrial level, is gaining and the food packaging field is experiencing new opportunities for the formulation of novel materials able to with extend the shelf life and control the quality of packaged foods. In the coming years a development of new and alternative food packaging technologies able to control oxidation of foods and microbial contamination is expected, in order to inhibit, limit, or delay microorganisms' proliferation and the rate of quality decay [1]. The simple blending of low molecular weight antimicrobials in polymer matrices has the disadvantage that the migration and the release of the active molecule cannot be easily predicted and controlled. Foods have peculiar physico-chemical characteristics that could alter the activity of antimicrobial substances; therefore, food components can significantly affect the efficiency of the antimicrobial substances and their release. For instance, the antimicrobial activity and chemical stability of incorporated active molecules could also be influenced by the water activity and the pH of food. The effect of the pH of the release medium has been studied with respect to the lysozyme activity [2] and rate of release from a silica carrier [3]. It was reported that the pH induced conformational changes in lysozyme. The increase of hydrodynamic radius of lysozyme at pH 2 suggested some expansion of the molecules caused by internal repulsive charge interactions. Such an expansion caused the jamming of lysozyme in the pores at acidic pHs [4]. The release of lysozyme from a porous structure can be, then, triggered by the pH, as the lysozyme molecules shrink to their original size at pH 7.

The release kinetics of antimicrobial agents must be able to ensure the concentration above the critical inhibitory dose with respect to the contaminating microorganisms [5]. The phase composition and the morphological organization of the manufacture used for active packaging also play a crucial role with respect to its physical properties and the molecules' release rate. For example, a possible chain orientation can determine the complex transport phenomena [6,7].

Lysozyme is a natural protein and a promising natural antimicrobial agent [8–10]. The European Union lists Llysozyme as a food additive (E 1105) with bacteriostatic, bacteriolytic, and bactericidal activity, and the Food and Drug Administration (FDA) considers this molecule as a GRAS (Generally Recognized as Safe). Chemically, it is characterized by a single polypeptide chain, and the antimicrobial activity is related to the capability to hydrolyze the beta 1–4 glycosidic bonds between *N*-acetylglucosamine and *N*-acetylmuramic acid present in peptidoglycans, which constitute 90% of the cell wall of Gram-positive bacteria [11]. Lysozyme has been immobilized on different supports using different methodologies, such as adsorption, entrapment, and surface conjugation [12–14], however, its use as a covalently-bonded antimicrobial agent is still very limited. Different strategies have also been used in the literature to modulate the release rate of lysozyme. They are focused on changing the packaging material morphology [15], degree of crosslinking [16–18], plasticizer loading [19], nature and amount of additive [20], and number of layers [21]. Some approaches are also focused on the control of the pH of the release medium to modulate its release [22,23]. All these studies show that the release ranges in relatively short times (few days). The goal is to control, as much as possible, the release of lysozyme without changing its chemical structure, and then its activity.

Very recently, halloysite nanotubes (HNTs) attracted increasing interest as inorganic fillers for polymers. They are cheap green materials and available in thousands of tons from natural deposits. They belong to the alumosilicate clays with a length of about 1000 nm, an internal diameter of 10–15 nm, and external diameters of about 50–80 nm. Their general chemical formula is $Al_2Si_2O_{54} \times nH_2O$, and the alumosilicate sheets are rolled into tubes [24,25]. HNTs can be dispersed in polymeric matrices without exfoliation, as required for a good dispersion of layered clays, due to the tubular shape. Polymeric materials have been widely filled with these tubular nano-containers [26–29] able to release active molecules (antimicrobial, drugs, essential oils, flame retardant, self-healing, anticorrosion, etc.) in particular environments [30–37].

In a previous paper we reported the preparation and analysis of novel composites based on poly(lactic acid) (PLA) and HNTs filled with lysozyme for potential in food packaging applications [38]. It is well known that PLA can be applied only in rigid packaging for its brittleness and low elongation at break. In recent years appear very promising biodegradable systems based on blends and copolymers of PLA and PCL to overcome the main drawbacks of both polymers (i.e., brittleness of PLA, and the low modulus and *Tg* of PCL). In this paper we filled PCL with 10% of a HNTs/lysozyme nano-hybrid (HNTs: lysozyme ratio equal to 1:1) and analyzed the physical and release properties. Being that PCL is a biodegradable polyester with a good elongation at break, we also analyzed the possibility to tune the physical and release properties of such active nano-composite through a uniaxially cold draw process. Films were drawn up to λ = 3, 4, and 5. The physical properties and the release analysis of lysozyme were studied and correlated to the phase composition and the drawing process. The possibility to apply the same drawing process to copolymers and blends of PLA and PCL, filled with reservoirs of active molecules, like halloysite nanotubes, can help to obtain novel biodegradable active materials with optimal thermal and mechanical properties, and tunable release for specific applications.

2. Experimental

2.1. Materials

Poly (ε-caprolactone) (PCL) Mn 50000 was supplied from Solvay (Solvay Chimica Italia S.p.A, Bollate, Italy). Halloysite nanoclay powders (CAS 1332-58-7), lysozyme powders (CAS 12650-88-3) and tetrahydrofuran (THF) (CAS: 109-99-9) were supplied from Sigma Aldrich (Milano MI, Italy) and used as received.

2.2. Preparation of HNTs-Lysozyme

The preparation of the nano-hybrid HNTs-lysozyme was conducted accordingly to a previously reported procedure [38]. Three grams of lysozyme were dissolved in 30 mL of distilled water at 50 °C for 20 min. The HNTs (3 g) were mixed with the lysozyme solution. Then, ultrasonic processing was performed for 10 min, at 40% of amplitude using a UP200S Ultrasonic Processor (Heilscher, Teltow, Germany) to sufficiently disperse the HNTs in the lysozyme solution. The solution was heated at 50 °C for 2 min, then reduced pressure ($\cong$0.085 MPa) was applied to remove the air between and within the hollow tubes. The vacuum was maintained for 15 min. The solution was then taken out from the vacuum and shaken for 5 min. Then vacuum was re-applied to remove the trapped air for 15 min. The HNTs loaded with lysozyme were filtered and dried in an oven for 16 h at 50 °C up to a constant weight. The cycle of vacuum vas then repeated for a second time. The content of lysozyme (wt %) in the HNT-lysozyme hybrid has been calculated according to following Equation, using the TGA analysis:

$$\alpha_3 = w \cdot \alpha_1 + (1 - w) \cdot \alpha_2$$

where α_1 is the mass loss of lysozyme (99.1%) at 413.2 °C; α_2 is the mass loss of HNTs (16.9%) at 457.9 °C; α_3 is the mass loss of lysozyme/HNTs (58.6%) at 416.0 °C (see TGA Figure 1). Therefore, the content of lysozyme (w) in HNTs-lysozyme was estimated to be 50.7%; the HNTs content was 49.3%.

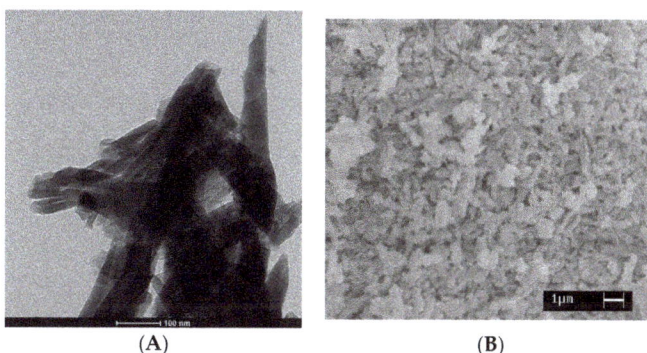

Figure 1.   (A) TEM image of the pristine HNTs and (B) SEM micrograph of the HNTs-lysozyme nanohybrid.

The evaluated lysozyme amount is too high if compared to the loading capacity of the halloysite nanotubes. This detected content is then relative either to the molecules that were filled into the nanotubes, or to the molecules external to the HNTs, which concurs in different ways to the release (see Section 3).

2.3. Nanocomposites Preparation and Draw Processing

Pure PCL and PCL/HNTs-lysozyme, in a weight ratio 90:10, were dissolved in THF at 40 °C for 20 min and sonicated for 10 min at 40% of amplitude. Cast films were hot pressed at 60 °C and cooled to ambient temperature. Strips 1 cm long, were obtained from the films. The strips were put between the clamps of an Instron Dynamometer (Mod 4301, INSTRON Ltd., Norwood, MA, USA), equipped with a temperature chamber. The temperature was fixed to 25 °C and the upper traverse of the dynamometric apparatus was moved at a speed rate of 5 mm/min. Films were drawn at different draw ratios $\lambda = l/l_0$, where l is the final length and l_0 the initial length. The selected values of the draw ratios for all the blends were $\lambda = 3$, 4, and 5. $\lambda = 1$ refers to the undeformed nanocomposite.

2.4. Methods of Investigation

Bright field transmission electron microscopy (TEM) experiments were performed on a FEI TECNAI G12 Spirit-Twin (120 kV, LaB6) microscope equipped with a FEI Eagle 4k CCD camera (Eindhoven, The Netherlands).

Scanning electron microscopy (SEM) analysis was performed with a LEO 1525 microscope (LEO Electron Microscopy Inc., Thornwood, NY, USA). Thermogravimetric analyses (TGA) were carried out from 30 to 900 °C (heating rate of 10 °C/min) under air flow, using a Mettler TC-10 thermo-balance (Mettler-Toledo GmbH, Greifensee, Switzerland).

Infrared spectra (IR) were recorded in attenuated total reflectance (ATR) mode (Bruker Italia, Milano, Italy) using a Bruker spectrometer, model Vertex 70 (average of 32 scans, at a resolution of 4 cm^{-1}).

Mechanical properties were evaluated on all the nanocomposites, using an INSTRON 4301 dynamometric apparatus. The experiments were conducted at room temperature with the deformation rate of 10 mm/min. The initial length of the samples was 10 mm. The mechanical properties, evaluated from stress (σ)/strain (ε) curves, were: elastic modulus E (MPa), evaluated by applying Hooke's law ($\sigma = E \times \varepsilon$) in the interval of deformation less than 0.2%; stress at the yield point σ_y (MPa); elongation at the yield point ε_y (mm/mm %); stress at the break point σ_b (MPa); and elongation at the break point ε_b (mm/mm %) [39]. Data were averaged on five samples.

Barrier properties of water vapor were evaluated using a McBain spring balance system. Samples were suspended from a helical quartz spring supplied by Ruska Industries (Houston, TX, USA) having a spring constant of 1.52 cm/mg. The temperature was controlled to 30 ± 0.1 °C by a constant temperature water bath. Samples were exposed to the water vapor at fixed pressures, P, giving different water activities $a = P/P_0$, where P_0 is the saturation water pressure at the experimental temperature. The spring position was recorded as a function of time using a cathetometer. The spring position data were converted to mass uptake data using the spring constant, and the process was followed to a constant value of sorption for at least 24 h.

Measuring the increase in weight as function of time, for the samples exposed to the vapor at a given partial pressure, it is possible to obtain the equilibrium value of sorbed vapor, $C_{eq, (gsolvent/100 \ gpolymer)}$. In the case of Fickian behavior, when the sorption follows a linear dependence on the square root of time, it is possible to derive the mean diffusion coefficients from the linear part of the reduced sorption curves, reported as C_t/C_{eq} versus square root of time, by the Equation (1) [40]:

$$\frac{C_t}{C_{eq}} = \frac{4}{d}\left(\frac{Dt}{\pi}\right)^{1/2} \tag{1}$$

where C_t is the penetrant concentration at the time t, C_{eq} the equilibrium value, d (cm) the thickness of the sample, and D (cm^2/s) is the average diffusion coefficient.

All the samples showed a Fickian at all the considered activities. Using Equation (1) the diffusion coefficient, D, at any vapor activity ($a = P/P_0$) in the range 0–0.6 was derived, as was the equilibrium concentration of solvent into the samples, $C_{eq, (gsolvent/100 \ gpolymer)}$. For polymer-solvent systems, the diffusion parameter is usually not constant, but depends on the vapor concentration, according to the empirical Equation (2):

$$D = D_0 \ exp(\gamma \ C_{eq}) \tag{2}$$

where D_0 (cm^2/s) is the zero concentration diffusion coefficient (related to the fractional free volume and to the morphology of the material), and γ is a coefficient which depends on the fractional free volume and on the effectiveness of the penetrant to plasticize the matrix.

The release kinetics of the lysozyme were followed using a UV-2401 PC Spectrometer (Shimadzu, Kyoto, Japan). The tests were performed using rectangular specimens of 4 cm^2 and the same thickness (150 μm), placed into 25 mL physiological solution and stirred at 100 rpm in an orbital shaker (VDRL MOD. 711+, Asal S.r.l., Milan, Italy). The release medium was withdrawn at fixed time intervals and replenished with fresh medium. The considered band was at 265 nm.

3. Results and Discussion

Figure 1A reports the TEM image of the pristine HNTs. It is evident a tubular structure where the average length of the tubes ranges between 0.5–1 μm, the inner diameter ranges between 10–15 nm, and outer diameters between 100–200 nm. Figure 1B shows SEM micrograph of the prepared halloysite nano-hybrid. In the used experimental conditions, it is evident a significant amount of lysozyme crystals outside the tubes.

The distribution of active molecules inside and/or around HNTs nano-hybrids is very complex to be accurately determined. The degradation temperatures of the single components and the nano-hybrid has been demonstrated to be a useful tool to have an indication about the molecules' intercalation efficiency [41]. Figure 2 reports the mass loss (%) (TG) and derivative thermogravimetry (DTG) evaluated on HNTs, lysozyme and HNTs-lysozyme. Halloysite shows one degradation step at about 500 °C, which is attributed to the dehydroxylation of the matrix [42]. The thermal degradation of lysozyme occurs in two main steps. The first one, centered at about 298 °C, is related to the breaking of the amide bonds. The second one, centered at about 542 °C is due to the formation and elimination of low molecular weight volatile oxygenated compounds. A shift toward higher values for the two degradation temperatures of lysozyme, centered at 316 and 566 °C is clearly evident in the nano-hybrid. The degradation of the HNTs does not change going from the crude material to the nano-hybrid. As evidenced from SEM analysis, a part of lysozyme molecules is external to the tubes and degrades at the degradation temperature of the free lysozyme. The lysozyme molecules that show a delay in the degradation were successfully entrapped into the nanotubes and their degradation occurs only after a spillage from the nanotubes.

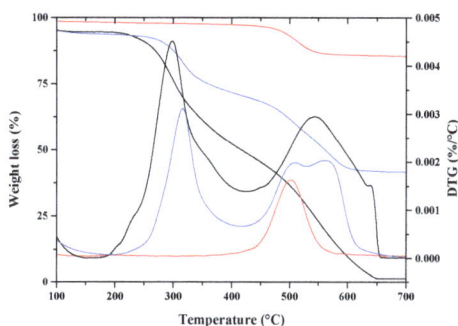

Figure 2. TGA/DTG of: HNTs (——), Lysozyme (——), nano-hybrid HNTs/Lysozyme (——).

Figure 3A reports the ATR spectra for all nanocomposites, undeformed and drawn at different draw ratios, in the range 2500–3200 cm^{-1}. In these regions no absorption bands are visible neither for HNTs nor for lysozyme. The bands at 2850 and 2930 cm^{-1} are related to the symmetrical and asymmetrical stretching from the bonds C–H of the PCL aliphatic chains [43]. It is evident that both bands increase with the draw ratio. Figure 3B reports the intensities of these bands in the drawn samples, divided by the intensity of the undeformed sample. It is evident a correlation between draw ratio and the symmetrical and asymmetrical stretching of the C–H bonds of the aliphatic PCL chains.

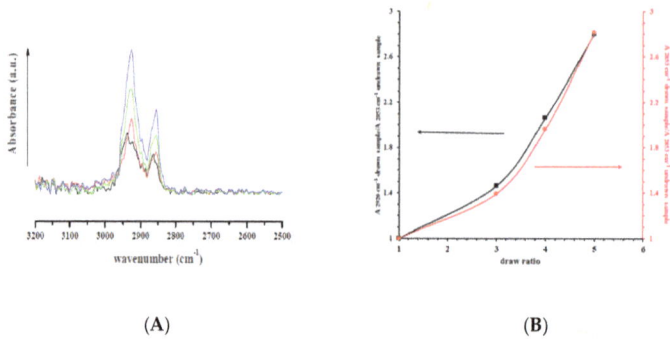

(A)   (B)

Figure 3. ATR spectra of (**A**): Undeformed nanocomposite (—), drawn at λ = 3 (—), drawn at λ = 4 (—), and drawn at λ = 5 (—); and (**B**) the ratio between absorbance bands of symmetrical (2850 cm^{-1}) and asymmetrical (2930 cm^{-1}) C–H stretching for drawn samples divided by the absorbance of undrawn samples.

Figure 4 shows the TGA analysis on the considered nanocomposites, either undrawn or drawn at different draw ratios. For comparison it is reported also the TGA of unfilled PCL and PCL drawn at λ = 5. The thermogram of neat PCL exhibits two peaks at about 385 and 480 °C, which evidenced a two-step mechanism for the PCL degradation [44]: (1) random chain scission induced by pyrolysis of the ester groups, with the release of hexanoic acid, CO_2 and H_2O, and (2) cyclic monomer of ε-caprolactone formation, as a result of unzipping depolymerization process. The drawing process does not significantly affect the thermal degradation of PCL (sample PCL drawn at λ = 5 in Table 1), while the incorporation of the nano-hybrid at 10 wt % of HNTs-lysozyme generates an improvement on thermal stability of PCL, which is reflected also in the drawn samples. Degradation temperatures, evaluated at 10%, 50%, and 95% of weight loss for all samples, are reported in Table 1. It has been demonstrated that HNTs in HNTs/polymer nanocomposites are responsible for an improvement in thermal stability [29,45]. As already found, it is hypothesized that the HNTs dispersed into the matrix, act as a barrier to the mass transport slowing down the escape of the volatile products during the degradation process [46].

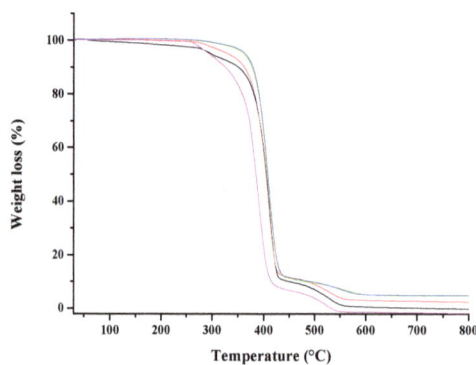

Figure 4. TGA of: PCL (—), undeformed nanocomposite (—), drawn at λ = 3 (—), drawn at λ = 4 (—), and drawn at λ = 5 (—).

Table 1. Thermal data for all nanocomposites and unfilled PCL, evaluated from TGA.

Sample	T_d (10% Weight Loss)	T_d (50% Weight Loss)	T_d (95% Weight Loss)
PCL	320 °C	385 °C	480 °C
PCL drawn at $\lambda = 5$	325 °C	390 °C	505 °C
PCL/10%HNTs-Lysozyme	350 °C	403 °C	518 °C
Nano-composite drawn at $\lambda = 3$	360 °C	405 °C	537 °C
Nano-composite drawn at $\lambda = 4$	373 °C	407 °C	573 °C
Nano-composite drawn at $\lambda = 5$	377 °C	409 °C	583 °C

Mechanical properties were conducted on all nanocomposites, either undeformed or monoaxially drawn samples. Table 2 reports the mechanical parameters for all samples, evaluated from the stress-strain curves here not reported. For comparison, the mechanical parameters for unfilled PCL are also reported, taken from a previous work [47]. It is also reported the mechanical behavior of unfilled PCL drawn at $\lambda = 5$. It is evident that either the introduction of the nano-hybrid, or the drawing process, determine a reinforcement of the PCL matrix. In the case of composite systems, the enhancement of the mechanical properties of nanocomposites requires a high degree of load transfer between continuous and the dispersed phases. If the interfacial adhesion between the phases is weak, the filler behaves as holes or nanostructured flaws, introducing local stress concentrations, and its benefits on the properties are lost [48]. The filler must be well dispersed because, in the case of poor dispersion, the strength is significantly reduced. Furthermore, solid state drawing of nanocomposites is largely dominated by regions of a lower HNTs volume fraction and this could certainly alter both orientation and distribution of HNTs along the length of the fiber. The improvement of the mechanical parameters is evident for the composite fibers, in particular at higher draw ratios. This could be attributed to highly-dispersed and well-aligned HNTs. The only parameter that results in a decrease, either with filler loading or with the draw ratio, is the elongation at break. This is quite expected because of the incompatibility between the two phases due to the different chemical nature of both, more evidenced at higher draw ratios where a stronger phase separation can occur for the mechanical treatment.

Table 2. Mechanical properties for all nanocomposites and unfilled PCL, evaluated from stress-strain curves.

Sample	E (MPa)	σ_y (MPa)	ε_y (%)	σ_b (MPa)	ε_b (%)
PCL [#]	185 ± 24	9.95 ± 0.23	11.50 ± 3.1	15.88 ± 0.23	616 ± 14.22
PCL drawn at $\lambda = 5$	530 ± 22	36.45 ± 0.67	13.57 ± 3.4	32.52 ± 0.34	320 ± 15.34
PCL/10%HNTs-Lysozyme	320 ± 16	10.24 ± 0.34	8.79 ± 3.7	18.14 ± 0.65	570 ± 12.31
Nano-composite drawn at $\lambda = 3$	335 ± 15	16.72 ± 0.46	9.31 ± 2.4	28.98 ± 0.47	157 ± 16.26
Nano-composite drawn at $\lambda = 4$	347 ± 23	23.60 ± 0.42	14.08 ± 3.6	36.10 ± 0.36	162 ± 8.420
Nano-composite drawn at $\lambda = 5$	447 ± 20	39.40 ± 0.57	14.06 ± 4.8	54.61 ± 0.74	102 ± 13.27

[#] Data from ref. [47].

Figure 5 reports the sorption isotherms of the undeformed nanocomposite and the ones drawn at different draw ratios. The mode of sorption, for all the samples, can be interpreted with a dual-mode mechanism. Sorption is visualized as a process in which there are dual modes: at low activity the penetrant molecules are normally dissolved and free to diffuse, or they are immobilized on polar sites of the polymer matrix. At higher activities the increase of sorption is due to the water plasticization effect on the material. It is evident that a decrement of sorption for all drawn samples in the whole investigated activity range. The lowering of water sorption is quite independent of the draw ratio. The morphological organization of the drawn samples, in terms of the amount of the permeable phase, is similar in the fibers. The drawing process either induces further crystallization in the PCL or allows an alignment in the polymeric chains that can represent impermeable domains to water sorption.

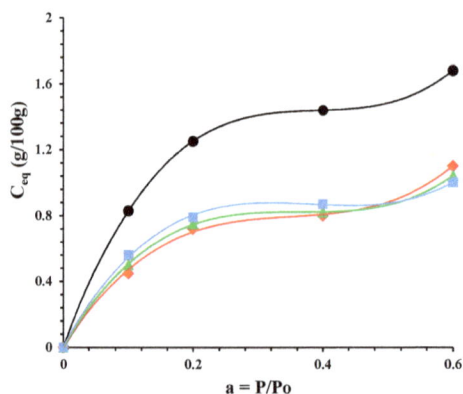

Figure 5. Sorption isotherms for: PCL/10%HNTs-lysozyme (•), a sample drawn at $\lambda = 3$ (♦), a sample drawn at $\lambda = 4$ (▲), and a sample drawn at $\lambda = 5$ (■).

The analysis of sorption and diffusion of low molecular weight molecules are of great importance in determining the applications of nanocomposites materials in the packaging field. Modulating the nanocomposite's composition, in terms of the nature of single components, loading, and morphology, it is possible to project novel and active materials for the required uses. In order to determine how the composition of the multiphase materials, drawn at different draw ratios, affects the transport properties of the PCL matrix, sorption (thermodynamic parameter) and diffusion (kinetic parameter) for all the samples using water vapor were evaluated.

Figure 6 reports the log of diffusion, D (cm^2/s), as a function of C_{eq} (g/100 g) of sorbed water vapor. It is evident a slight increase of the diffusion coefficient with the draw ratio. This can be due to micro-voids formed during the drawing process. It is worth noting, as for the mechanical properties, that the two phases (organic-inorganic) are not perfectly miscible. The drawing process, at ambient temperature, can generate a slight disconnection between the two phases, evidenced by the diffusion. The small molecules can diffuse through mechanically-induced microvoids.

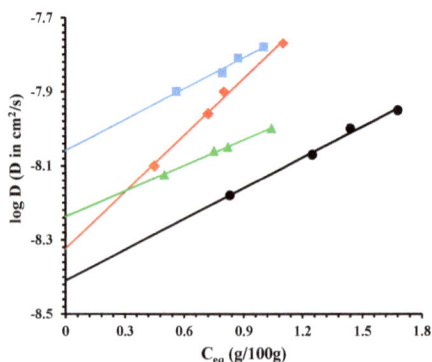

Figure 6. Log diffusion for: PCL/10%HNTs-lysozyme (●), a sample drawn at $\lambda = 3$ (♦), a sample drawn at $\lambda = 4$ (▲), and a sample drawn at $\lambda = 5$ (■).

Figure 7 reports the release of lysozyme (in wt %) in physiological solution, as a function of time (hours). In order to demonstrate the capability of HNTs to slow down the lysozyme release, we prepared also a sample in which the same amount of lysozyme (5 wt %) is free dispersed in PCL film. We observed that in the first 24 h about the 90% of lysozyme is released, while it is evident from

Figure 7 that only the 25% is released from the nano-composite in which the molecule is entrapped into the HNTs. After 720 h the lysozyme release from the nano-composite is around 50%. The release of Lysozyme from the nano-composites is a complex phenomenon. It is evident a first stage of fast release in the first 24 h could be due to the free lysozyme molecules located on the surface of the sample of free to diffuse through the microvoids mechanically inducted (see discussion on the diffusion). The second stage, from 24 to about 60 h, could be related to the diffusion of the molecules from the bulk and/or the HNTs, the third stage represents the spillage of the most entrapped lysozyme molecules. The higher is the draw ratio, the higher the lysozyme (wt %) capable to exit from the nanocomposites. After one month (720 h) the lysozyme is released for 100% in the case of a sample drawn at $\lambda = 5$, for 94% and 77% for the nanocomposites drawn at $\lambda = 4$ and 3, respectively, and only for 50% for the undeformed sample. Figure 8 reports the Lysozyme release (wt %) at two different times: 24 h and one month. It is evident a linear increase of the active molecule released, as function of the draw ratio, for both the chosen times. In the case of food packaging application 1 month is a long time of storage for fresh foods, and in this time it is desirable that 100% of the antimicrobial molecule is released. In particular, in the case of meat, it has been found that a lower burst and a slower release is preferable, like for the sample drawn at $\lambda = 3$ [19], while for foods rich of water, like tomatoes, strawberries, or grapefruit [50], an initial fast release around 50%, followed by a slower step of release is better, like for the samples drawn at $\lambda = 4$ and 5.

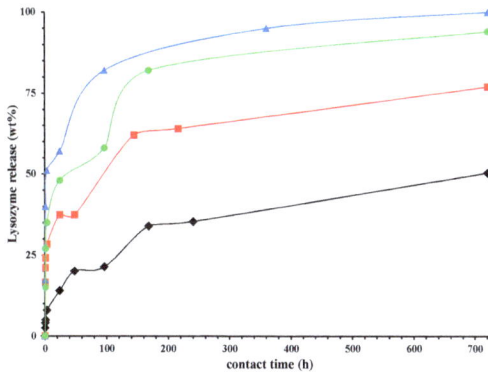

Figure 7. Lysozyme released (wt %), as function of time (hours) for: PCL/10%HNTs-lysozyme ($\blacklozenge$), a sample drawn at $\lambda = 3$ ($\blacksquare$), a sample drawn at $\lambda = 4$ ($\bullet$), and a sample drawn at $\lambda = 5$ ($\blacktriangle$).

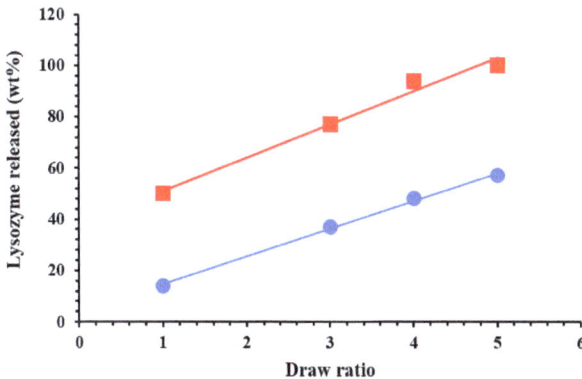

Figure 8. Lysozyme released (wt %) after: 24 h ($\bullet$) and one month ($\blacksquare$).

4. Concluding Remarks

In this work we investigated the influence of the mechanical drawing on nanocomposites composed of poly (ε-caprolactone) (PCL) filled with 10% of HNTs/Lysozyme nano-hybrid. Films were uniaxially cold drawn up to $\lambda = 3$, 4, and 5. The physical properties and the release analysis of lysozyme were studied and correlated to the phase composition and the drawn ratio.

From ATR measurements it was demonstrated that the bands at 2850 and 2930 cm^{-1}, related to the symmetrical and asymmetrical stretching from the bonds C–H of the PCL aliphatic chains, increased with the draw ratio.

The incorporation of the nano-hybrid at 10 wt % of HNTs-lysozyme allowed an improvement of thermal stability of PCL, which was also reflected in the drawn samples.

Mechanical properties demonstrated that the strength of the samples were improved, in particular, at higher draw ratios. This was attributed to highly-dispersed and well-aligned HNTs. The elongation at break resultedly decreased because of the incompatibility between the two phases due to the different chemical nature of both, more evidenced at higher draw ratios where a stronger phase separation occurs for the mechanical treatment.

Barrier properties to water vapor showed that sorption decreased for all drawn samples, while the diffusion slightly increased with the draw ratio. This was attributed to possible micro-voids formed during the drawing process.

The release kinetics of lysozyme in physiological solution showed, in accordance with diffusion data, an increasing of the release with increasing the draw ratio. It was demonstrated that with a mechanical treatment it is possible to obtain nanocomposites at controlled release for specific active packaging requirements.

Work is in progress in order to test the efficiency of such systems with respect to *Bacillus sp.* considering the effect of the solvent treatment, the temperature used for hot pressing the material, and the drawing process.

Author Contributions: Giuliana Gorrasi conceived the paper and designed the experiments. Gianluca Viscusi, Carlo Naddeo and Valeria Bugatti performed the experiments. Giuliana Gorrasi and Valeria Bugatti analyzed the data and also wrote this paper.

Conflicts of Interest: The authors declare no conflict of interest.

References

1. Lagaròn, J. *Multifunctional and Nanoreinforced Polymers for Food Packaging*; Woodhead Publishing: Sawston, Cambridge, UK, 2011.
2. Benelhadj, S.; Fejji, N.; Degraeve, P.; Attia, H.; Ghorbel, D.; Gharsallaoui, A. Properties of lysozyme/Arthrospira platensis (Spirulina) protein complexes for antimicrobial edible food packaging. *Algal Res.* **2016**, *15*, 43–49. [CrossRef]
3. Xue, M.; Findenegg, G.H. Lysozyme as a pH-Responsive Valve for the Controlled Release of Guest Molecules from Mesoporous Silica. *Langmuir* **2012**, *28*, 17578–17584. [CrossRef] [PubMed]
4. Bonincontro, A.; de Francesco, A.; Onori, G. Influence of pH on Lysozyme Conformation Revealed by Dielectric Spectroscopy. *Colloids Surf. B* **1998**, *12*, 1–5. [CrossRef]
5. Mastromatteo, M.; Conte, A.; Del Nobile, M.A. Advances in controlled release devices for food packaging applications. *Trends Food Sci. Technol.* **2010**, *21*, 591–598. [CrossRef]
6. Bugatti, V.; Acocella, M.; Maggio, M.; Pantani, R.; Gorrasi, G. Release of Lysozyme from cold drawn Poly (ε-caprolactone) at different draw ratios. *Marcomol. Mater. Eng.* **2017**, in press.
7. Zhang, W.; Ronca, S.; Mele, E. Electrospun Nanofibres Containing Antimicrobial Plant Extracts. *Nanomaterials* **2017**, *7*, 42. [CrossRef] [PubMed]
8. James, S.; McManus, J.J. Thermal and solution stability of lysozyme in the presence of sucrose, glucose, and trehalose. *J. Phys. Chem. B* **2012**, *116*, 10182–10188. [CrossRef] [PubMed]
9. Venkataramani, S.; Truntzer, J.; Coleman, D.R. Thermal stability of high concentration lysozyme across varying pH: A fourier transform infrared study. *J. Pharm. Bioallied Sci.* **2013**, *5*, 148–153. [CrossRef] [PubMed]

10. Al Meslmani, B.M.; Mahmoud, G.F.; Leichtweiß, T.; Strehlow, B.; Sommer, F.O.; Lohoff, M.D.; Bakowsky, U. Covalent immobilization of lysozyme onto woven and knitted crimped polyethylene terephthalate grafts to minimize the adhesion of broad spectrum pathogens. *Mater. Sci. Eng. C* **2016**, *58*, 78–87. [CrossRef] [PubMed]

11. Losso, J.N.; Nakai, S.; Charter, E.A. Lysozyme. In *Natural Food Antimicrobial Systems*; Naidu, A.S., Ed.; CRC Press LLC: Boca Raton, FL, USA, 2000; pp. 185–210.

12. Conte, A.; Buonocore, G.G.; Bevilacqua, A.; Sinigaglia, M.; Del Nobile, M.A. Immobilization of lysozyme on polyvinylalcohol films for active packaging applications. *J. Food Prot.* **2006**, *69*, 866–870. [CrossRef] [PubMed]

13. Edwards, J.V.; Prevost, N.; Condon, B.; Sethumadhavan, K.; Ullah, J.; Bopp, A. Immobilization of lysozyme on cotton fabrics: Synthesis, characterization, and activity. *AATCC Rev.* **2011**, *11*, 73–79.

14. Muriel-Galet, V.; Talbert, J.N.; Hernandez-Munoz, P.; Gavara, R.; Goddard, J.M. Covalent Immobilization of Lysozyme on Ethylene Vinyl Alcohol Films for Nonmigrating Antimicrobial Packaging Applications. *J. Agric. Food Chem.* **2013**, *61*, 6720–6727. [CrossRef] [PubMed]

15. Farhoodi, M. Nanocomposite Materials for Food Packaging Applications: Characterization and Safety Evaluation. *Food Eng. Rev.* **2016**, *8*, 35–51. [CrossRef]

16. Buonocore, G.G.; Del Nobile, M.A.; Panizza, A.; Corbo, M.R.; Nicolais, L. A general approach to describe the antimicrobial agent release from highly swellable films intended for food packaging applications. *J. Control. Release* **2003**, *90*, 97–107. [CrossRef]

17. Buonocore, G.G.; Sinigaglia, M.; Corbo, M.R.; Bevilacqua, A.; La Notte, E.; Del Nobile, M.A. Controlled release of antimicrobial compounds from highly swellable polymers. *J. Food Prot.* **2004**, *67*, 1190–1194. [CrossRef] [PubMed]

18. Fajardo, P.; Balaguer, M.P.; Gomez-Estaca, J.; Gavara, R.; Hernandez-Munoz, P. Chemically modified gliadins as sustained release systems for lysozyme. *Food Hydrocoll.* **2014**, *41*, 53–59. [CrossRef]

19. Min, S.; Rumsey, T.R.; Krochta, J.M. Diffusion of the antimicrobial lysozyme from a whey protein coating on smoked salmon. *J. Food Eng.* **2008**, *84*, 39–47. [CrossRef]

20. Mastromatteo, M.; Lecce, L.; De Vietro, N.; Favia, P.; Del Nobile, M.A. Plasma deposition processes from acrylic/methane on natural fibres to control the kinetic release of lysozyme from PVOH monolayer film. *J. Food Eng.* **2011**, *104*, 373–379. [CrossRef]

21. Conte, A.; Buonocore, G.G.; Nicolais, L.; Del Nobile, M.A. Controlled release of active compounds from antimicrobial films intended for food packaging applications. *Ital. J. Food Sci.* **2003**, *15*, 216–218.

22. Lu, J.R.; Su, T.J.; Howlin, B.J. The Effect of Solution pH on the Structural Conformation of Lysozyme Layers Adsorbed on the Surface of Water. *J. Phys. Chem. B* **1999**, *103*, 5903–5909. [CrossRef]

23. Mendes de Souza, P.; Fernandez, A.; Lopez-Carballo, G.; Gavara, R.; Hernandez-Munoz, P. Modified sodium caseinate films as releasing carriers of lysozyme. *Food Hydrocoll.* **2010**, *24*, 300–306. [CrossRef]

24. Yuan, P.; Tan, D.; Annabi-Bergaya, F. Properties and applications of halloysite nanotubes: Recent research advances and future prospects. *Appl. Clay Sci.* **2015**, *112–113*, 75–93. [CrossRef]

25. Lvov, Y.; Abdullayev, E. Green and functional polymer-clay nanotube composites with sustained release of chemical agents. *Prog. Polym. Sci.* **2013**. [CrossRef]

26. Lvov, Y.; Shchukin, D.; Möhwald, H.; Price, R. Clay Nanotubes for Controlled Release of Protective Agents—Perspectives. *ACS Nano* **2008**, *2*, 814–820. [CrossRef] [PubMed]

27. Arcudi, F.; Cavallaro, G.; Lazzara, G.; Massaro, M.; Milioto, S.; Noto, R.; Riela, S. Selective Functionalization of Halloysite Cavity by Click Reaction: Structured Filler for Enhancing Mechanical Properties of Bionanocomposite Films. *J. Phys. Chem.* **2014**, *118*, 15095–15101. [CrossRef]

28. Gorrasi, G.; Pantani, R.; Murariu, M.; Dubois, P. PLA/Halloysite Nanocomposite Films: Water Vapor Barrier Properties and Specific Key Characteristics. *Macromol. Mater. Eng.* **2014**, *299*, 104–115. [CrossRef]

29. Liu, M.; Jia, Z.; Jia, D.; Zhou, C. Recent advance in research on halloysite-nanotubes polymer nanocomposites. *Prog. Polym. Sci.* **2014**, *39*, 1498–1525. [CrossRef]

30. Abdullayev, E.; Lvov, Y. Clay Nanotubes for Corrosion Inhibitor Encapsulation: Release Control with End Stoppers. *J. Mater. Chem.* **2010**, *20*, 6681–6687. [CrossRef]

31. Abdullayev, E.; Lvov, Y. Clay Nanotubes for Controlled Release of Protective Agents—A Review. *J. Nanosci. Nanotechnol* **2011**, *11*, 10007–10026. [CrossRef] [PubMed]

32. Abdullayev, E.; Price, R.; Shchukin, D.; Lvov, Y. Halloysite Tubes as Nanocontainers for Anticorrosion Coating with Benzotriazole. *Appl. Mater. Int.* **2009**, *2*, 1437–1442. [CrossRef] [PubMed]

33. Abdullayev, E.; Shchukin, D.; Lvov, Y. Halloysite Clay Nanotubes as a Reservoir for Corrosion Inhibitors and Template for Layer-by-Layer Encapsulation. *Mater. Sci. Eng.* **2008**, *99*, 331–332.

34. Gorrasi, G.; Vertuccio, L. Evaluation of zein/halloysite nano-containers as reservoirs of active molecules for packaging applications: Preparation and analysis of physical properties. *J. Cereal Sci.* **2016**, *70*, 66–71. [CrossRef]

35. Veerabadran, N.; Price, R.; Lvov, Y. Clay nanotubes for encapsulation and sustained release of drugs. *Nano* **2007**, *2*, 215–222. [CrossRef]

36. Lvov, Y.; Price, R. Halloysite Nanotubules a Novel Substrate for the Controlled Delivery of Bioactive Molecules. In *Bio-Inorganic Hybrid Nanomaterials*; Ruiz-Hitzky, E., Ariga, K., Lvov, Y., Eds.; Wiley: London, UK, 2008; pp. 440–478.

37. Liu, M.; Dai, L.; Shi, H.; Xiong, S.; Zhou, C. In vitro evaluation of alginate/halloysite nanotube composite scaffolds for tissue engineering. *Mater. Sci. Eng. C* **2015**, *49*, 700–712. [CrossRef] [PubMed]

38. Bugatti, V.; Sorrentino, A.; Gorrasi, G. Encapsulation of Lysozyme into halloysite nanotubes and dispersion in PLA: Structural and physical properties and controlled release analysis. *Eur. Polym. J.* **2017**, *93*, 495–506. [CrossRef]

39. *Mechanical Properties of Solid Polymers*, 3rd ed.; Ward, I.M.; Sweeney, J. (Eds.) Wiley: Hoboken, NJ, USA, 2012.

40. Koros, W.J.; Burgess, S.K.; Chen, Z. *Encyclopedia of Polymer Science and Technology*; Wiley: Hoboken, NJ, USA, 2015; pp. 1–96. [CrossRef]

41. Gorrasi, G. Dispersion of halloysite loaded with natural antimicrobials into pectins: Characterization and controlled release analysis. *Carbohydr. Polym.* **2015**, *127*, 47–53. [CrossRef] [PubMed]

42. Horvath, E.; Frost, R.L.; Mako, E.; Kristof, J.; Cseh, T. Thermal treatment of mechano-chemically activated kaolinite. *Thermochim. Acta* **2003**, *404*, 227–235. [CrossRef]

43. Ciardelli, G.; Chiono, V.; Vozzi, G.; Pracella, M.; Ahluwalia, A.; Barbani, N.; Cristallini, C.; Giusti, P. Blends of Poly(ε-caprolactone) and Polysaccharides in Tissue Engineering Applications. *Biomacromolecules* **2005**, *6*, 1961–1976. [CrossRef] [PubMed]

44. Pantoustier, N.; Alexandre, M.; Degee, P.; Calberg, C.; Jerome, R.; Henrist, C.; Cloots, R.; Rulmont, A.; Dubois, P. Poly(ε-caprolactone)/clay nanocomposites prepared by melt intercalation. Mechanical, thermal and rheological properties. *Polymer* **2002**, *43*, 4017–4023.

45. Du, M.; Guo, B.; Jia, D. Thermal stability and flame retardant effects of halloysite nanotubes on poly(propylene). *Eur. Polym. J.* **2006**, *42*, 1362–1369. [CrossRef]

46. Gorrasi, G.; Senatore, V.; Vigliotta, G.; Belviso, S.; Pucciariello, R. PET-halloysite nanotubes composites for packaging application: Preparation, characterization and analysis of physical properties. *Eur. Polym. J.* **2014**, *61*, 145–156. [CrossRef]

47. Bugatti, V.; Costantino, U.; Gorrasi, G.; Nocchetti, M.; Tammaro, L.; Vittoria, V. Nano-hybrids incorporation into poly(ε-caprolactone) for multifunctional applications: Mechanical and barrier properties. *Eur. Polym. J.* **2010**, *46*, 418–427. [CrossRef]

48. Koo, J.H. Mechanical Properties of Polymer Nanocomposites. In *Fundamentals, Properties, and Applications of Polymer Nanocomposites*; Cambridge University Press: Cambridge, UK, 2016; Chapter 7; pp. 273–331.

49. Quintavalla, S.; Vicini, L. Antimicrobial food packaging in meat industry. *Meat Sci.* **2002**, *62*, 373–380. [CrossRef]

50. Bugatti, V. Dispersion of Inorganic Fillers in Polymeric Matrices for Food Packaging Applications. Ph.D. Thesis, University of Salerno, Italy, 2012. Available online: http://hdl.handle.net/10556/274 (accessed on 18 May 2012).

nanomaterials

MDPI

Article

3D-Hydrogel Based Polymeric Nanoreactors for Silver Nano-Antimicrobial Composites Generation

Albanelly Soto-Quintero [1], Ángel Romo-Uribe [2], Víctor H. Bermúdez-Morales [3], Isabel Quijada-Garrido [4,*] and Nekane Guarrotxena [4,*]

1 Centro de Investigación en Ingeniería y Ciencias Aplicadas, Universidad Autónoma del Estado de Morelos, Cuernavaca 62209, Morelos, Mexico; soquia_17@hotmail.com
2 Research & Development, Advanced Science & Technology Division, Johnson & Johnson Vision, Jacksonville, FL 32256, USA; aromouribe@gmail.com
3 Centro de Investigación sobre Enfermedades Infecciosas, Instituto Nacional de Salud Pública, Dirección de Infecciones Crónicas y Cáncer, Avenida Universidad No. 655, Cerrada los Pinos y Caminera, Colonia Santa María Ahuacatitlán, Cuernavaca 62100, Morelos, Mexico; vbermudez@insp.mx
4 Instituto de Ciencia y Tecnología de Polímeros, Consejo Superior de Investigaciones Científicas (ICTP-CSIC), c/ Juan de la Cierva, 3. E-28006 Madrid, Spain
* Correspondence: iquijada@ictp.csic.es (I.Q.-G.); nekane@ictp.csic.es (N.G.); Tel.: +34-915-622-900 (I.Q.-G. & N.G.)

Received: 26 June 2017; Accepted: 25 July 2017; Published: 1 August 2017

Abstract: This study underscores the development of Ag hydrogel nanocomposites, as smart substrates for antibacterial uses, via innovative in situ reactive and reduction pathways. To this end, two different synthetic strategies were used. Firstly thiol-acrylate (PSA) based hydrogels were attained via thiol-ene and radical polymerization of polyethylene glycol (PEG) and polycaprolactone (PCL). As a second approach, polyurethane (PU) based hydrogels were achieved by condensation polymerization from diisocyanates and PCL and PEG diols. In fact, these syntheses rendered active three-dimensional (3D) hydrogel matrices which were used as nanoreactors for in situ reduction of $AgNO_3$ to silver nanoparticles. A redox chemistry of stannous catalyst in PU hydrogel yielded spherical AgNPs formation, even at 4 °C in the absence of external reductant; and an appropriate thiol-functionalized polymeric network promoted spherical AgNPs well dispersed through PSA hydrogel network, after heating up the swollen hydrogel at 103 °C in the presence of citrate-reductant. Optical and swelling behaviors of both series of hydrogel nanocomposites were investigated as key factors involved in their antimicrobial efficacy over time. Lastly, in vitro antibacterial activity of Ag loaded hydrogels exposed to *Pseudomona aeruginosa* and *Escherichia coli* strains indicated a noticeable sustained inhibitory effect, especially for Ag–PU hydrogel nanocomposites with bacterial inhibition growth capabilities up to 120 h cultivation.

Keywords: smart-hydrogel; polyurethane networks; thiol-acrylate networks; silver-hydrogel nanocomposites; Ag nanoparticles; antibacterial materials

1. Introduction

To date, many advanced infection treatments have been developed based on the discovery of new families of broad spectrum antibiotics. Nevertheless, the intense abuse and misuse of them makes unfeasible so far to ward off the increasing resistance of bacteria strains. Additionally, the declining interest in antibiotics research from the major pharmaceutical corporations [1] endangers the successful control of bacterial infections to address the challenge. Thus, efficient and innovative therapies against the emerging threat of resistant bacteria are highly desirable. In this respect, silver nanoparticles-based integrative solutions, due to their potential antimicrobial effects, could provide a compelling alternative [2–5]. Since ancient times, silver, in different forms, has been used as an antimicrobial agent against infections; thus, for more than 100 years, silver nanoparticles (AgNPs) have been used before recognizing their nanometric dimension [6,7]. More recently, the development of biomaterials based on polymeric hydrogels containing AgNPs is attracting significant interest. In particular, hydrophilic hydrogels are good matrices for the sustained dosage of silver ions (Ag$^+$) also decreasing the toxicity of AgNPs [8,9].

Hydrogels are three-dimensional (3D) networks of hydrophilic polymers that swell in the presence of water; the versatility of these systems lies on the fact that by controlling chemical composition and crosslinking ratio, a wide range of materials can be modulated as function of the desired application [10,11]. Moreover, properties such as swelling degree and kinetics, stimuli responsiveness, or degradability can be tuned. In addition, their ability to absorb other substances makes them invaluable tools for the controlled dosage of drugs and other active substances [12]. Thus, nanohybrids generated by embedding AgNPs into hydrogels are of particular interest in medical treatments, for example preventing infection of topical wounds [8,13–15], because of their low toxicity, transparency to observe the wound, potential for extended release of drugs, and ability to keep the wound hydrated. Similarly, AgNP-hydrogel nanocomposites are being used as therapeutic contact lenses [16,17] for the treatment of fungal keratitis [16], as tissue engineering scaffolds [18] for bone replacement [19], and as colored cornea substitutes due to their anti-infective properties and potential for color modulation [20].

Several approaches for the synthesis of hydrogel nanocomposite materials with nanoparticles embedded in the hydrogel scaffold have been developed. Most of them involve a physical or chemical combination of independently generated NPs and hydrogels [18,21,22], or gel formation by mixing pre-synthesized NPs with polymers and gelator precursors [16,23,24]. Other strategies involved the in situ radical polymerization and reduction of AgNO$_3$ during the hydrogel formation, mainly via the photochemical process [25,26]. Potential drawbacks of using these protocols are their synthetic complexity and the low control of the amount and distribution of particles formed within the 3D network.

As an alternative, in this study, we report a simple strategy to design two structurally diverse nanoparticle-hydrogel composites based on in situ reactive nanoparticle generation within a previously built hydrogels, by using two different reactive synthetic routes. So, while Ag-Thiol acrylate (Ag–PSA) nanocomposite generation involves a thiol-functionalized hydrogel matrix able to modulate the AgNP formation in the presence of a reducing agent; a new and improved protocol to synthesis nanosilver-loaded PU hydrogel (Ag–PU), with no presence of thiol functionalities neither external reducing agent, is performed. We tactically selected two series of hydrogels (thiol-acrylate (PSA) and polyurethane (PU)), as templates in the reactive synthetic approaches, and whose specific compositional structures satisfy the restrictive demand of biocompatibility and biodegradability; fundamental features for innovative biological applications [27]. In fact, both systems are mainly composed by poly(ethylene glycol) (PEG), a biocompatible hydrophilic polymer, and poly(ε-caprolactone) (PCL) which enables to modulate properties as viscoelasticity, hydrophobicity, shape memory effect, and degradability [28]. Furthermore, we took advantage of the synthetic protocols used in performing the 3D hydrogel networks to develop a novel and advantageous nanoparticle-formation procedure within the hydrogel. The yielded hydrogel matrices exhibit optical and swelling properties influenced by temperature and AgNP presence which last determine the Ag-nanocomposite efficiency on competitive and sustained antimicrobial properties. The effect of synthetic conditions and polymer composition in AgNPs uniformity and distribution together with the antimicrobial properties of the nanocomposites, on the basis of swelling-shrinking equilibrium and NP surface charge induced effect will be discussed in detail.

2. Materials and Methods

2.1. Materials

For thiol-acrylate (PSA) hydrogels preparation, poly(ethylene glycol) of average molecular weight (M_n) of 3350 g·mol^{-1} (PEG3.3k), poly(ε-caprolactone) of M_n 10,000 g·mol^{-1} (PCL10k), triethylamine (99%), acryloyl chloride (97%), pentaerythritol tetrakis(3-mercaptopropionate) (>95%), and 2,2-dimethoxy-2-phenylacetophenone (DMPA, 99%) were purchased from Sigma-Aldrich (St. Louis, MO, USA).

For polyurethane (PU) hydrogel synthesis, polycaprolactone diol (PCL$_{530}$) M_n (530 g·mol^{-1}), poly(ethylene glycol) diol (PCL$_{600}$) of M_n (600 g·mol^{-1}), and pentaerythritol ethoxylate crosslinker (PEG$_{300}$) of M_n (270 g·mol^{-1})—purchased from Sigma-Aldrich (St. Louis, MO, USA)—were vacuum dried at 80 °C for 24 hours before use. Stannous 2-ethylhexanoate (Aldrich, St. Louis, MO, USA, 92%) was used as received. Hexamethylene diisocyanate (HDI, Aldrich, St. Louis, MO, USA, 98%) was distilled before use, and 1, 2-dichloroethane (DCE, Aldrich, St. Louis, MO, USA, 99%) was dried by distillation from phosphorus pentoxide (Sigma-Aldrich, St. Louis, MO, USA, ≥98%).

Silver nitrate (AgNO$_3$) and trisodium citrate dihydrate (NaC$_6$H$_5$O$_7$·2H$_2$O) were purchased from Sigma-Aldrich (St. Louis, MO, USA). The solvents chloroform (≥99.8%), ethanol (99.8%), *n*-hexane (≥99%), dichloromethane (DCM, ≥99%) from Sigma-Aldrich (St. Louis, MO, USA) and 2-propanone (Merck, Darmstadt, Germany, 99.8%) were used without previous purification. All aqueous solutions were prepared with ultrapure water purified with a Milli Q-POD water purification system (Millipore, Bedford, MA, USA).

2.2. Hydrogels Synthesis and Characterization

Both types of hydrogels used in this work were synthesized following existing protocols [29–32].

2.2.1. Thiol-Acrylate Hydrogels (PSAs)

PSA molecular networks were synthesized by a two-step sequential procedure based on UV-initiated free radical polymerization of previously functionalized macromonomers (Scheme 1a). The macromonomer functionalization consisted of exchanging the end hydroxyl groups of PEG and PCL by alkene groups, as described by Kweon et al. [30] and Alvarado-Tenorio et al. [29]. Afterward, the chemically crosslinked PEG–PCL molecular networks were prepared via radical photopolymerization of thiols and acrylate alkene groups in the presence of the photo-initiator DMPA, under UV-irradiation at a wavelength of 365 nm and at room temperature for 1 h, using glass molds to obtain hydrogel films of 1 mm thickness.

As reported by Anseth and coworkers [32] these thiol-acrylate photopolymers polymerize using mixed-mode chain (acrylate homopolymerization) and step-growth (thiol-acrylate) competitive mechanisms, and their final network structures depend on the thiol/acrylate ratio. In the present work, all photopolymerizations were performed at a thiol/acrylate functional group ratio of 1/5 mol. Subsequently, a series of hydrogels were prepared by varying the concentration of PCL10k from 0 to 15%-mol. In addition, PEG3.3k and PCL10k homo-networks were also prepared using the respective acrylate macromonomers and tetrathiol, following the same procedure. The resulting gel sheets were removed from the glass plate, and 6 mm diameter uniform disks were punched out from each gel sheet using a stainless steel cork-borer. These hydrogel disks were immersed in DCM at room temperature for two days, in order to remove residual monomers. Finally, hydrogels were dried at room temperature until they achieved a constant weight.

2.2.2. Polyurethane Based Hydrogels (PUs)

PU based hydrogels were synthesized by condensation polymerization between PEG and PCL diols, the diisocyanate HDI and PEG$_{300}$ tetraol as crosslinker (Scheme 1b). All polymerizations were carried out using 5 g of macromonomers and 10%-mol of PEG$_{300}$ crosslinker, the solvent

DCE (40 wt %), the catalyst stannous 2-ethylhexanoate (0.3 wt %), and the diisocyanate HDI with a isocyanate/hydroxyl molar ratio of 1.05/1. The synthetic procedure was as follows: the oligomeric compounds were dried under vacuum at 80 °C for at least 6 h. Then, polyols were solved in DCE at room temperature and the diisocyanate and the catalyst were added under magnetic stirring. To obtain sheet shaped gels, the mixture solution was cast on a glass plate enclosed by a rubber framework-spacer with 1 mm thickness and sealed off with other glass plate in order to avoid air contact during the polymerization (24 h, 40 °C). Afterward, the gel sheet was removed from the glass plate and uniform disks 6 mm in diameter were punched out from the gel sheet using a cork-borer. In any case, disk-shaped hydrogels were immersed in ethanol for, at least, three days to remove the unreacted chemicals. During this time the solvent was replaced several times. After that, the hydrogels were dried at room temperature until constant weight was reached.

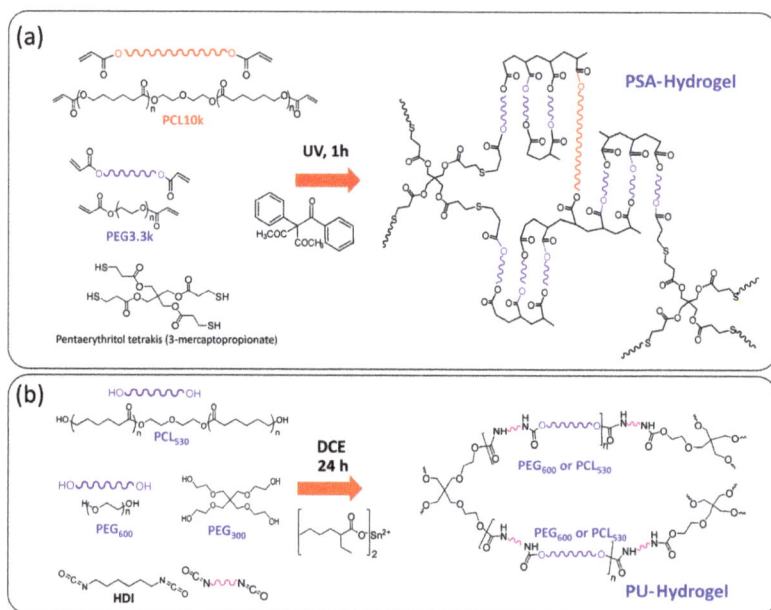

Scheme 1. Schematic representation displaying the synthesis of thiol-acrylate (PSA) (**a**) and polyurethane (PU) (**b**) hydrogels, providing the polymeric chains throughout the hydrogel matrix for AgNP coordination and stabilization. The left-hand side of the scheme depicts more schematic versions of the polymeric chains, crosslinking agents, and catalysts used in the 3D-hydrogel matrix generation.

Both hydrogel types were characterized using a combination of attenuated total reflectance Fourier transform infrared spectroscopy (ATR-FTIR), thermogravimetric analysis (TGA), and differential scanning calorimetry (DSC). FTIR spectra were recorded on a Perkin Elmer Spectrum One spectrophotometer (Waltham, MA, USA) with the ATR technique (ATR-FTIR) and with a resolution of 4 cm^{-1}. The thermal stability was determined from thermogravimetric analysis (TGA) with a TA Hi-Res TGA 2950 instrument (New Castle, DE, USA), at 10 °C·min^{-1}, under 20 mL·min^{-1} of dry helium. The temperature at which the weight loss rate is maximum (T_{max}) was determined from the peak maximum of the first derivative of the weight lost. The thermal transitions (glass transition (T_g), endothermic crystallization (T_c), and exothermic melting (T_m) temperatures) were measured by means of differential scanning calorimetry (DSC) using a TA DSC Q100 apparatus (New Castle, DE, USA) connected to a cooling system to work at low temperatures. Samples were scanned at 20 °C·min^{-1}

under 20 mL·min^{-1} of dry nitrogen from -75 to 150 °C. T_g values were determined in the second heating run cycle. Morphology of the lyophilized hydrogel samples was examined using a scanning electron microscope (SEM) Philips XL 30 (Eindhoven, The Netherlands) operated at 25 kV after coating the sample with Au-Pd (80/20) 10 nm film.

2.3. Ag-Hydrogel Nanocomposite Synthesis and Characterization

The loading of silver nanoparticles into the hydrogel networks was performed according to the following procedures.

2.3.1. AgNPs within Thiol-Acrylate Hydrogels (Ag–PSA)

Dry pure hydrogel disks were swollen in an aqueous solution of AgNO$_3$ (1×10^{-3} M) and sodium citrate (6.4×10^{-4} M) for 24 h at 4 °C. The aqueous solution containing silver salt loaded hydrogel disks was then placed into a silicon bath at 103 °C for 30 min to induce the reduction of the silver ions and the formation of AgNPs in the hydrogel structure. After 5 min, the colorless reaction mixture started turning yellowish, which further turned brown within 15 min, indicating the formation of AgNPs. The solution was then cooled down in a water bath.

2.3.2. AgNPs within PU Hydrogels (Ag–PU)

AgNPs were incorporated into the structure of hydrogel after swelling pure hydrogel disks in an aqueous solution of AgNO$_3$ (1×10^{-3} M) at 4 °C. During this stage, the silver ions were quickly absorbed and AgNPs were progressively formed. The hydrogel disk color changed nearly instantly from transparent to light yellow, which turned into brown color within 3 h, and finally evolved to dark brown within 10 h. Then, no color change was observed and reaction was stopped by replacing the solution by fresh Milli Q water. To study the influence of Ag precursor concentration on AgNPs formation in PU hydrogels, the above procedure was employed, embedding hydrogel disks in solutions of 5×10^{-4}, 1×10^{-4}, 5×10^{-5}, and 1×10^{-5} M AgNO$_3$.

Both Ag-hydrogel nanocomposite types were characterized using a combination of transmission electron microscopy (TEM) and UV-Vis extinction spectroscopy. UV-visible extinction spectroscopy analysis of the localized surface plasmon resonance (LSPR) and cloudy point measurements were carried out in a Cary 3 BIO-Varian UV-visible spectrophotometer (Palo Alto, CA, USA) equipped with a Peltier temperature control device. The transmission electron microscopy (TEM) was used to find out the size of AgNPs inside the hydrogel nanocomposite. To image the AgNPs, the hydrogel was cut into a number of very small pieces and incorporated within LR-White resin (London Resin Co. Ltd., Basingstoke, Hampshire, UK) for posterior ultrathin sectioning (Ultracut Reichert-Jung, Austria). 90 nm ultrathin sections were finally collected on Formvar-coated copper grids and observed with a field emission scanning electron microscope (FE-SEM) (Hitachi, SU 8000, Tokyo, Japan) at 30 kV using scanning transmission electron microscope (S-TEM) detector to detect the transmitted electrons.

2.4. Equilibrium Swelling Values as Function of Temperature

Equilibrium swelling values at different temperatures were determined gravimetrically. Gel disks were left to swell in water solutions for 24 h to achieve equilibrium at each temperature. At regular intervals, the swollen gels were taken out, wiped superficially with filter paper, weighed until equilibrium was attained, and placed again in the same immersion bath. Data from the swelling studies are usually expressed in terms of water uptake, defined as the weight of water absorbed by the sample per unit weight of dry polymer [33]. All measurements were performed in triplicate and averaged.

2.5. Antibacterial Activity

Sensitivity of *Pseudomonas aeruginosa (ATCC 25922)* and *Escherichia coli (ATCC 2785)* to the different silver-hydrogel nanocomposites was tested using disk diffusion test. The cells were reactivated from stocks culture in Mueller & Hinton (Difco Laboratories, Detroit, MI, USA), after 24 h, the samples of *P. aeruginosa* and *E. coli* were placed in 0.9% saline solution. The concentration of the bacteria was controlled to 10^6 CFU/mL and they were seeded in Mueller & Hinton culture medium using a sterile spreader under sterile conditions. Prior to the experiment, uniform 6 mm diameter disks were washed by soaking each of them in 3 mL of MilliQ water. Afterward, they were placed on the culture plate. The plates were then incubated at 37 °C for 24 h. To study the sustained effect, every 24 h, the disks were changed to a new agar plate in which the bacterial suspension was previously spread. Inhibition zone diameters were measured in millimeters for each specimen. Commercially available, antimicrobial susceptibility test disks containing ceftazidime (30 µg, Oxoid, Columbia, MD, USA)) were used for positive control experiments of the antimicrobial effect. Assays were performed in triplicate and the data are shown as the mean ± standard deviation (SD).

3. Results and Discussion

3.1. Hydrogel Characterization

ATR-FTIR measurements were realized to corroborate the chemical structure of the hydrogel networks, via the identification of specific functional groups; and the typical spectra for dried thiol-acrylate (PSA) and polyurethane (PU) hydrogels, (Figures S1 and S2 of Supplementary Information SI, respectively), shows their achievement. In fact, Figure S1 exhibits the vibration bands associated with the expected PSA-hydrogel functional groups generated via thiol-ene radical polymerization (Scheme 1a). However, it is worth noting that even when no free thiol group signals can be observed, the presence of acrylate double bond vibration band at 1640 cm^{-1} denotes that not all the acrylate groups reacted. A plausible explanation of such behavior can be found in the known competitive chain- and step-growth reaction mechanism directly related to the thiol/acrylate ratio [32]. Typical morphology of pore structure of the formed hydrogels is presented in Figure S3.

With regard to the other hydrogel system, the absence of the band at 2270–2285 cm^{-1}, associated to the isocyanate groups (Figure S2), supports the successful formation of the PU hydrogel network via policondensation [31] (Scheme 1b).

The thermal properties of these hydrogels, in terms of degradation temperature (T_{max}) and glass transition (T_g), were evaluated by TGA and DSC, respectively. The comparative data concerning thermal degradation values are collected in Table 1. From Table 1 it can be seen that each hydrogel system presents a unique degradation process, T_{max} , in the range of 350 to 400 °C with quite similar thermal stability in all cases and also similar to PEG linear polymers [34]. However, the slightly higher values for thiol-acrylate polymer based hydrogels compared with the data corresponding of polyurethane polymer based ones indicate that PSA hydrogel-related structure could still provide better thermal stability. Quite similar T_g values for hydrogels containing 100%-mol PEG (PSA100 and PU100, Table 1) were observed, in accordance to the values (about −45 °C) reported in the literature for linear PEG$_{600}$/HDI Pus [31,35]. However, increases of PCL content confer a T_g diminution in the thiol-acrylate hydrogels (PSAs), that can be attributed to the molecular weight differences between the two polymers (PCL, 10 kDa; PEG, 3.35 kDa) involved in the polymerization. Basically, a higher PCL molecular weight would exert an increase of the length between crosslinking points. No similar tendency was observed for the PU-based hydrogels. Actually, glass transition exhibits no dependence on composition since PEG and PCL segments, involved in these hydrogels structure, have similar lengths. The glass transition and thermal stability values for Ag–PSA and Ag–PU nanocomposites show that these properties are not affected by the presence of AgNPs through the polymer matrix.

Table 1. Thermal properties of (PSA and PU) polymer hydrogels and (Ag–PSA and Ag–PU) silver hydrogel nanocomposites: maximum degradation temperature (T_{max}), glass transition temperature (T_g) and melting temperature (T_m).

Hydrogel [a]	T_{max} (°C)	T_g (°C)	T_m (°C)	Ag-Nanocomposites	T_{max} (°C)	T_g (°C)	T_m (°C)
PSA100	376	−45	43	Ag–PSA PEG 100	378	−46	44
PSA95	367	−47	44	Ag–PSA PEG 95	365	−48	46
PSA90	394	−51	42	Ag–PSA PEG 90	383	−52	44
PSA85	389	−53	41	Ag–PSA PEG 85	385	−52	43
PSA PCL	403	−58	54	Ag–PSA PCL	-	-	-
PU100	351	−44	-	Ag–PU PEG 100	376	−44	-
PU95	341	−42	-	Ag–PU PEG 95	-	-	-
PU90	360	−43	-	Ag–PU PEG 90	382	−45	-
PU85	356	−43	-	Ag–PU PEG 85	350	−43	-
PU80	364	−42	-	Ag–PU PEG 80	-	-	-
PU75	356	−43	-	Ag–PU PEG 75	351	−43	-
PU70	357	−41	-	Ag–PU PEG 70	-	-	-
PU60	357	−41	-	Ag–PU PEG 60	358	−43	-
PU–PCL	330	−35	-	Ag–PU PCL	-	-	-

[a] PSA- and PU-hydrogel nomenclature is referred to the PEG %-mol content. Ag–PSA PCL and Ag–PU PCL denote to PCL 100%-mol content (PEG 0%-mol content).

3.2. Swelling as a Function of Temperature

To understand and control the properties of hydrogel, structural parameters, such as crosslinking density and hydrophilicity of the polymer network that form the hydrogel, need to be taken into consideration. One important characteristic of hydrogels is the swelling response upon contact with water to an equilibrium disk thickness [33].

Consequently, volume phase transition is caused by a change in the hydrogel's water content. In this regard, Figure 1 depicts the swelling ratio of PSA and PU hydrogels plotted as function of PEG composition and temperature. As can be extracted from Figure 1 comparison, the PSA hydrogels (Figure 1a) lead to a higher water uptake and larger swelling ratios. As far as the hydrophilicity is concerned, a higher hydrophobic character of PU hydrogels due to HDI segments content (Scheme 1b) leads to a lower swelling ratio (Figure 1b). Therefore, a clear dependence of the equilibrium swelling on the composition of the hydrogel is stated.

Another important behavior extracted from Figure 1 is the negative thermo-sensitivity of both hydrogels, since the equilibrium of swelling clearly depends on temperature. Hydrogels composed of PEG can abruptly change their physico-chemical properties in response to external stimuli, such as temperature [36,37], undergoing a phase separation at a critical temperature. This volume phase transition temperature (VPTT) is intrinsically linked to the chemical nature of the polymer backbone and can be tuned by polymer characteristics, such as molecular weight [36] and the PEG end groups [37]. In this sense, Figure 1 shows that in all cases, the higher the temperature of solution is, the lower the degree of swelling and solubility is. Figure 1 also reveals a more noticeable volume transition phase decrease with increasing temperature for PU hydrogels than for PSA hydrogels. Again, the balance between hydrophobic and hydrophilic polymer chains inside hydrogel networks must be invoked. Actually, Figure S4 of SI shows a change from transparent (swollen state) to opaque when polymer chains collapse due to hydrophobic chain dominance. Similar behaviors were reported for linear [35] and crosslinked [31] polyurethanes comprised by PEG and HDI. Actually, both temperature sensitive hydrogels exhibited a tunable cloud point or volume transition temperature via control of hydrophilic/hydrophobic balance. Thus, increases of hydrophilic PEG molecular weight [35] or incorporation of hydrophobic PCL chains [31] raise cloud point and volume transition temperature, respectively.

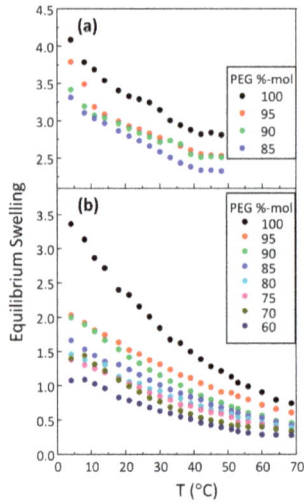

Figure 1. Equilibrium swelling degree of thiol-acrylate (PSA) (**a**) and polyurethane (PU) (**b**) hydrogels as a function of temperature and of composition.

3.3. Silver/Hydrogel Nanocomposites

To get a controlled nanostructured dispersion of Ag inside the two hydrogels, an in situ reactive nanoparticle formation within a preformed hydrogel-approach was considered. A schematic illustration of the two procedures is given in Scheme 2.

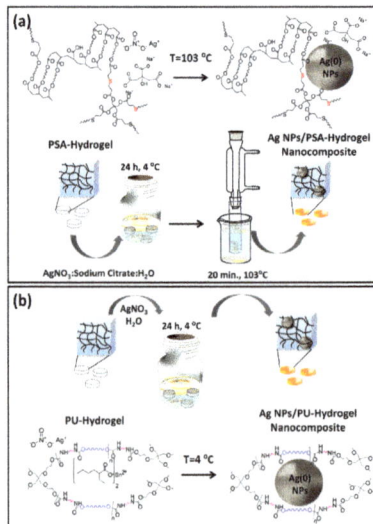

Scheme 2. Schematic illustration portraying the formation of silver nanocomposites: AgNPs/ PSA-hydrogel (**a**) and AgNPs/PU-hydrogel nanocomposites (**b**). Amphiphilic AgNPs were formed by in situ reactive-reduction process within preformed 3D-hydrogel matrices: thiol-acrylate (PSA) (**a**) and polyurethane (PU) (**b**) hydrogels.

3.3.1. AgNPs within Thiol-Acrylate Hydrogels (Ag–PSA)

3D PSA-hydrogel network, containing thiol reactive groups (Scheme 1a), was set up via crosslinking radical photo-polymerization (see Section 2.2 in Experimental) to act as template for the well-spacing growth of AgNPs into its structure (Scheme 2a). Then, nanoscale silver formation was conducted by soaking accurately weighed PSA hydrogel disks in an aqueous solution of $AgNO_3$ and sodium citrate for a period of 24 h at 4 °C. During this incubation time, the hydrogel underwent appreciable swelling but no AgNP formation was observed. Afterward, glass tubes containing the silver salt absorbed hydrogels absorbed in the precursor solution were transferred into a silicon oil bath at 103 °C and allowed for 30 min to reduce the silver ions to AgNPs. A dark brown color in Figure S5 of SI indicates formation of AgNPs in hydrogel matrix. It is noteworthy that, during the exchange process of the silver ions from solution to hydrogel, the occupancy of free-network spaces of hydrogel is modulated by thiol-silver complexes formation in the presence of sodium citrate reductant. In this way, after temperature activation, an almost uniform distribution of AgNPs might be achieved within the hydrogel. Note that thiol groups exhibit high affinity to both Ag^+ ions and colloidal Ag. Nevertheless, Figure S5 also reveals AgNP formation in the water outside the hydrogel. The explanation might be found in the key role played by hydrophobic/hydrophilic balance on the swelling degree of hydrogel over the temperature.

Figure 2a shows the photographic images of the resulting Ag–PSA hydrogel nanocomposite disks. As it can be observed, the color intensity strongly depends on polymer composition, decreasing color intensity for copolymers with higher PCL content. This behavior could be explained by considering that the increased presence of PCL chains in the hydrogel copolymeric architecture leads to an increment of the hydrophobicity of the network, thus decreasing the swelling degree. In this context, another important factor to take into consideration is the swelling degree diminution with temperature, which undoubtedly leads to lower particle formation inside the hydrogels. In fact, the strong color of the soaking gel solution indicates that most of the AgNPs are formed outside the hydrogel network disks (Figure S5).

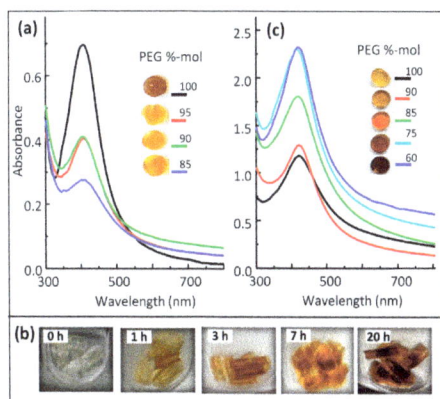

Figure 2. UV-Vis spectra of (**a**) Ag–PSA and (**c**) Ag–PU hydrogel nanocomposite disks for several PEG polymer compositions (%-mol), (Table 2). Photographs of the resulting Ag–PSA and Ag–PU hydrogel nanocomposite disks (Inset of (**a**) and (**c**) respectively). Photographic images following the AgNPs formation inside PU hydrogel matrices during the course of the reaction (**b**). The PU hydrogel disks were soaked in $AgNO_3$ aqueous solutions at 4 °C.

3.3.2. AgNPs within Polyurethane Hydrogels (Ag–PU)

As mentioned in the introduction, an improved approach from the loading nanoparticle precursors into a gel was considered for building up Ag–PU hydrogel nanocomposites. Interestingly, AgNPs were formed in water solution not only in the absence of sodium citrate but even with a reaction temperature of 4 °C. The strategy based on the redox chemistry of the catalyst was involved in the PU hydrogel network-assembly. Residual stannous ethylhexanoate amount strategically acts as a reducing agent of silver precursor (AgNO$_3$) to nanoparticles, so that nanoparticle-hydrogel composites are formed in the absence of additional external reducing agent.

Therefore, copolymer PU hydrogel was previously generated by condensation polymerization, and subsequently used as nanoreactor for silver nanoparticle formation (Scheme 2b). The formation of the silver nanoparticles into the hydrogel structure was then performed by contacting hydrogel with AgNO$_3$ in deionized water solution at 4 °C. During this stage, the fast silver solution absorption derived on progressive AgNP formation, due to the oxidation/reduction reaction between the silver salt and stannous ion in the hydrogel copolymer-network (disk). The color of the hydrogel evolved from light yellow to dark brown. The snapshots, displayed in Figure 2b, demonstrate this nanoparticle generation. As it can be observed, a fast increase of yellowish color indicates almost immediate nanoparticle formation (0–30 min), which turned then to deep brown color (around 3–7 h) and remained stable after 12 h.

It is worth noting that, contrary to what happens to PSA hydrogels—where hydrogel surrounding the solution exhibits AgNP presence (Figure S5)—PU hydrogels only have particles forming inside the hydrogel disks. In analogy to the reported in the literature [38–40], stannous compound acts as reducing catalyst, which promotes AgNPs formation through AgNO$_3$ reduction. So, during the reduction process, Sn^{2+} oxidizes into Sn^{4+} which reduces Ag$^+$ into Ag$°$, and AgNPs are formed in a high yield. To the best of our knowledge, this is the first report about the in situ formation of AgNPs in polyurethane gels, taking advantage of remaining stannous ethylhexanoate catalyst. The reported references [38–40] involve AgNPs formation in organic solvents using stannous acetate as reducing agent and silver acetate as Ag precursor at higher temperature. Therefore, the low temperature reaction (4 °C) and the aqueous solution as reaction media make the main advantages of our system evident.

3.4. Optical Properties of Silver/Hydrogel Nanocomposites

UV-Vis absorption spectroscopy was carried out to monitor the AgNPs formation and gain insight into the particle size and shape. All UV-Vis traces (Figure 2a,c) showed an intense absorption band near 400–420 nm, depending on sample composition, characteristic of spherical AgNPs. This band, known as 'surface plasmon resonance' (SPR), is sensitive to geometric parameters (size and shape), aggregation, and the surrounding matrix of metal NPs. Hence, the plasmon band intensity is related to nanoparticle concentration, whereas the absorption maximum wavelength and shape is associated with nanoparticle size and aggregation.

Figure 2a,c displays the SPR band for our hydrogel nanocomposite systems (Ag–PSA and Ag–PU, respectively) synthesized. From mere inspection of figures, it follows that the peak intensity clearly decreases (Figure 2a) and increases (Figure 2c) with increasing PCL content; as result of the opposite contribution that the hydrophobic PCL segments exert on the silver nanoparticles genesis in both hydrogel systems. Moreover, this chemical composition dependence seems to be more noticeable for PSA hydrogels (Figure 2a); since approximately three-fold higher peak intensity diminutions were obtained as compared to the found intensity variations for PU hydrogels (Figure 2c). In fact, 50% and 75% peak intensity reductions were observed by simple increment of PCL content from 5% up to 10%-mol; whereas increased values of only 15% and 20% peak intensity were reported for similar PCL %-mol contribution in PU hydrogels (Table 2). These results demonstrate a clear shrinkage contribution of hydrophobic PCL chains in the swelling behavior of Ag-PSA hydrogel nanocomposites, which leads to a lower AgNP generation at 103 °C.

Furthermore, if we compare the absolute values of the peak intensities for equivalent PCL %-mol contributions of the both types of hydrogel nanocomposites (Ag–PSA and Ag–PU, Table 2), it results in the remarkable formation of higher AgNP concentration for PU hydrogels (Figure 2c), almost independently of their polymer composition. The explanation of these differences might be found in the two different approaches used by in situ reducing of silver nitrate in the hydrogel network. So, while PU hydrogel absolutely contains reduction agent inside hydrogel network, PSA hydrogel guarantees its presence through its dynamic diffusion equilibrium between the surrounding media and the inside of hydrogel. Therefore, NPs in the PSA system will be formed inside and outside the hydrogel framework (disk), and subsequently lower NP concentration (peak intensity) will be observed by UV measurement (Figure 2a and Table 2).

With respect to PU system, the NPs formation follows a process markedly driven by the stannous catalyst; so the chemical composition of the PU hydrogel should preferably influence the NP formation rate rather than the final NP concentration. In spite of this fact, PCL chains contribute, at some extent, in the process. Indeed, they act as: (i) stannous catalyst traps on their hydrophobic domains, leading to an improved SPR intensity effect with PCL content increase; and (ii) a lowered water uptake promoter, as a result of the much lowest swelling capacity of PU hydrogels (Figure 1), which ends up as a fictional SPR intensity increase appearance. These effects might explain the increased tendency of SPR intensity with increasing hydrophobic chain (PCL) content (Figure 2c); somehow inferred from dissimilarities on swelling degree capacities of hydrogels.

Table 2. Characteristics of the surface plasmon resonance (SPR) band for Ag–PSA and Ag–PU hydrogel nanocomposites.

Sample	Temperature of Synthesis (°C)	SPR Maximum Wavelength (nm)	SPR Maximum Absorbance
Ag–PSA PEG 100		405.7 ± 0.6	0.83 ± 0.05
Ag–PSA PEG 95		409.2 ± 0.7	0.41 ± 0.03
Ag–PSA PEG 90	103	406.2 ± 0.7	0.25 ± 0.02
Ag–PSA PEG 85		406.5 ± 0.8	0.37 ± 0.02
Ag–PU PEG 100		419.9 ± 0.8	1.02 ± 0.05
Ag–PU PEG 90		421.3 ± 1.1	1.22 ± 0.04
Ag–PU PEG 85	4	416.8 ± 0.6	1.61 ± 0.06
Ag–PU PEG 75		416.1 ± 0.9	1.60 ± 0.05
Ag–PU PEG 60		418.7 ± 0.9	1.75 ± 0.05

The optical properties of both hydrogel nanocomposites were further investigated by variable SPR band shape. The band shape symmetry and the lack of peak shift (Figure 2a,c) indicate that the spherical AgNPs were well dispersed and not forming aggregated structures. S-TEM image of Ag–PSA and Ag–PU hydrogel nanocomposites (Ag–PSA PEG 100 and Ag–PU PEG 90, respectively in Table 2) mostly reveals AgNPs with uniform dimension and morphology (Figure 3). Sizes of 97 ± 8 nm for Ag–PSA PEG 100 were estimated for an overall of 50 nanoparticles counted (Figure 3a). However, direct observation of Ag–PU hydrogel nanocomposite samples (Table 2) by means of several S-TEM photographs evidenced two populations of nanoparticles: one of small size (25 ± 8 nm) and another with a different size around 73 ± 13 nm. Figure 3b displays representative S-TEM images of Ag–PU PEG 90 hydrogel nanocomposite to visualize these two different NP size-populations.

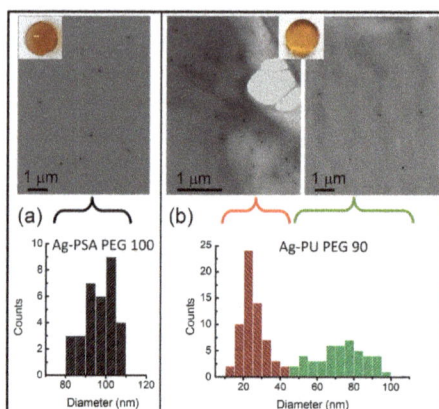

Figure 3. Representative bright field S-TEM images (upper) and respective five-bin histograms showing the size-distribution of AgNPs (below) prepared in Ag–PSA PEG 100 (**a**) and Ag–PU PEG 90 (**b**) hydrogel nanocomposites. Insets reproduce the light photograph of each Ag-hydrogel nanocomposite. Polyurethane-based nanocomposites show two different NP size-populations (25 ± 8 nm and 73 ± 13 nm right representations).

Interestingly, from Figure 2c slight asymmetric broadening of the SPR band toward shorter wavelengths with increasing hydrophobic chain content can be inferred. Complementary information to that in Figure 2c is obtained by inspection of the signal intensity and position collected from SPR bands of a silver-polymer hydrogel with PEG 60%-mol (Ag-P 60*) nanocomposite sample in Figure S6. Basically, the Ag-P 60* hydrogel matrix sample was performed in the absence of PEG_{300} crosslinker and with similar PCL and PEG %-mol contents (Scheme 1b) as those of Ag–PU PEG 60 (see Section 2.2 in Experimental). By this protocol, the potential to dissolve a physically crosslinked 3D-hydrogel structure under basic conditions in ethanol was obtained. The resulting un-crosslinked 3D-polyurethane hydrogel matrix (P 60*) was then used to prepare AgNPs within, following the procedure described in Section 2.3 of the experimental part. Interestingly, UV-Vis spectra (Figure S6) display a unique SPR band for the both Ag–PU nanocomposites (gel-disk (Ag–PU PEG 60, red line) and gel-solution (Ag-P 60*, green line) sample) evidencing the cooperative behavior of the two NP size populations; which only splits into two SPR bands when the enriched fraction on the smallest NPs (supernatant SN-Ag-P 60*)—as obtained by low speed centrifugation of Ag-P 60* (gel-solution)—is measured (Figure S6, black line). Indeed, the appearance of a second peak toward blue end of the spectrum can be attributed to a different NP size-population, formed during the reduction process within the hydrogel, attesting to the two observed NP sizes in the S-TEM image of Figure 3b. Moreover, considering the pivotal NP surface charge effect in the osmotic pressure of hydrogel nanocomposites [41], we hypothesized that the initial content of immobilized charge (due to small silver colloids generated along the in situ reactive nanosilver-loading of PU hydrogels) would induce a hydrogel osmotic swelling, despite the opposite PCL effect. Then, at a higher surface charge, more nuclei are formed and result in the formation of another NP population with a different range of sizes inside the hydrogel scaffold. Note that small NPs have a greater surface area at a fixed NP concentration; and a more effective surface charge, subsequently. It is worth pointing out though that longer SPR wavelengths are obtained over PU-based Ag-nanocomposites than over PSA-based ones, as depicted from Figure 2a,c comparison; mainly attributed to differences on the surrounding refractive index of both systems. Note that AgNPs within PSAs are protected by sodium citrate, whereas within PUs must be covered by the polymer.

Further investigations of the SPR band evolution at various temperatures (Figure 4) for Ag–PU nanocomposites containing 85%-mol PEG (Ag–PU PEG 85, Table 1) confirmed the effective dispersion of NPs inside the hydrogel.

Figure 4 unequivocally shows a strong plasmon intensity increase (about 50%), and no shift in the absorption profile, as a result of the volume phase transition of the hydrogel when the temperature rises to 50 °C; which may be ascribed to the surrounding polymer collapse on particles core surface. Indeed, PU hydrogel with 85% PEG exhibits a quite similar decrease of swelling equilibrium (around 57%) in between 25 and 50 °C, as displayed in Figure 1.

An additional experiment concerning, now, the influence of Ag precursor concentration on AgNPs formation in the PU hydrogel was evaluated. UV-Vis spectra were recorded for Ag–PU hydrogel nanocomposites with a constant PEG content (85%-mol, Ag–PU PEG 85) but varying $AgNO_3$ content (Table 3 and Figure 5). As expected, the number of AgNPs created was proportional to the $AgNO_3$ content; a linear trend of increase of SPR maximum intensity with the increase of $AgNO_3$ concentration was observed (Figure 5). Interestingly, a blueshift of SPR band and a yellowish color with Ag precursor content diminution (inset, Figure 5) argue favorably for the initial formation of small and discrete AgNPs. For 10^{-5} M $AgNO_3$ concentration, no silver nanoparticle formation was observed.

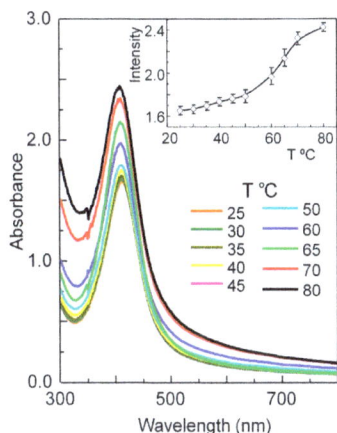

Figure 4. UV-Vis spectra of Ag–PU PEG 85 hydrogel nanocomposites when temperature increases from 25 to 80 °C. Inset: SPR band maximum intensity of Ag–PU PEG 85 as a function of the temperature.

Table 3. Influence of $AgNO_3$ concentration on the SPR band (position and intensity) of PU hydrogel with 85%-mol of PEG (Ag–PU PEG 85, Table 2).

Sample	$AgNO_3$ (M)	SPR Maximum Wavelength (nm)	SPR Maximum Absorbance
Ag–PU PEG 85-1	1×10^{-3}	416.8 ± 0.6	1.61 ± 0.06
Ag–PU PEG 85-2	5×10^{-4}	412.9 ± 0.4	1.15 ± 0.05
Ag–PU PEG 85-3	1×10^{-4}	400.8 ± 0.7	0.64 ± 0.03
Ag–PU PEG 85-3	5×10^{-5}	397.7 ± 0.5	0.52 ± 0.02
Ag–PU PEG 85-3	1×10^{-5}	-	0

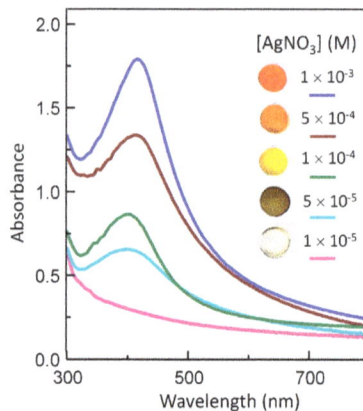

Figure 5. UV-Vis of spectra of Ag–PU PEG 85 hydrogel nanocomposite disks, while increasing the AgNO$_3$ concentration added to the PU hydrogel disk (Table 3). Inset: Pictures of the resulting Ag–PU hydrogel nanocomposite disks (error bars represent standard deviation for studies conducted in triplicate).

3.5. Antimicrobial Properties

Antimicrobial properties of Ag–PSA and Ag–PU hydrogel nanocomposites were evaluated against Gram-negative bacteria, *P. aeruginosa* and *E. coli*, using the disk diffusion test method. In all cases, these properties were measured by evaluating the zone of inhibition around the disk after incubation at 37 °C. With this aim, uniform disks with a 6 mm diameter and 1 mm thickness obtained from Ag–PSA and Ag–PU hydrogel nanocomposites, were placed on agar Petri dishes, where the bacterial suspension was previously spread (Figures S7 and S8, respectively). A simple observation of the photographs (Figures S7 and S8) revealed that the Ag-hydrogel nanocomposites exhibited antimicrobial properties. However, the hydrogel systems without Ag showed no zone of inhibition against *E. coli* and *P. aeruginosa*. A quantitative analysis of these results is reported in Figure 6a,b. Figure 6a shows measurements of the inhibition zone for Ag–PSA hybrids over time for *E. coli* (Figure 6a-i) and *P. aeruginosa* (Figure 6a-ii). Comparing the inhibition results after the first 24 h (Figure 6a), it appears that all Ag-nanocomposites were able to inhibit bacterial growth without copolymer composition dependence. A similar behavior was found for Ag–PU hydrogel nanocomposites in Figure 6b. Moreover, both hydrogel nanocomposite systems exhibited a more effective antibacterial activity against *P. aeruginosa* (Figure 6a-ii and 6b-ii) than against *E. coli* (Figure 6b-i), as proven with the higher inhibition halo. The explanation of this fact could lie on the ability of *E. coli* to develop heavy metal resistance, particularly for silver [1,42]. Remarkably, the inhibition halos after 24 h cultivation were slightly higher for PU hydrogel nanocomposites. Although the most interesting feature of our systems is the sustained antimicrobial activity against *P. aeruginosa* (Figure 6a-ii and 6b-ii). In fact, after 48 h incubation, the bacterial inhibition effect strongly increases, almost doubling for Ag–PU hydrogel nanocomposites. Polyurethane-based nanocomposites, possibly due to their higher AgNP content (Figure 2c and Table 2), exhibit a longer sustained effect (Figure 6b-ii). Values up to above 120 h incubation can be observed for samples containing 90%-mol PEG. We speculate that prolonged inhibition may be associated with the position and the accumulation of generated AgNPs through the hydrogel network (disk). So, a closer AgNP location to the contact surface between 3D hydrogel nanocomposites and culture medium, where the bacteria is spread, will induce an increased and sustained inhibition effect. The antibacterial efficacy of the hydrogel nanocomposites is determined by the size of nanoparticles within the polymeric structure. A bigger size of AgNPs in PSA hydrogels decreases the antibacterial effect, whereas the smaller size of AgNPs would increase antibacterial activity. The reason is because, with decreasing size, the total surface area per unit volume is greater; this phenomenon facilitates nucleation of a higher number of nanoparticles within the polymeric network [43,44]. The degree of swelling and cross-linking is also related to this effect. In the specific case of the PU hydrogels, the diffusion rate is slower than in the case of PSA hydrogels, whose rate promotes a faster diffusion, having a higher degree of swelling. This analysis demonstrates the sustained antibacterial effect until 120 h incubation for the PU hydrogel nanocomposites.

In addition, the presence of different NP sizes with larger antibacterial reactive surface areas as attested by UV-Vis (Figure S6) and confirmed by S-TEM (Figure 3) correlates well with an increased sustained effect of PU based Ag-nanocomposites.

Resistance—more by *E. coli* than by *P. aeruginosa*—may be due to the features of its cell membrane, which has properties of altering the permeability [45,46]. This defense mechanism is the most likely that limits the sustained antibacterial activity against *E. coli* strain. The analysis of antimicrobial activity by inhibition zone shows that the effect against studied bacterial strains is effective, which represents, in the future, great advantages in biomedical application.

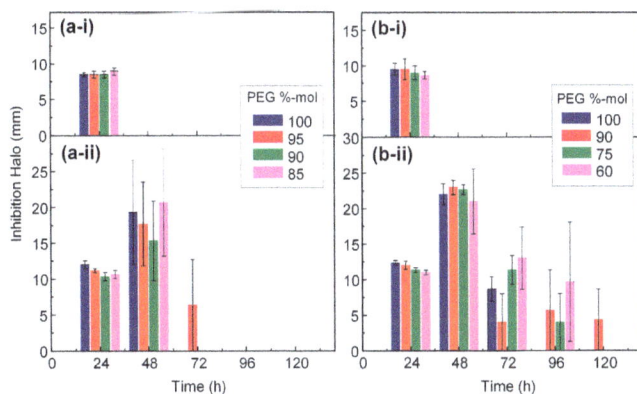

Figure 6. Graphical representation showing the antimicrobial efficacy of (**a**) Ag–PSA and (**b**) Ag–PU hydrogel nanocomposites: Inhibition halo distance (mm) over time against (i) *E. coli* and (ii) *P. aeruginosa*. Error bars represent standard deviation for studies conducted in triplicate.

4. Conclusions

In this work, we reported an easy and effective strategy to produce Ag hydrogel nanocomposites by in situ reactive-reduction of AgNPs within polymeric-hydrogel matrices. The system is based on biocompatible and biodegradable PSA and PU hydrogels, previously synthesized by thiol-acrylate radical polymerization and polyurethane chemistry. Then subsequently used as templates for the growth of AgNPs. A good dispersion and sharp-particle-size distribution of AgNPs is followed by simple loading of nanoparticle precursor (AgNO$_3$)—in the presence and absence of an external reducing agent—within PSA and PU hydrogels respectively. Interestingly, Ag–PU hydrogel nanocomposites were newly obtained via redox chemistry of residual stannous catalyst from the polyurethane synthesis in the absence of temperature (4 °C); while thiol-silver ion affinities would facilitate some control over NP size and distribution in Ag–PSA hydrogel nanocomposites. We systematically investigated the swelling and optical behaviors of the both series of thermosensitive hydrogels on their in situ nanoparticle generation. It was shown that a higher contribution of hydrophobic segments in the hydrogel network induced a lower swelling response, which led to a lower NP content. Also, SPR shape and intensity as a function of the hydrogel chemical-composition revealed spherical and well-spaced AgNPs with non-aggregated nanoparticle-assembly evidence. Nevertheless, certain SPR asymmetry of Ag–PU hydrogel nanocomposites demonstrates the formation of two NP-populations, probably due to the increase of nucleation/growth rates generated by changes in osmotic swelling induced by surface charge on small NPs. S-TEM images confirm these data. Furthermore, the antimicrobial behavior of both Ag-loaded hydrogels showed growth inhibition of Gram-negative bacteria as *P. aeruginosa* and *E. coli*; suggesting a more effective Ag^+ releasing hydrogel against *P. aeruginosa* strain. Remarkably, Ag–PU hydrogel nanocomposites exhibited sustained inhibition of *P. aeruginosa*, detectable even after 120 h cultivation. Therefore, these mechanisms emphasize the pivotal role of AgNP size and location in hydrogel matrix, in the efficacy and sustainability of antibacterial activity. This illustrates the potential of these hydrogels for application biomedical fields, such as smart healing dressings for burns and surgical wounds, tissue engineering scaffolds, and ophthalmic therapies.

Supplementary Materials: The following are available online at http://www.mdpi.com/2079-4991/7/8/209/s1, Thiol-acrylate Hydrogel characterization by ATR-FTIR (Figure S1) and SEM (Figure S2); Polyurethane Hydrogel characterization by ATR-FTIR (Figure S3); UV-Vis spectra of different AgNP sizes formed within PU hydrogels (Figure S4); AgNPs generation outside-inside PSA hydrogels (Figure S5); Comparative UV–Vis spectra of Ag-PU PEG 60 (gel-disk, dissolved gel solution and supernatant solution obtained by centrifugation) (Figure S6); and Photographs of sustained antibacterial activity of Ag–PSA and Ag–PU hydrogel nanocomposites (Figures S7–S9).

Acknowledgments: A.S.-Q. thanks the Mexican Council for Science and Technology, CONACyT, for a graduate scholarship. Authors thank the financial support of the Spanish Ministerio de Economía y Competitividad through the Projects MAT 2011-25513 and MAT 2014-57429-R. The authors also acknowledge Virginia Souza-Egipsy for the preparation of ultrathin samples for microscopy observation and David Gómez Varga for S-TEM measurement.

Author Contributions: Nekane Guarrotxena and Isabel Quijada-Garrido conceived the idea and designed the experiments; Albanelly Soto-Quintero performed the hydrogel synthesis, swelling experiments, and nanocomposite generation. Albanelly Soto-Quintero and Victor Bermúdez-Morales performed antimicrobial experiments; Nekane Guarrotxena and Isabel Quijada-Garrido performed the optical experiments and nanocomposite generation and processed the experimental data; Ángel Romo-Uribe contributed to the hydrogel design strategy and reagents/materials/analysis tools. All authors contributed to the scientific discussion and Nekane Guarrotxena and Isabel Quijada-Garrido wrote the paper.

Conflicts of Interest: The authors declare no conflict of interest.

References

1. Rizzello, L.; Pompa, P.P. Nanosilver-based antibacterial drugs and devices: Mechanisms, methodological drawbacks, and guidelines. *Chem. Soc. Rev.* **2014**, *43*, 1501–1518. [CrossRef] [PubMed]

2. Rai, M.K.; Deshmukh, S.D.; Ingle, A.P.; Gade, A.K. Silver nanoparticles: the powerful nanoweapon against multidrug-resistant bacteria. *J. Appl. Microbiol.* **2012**, *112*, 841–852. [CrossRef] [PubMed]

3. Hajipour, M.J.; Fromm, K.M.; Akbar Ashkarran, A.; Jimenez de Aberasturi, D.; Ruiz de Larramendi, I.; Rojo, T.; Serpooshan, V.; Parak, W.J.; Mahmoudi, M. Antibacterial properties of nanoparticles. *Trends Biotechnol.* **2012**, *30*, 499–511. [CrossRef] [PubMed]

4. Eckhardt, S.; Brunetto, P.S.; Gagnon, J.; Priebe, M.; Giese, B.; Fromm, K.M. Nanobio Silver: Its Interactions with Peptides and Bacteria, and Its Uses in Medicine. *Chem. Rev.* **2013**, *113*, 4708–4754. [CrossRef] [PubMed]

5. Wang, L.; Hu, C.; Shao, L. The antimicrobial activity of nanoparticles: Present situation and prospects for the future. *Int. J. Nanomed.* **2017**, *12*, 1227–1249. [CrossRef] [PubMed]

6. Nowack, B.; Krug, H.F.; Height, M. 120 years of nanosilver history: Implications for policy makers. *Environ. Sci. Technol.* **2011**, *45*, 1177–1183. [CrossRef] [PubMed]

7. Brett, D.W. A discussion of silver as an antimicrobial agent: Alleviating the confusion. *Ostomy Wound Manag.* **2006**, *52*, 34–41.

8. Reithofer, M.R.; Lakshmanan, A.; Ping, A.T.K.; Chin, J.M.; Hauser, C.A.E. In situ synthesis of size-controlled, stable silver nanoparticles within ultrashort peptide hydrogels and their anti-bacterial properties. *Biomaterials* **2014**, *35*, 7535–7542. [CrossRef] [PubMed]

9. Leawhiran, N.; Pavasant, P.; Soontornvipart, K.; Supaphol, P. Gamma irradiation synthesis and characterization of AgNP/gelatin/PVA hydrogels for antibacterial wound dressings. *J. Appl. Polym. Sci.* **2014**, *131*. [CrossRef]

10. Gil, E.S.; Hudson, S.M. Stimuli-reponsive polymers and their bioconjugates. *Prog. Polym. Sci.* **2004**, *29*, 1173–1222. [CrossRef]

11. Hoffman, A.S. Hydrogels for biomedical applications. *Adv. Drug Deliv. Rev.* **2002**, *54*, 3–12. [CrossRef]
12. Peppas, N.A.; Bures, P.; Leobandung, W.; Ichikawa, H. Hydrogels in pharmaceutical formulations. *Eur. J. Pharm. Biopharm.* **2000**, *50*, 27–46. [CrossRef]
13. Castellano, J.J.; Shafii, S.M.; Ko, F.; Donate, G.; Wright, T.E.; Mannari, R.J.; Payne, W.G.; Smith, D.J.; Robson, M.C. Comparative evaluation of silver-containing antimicrobial dressings and drugs. *Int. Wound J.* **2007**, *4*, 114–122. [CrossRef] [PubMed]
14. Travan, A.; Pelillo, C.; Donati, I.; Marsich, E.; Benincasa, M.; Scarpa, T.; Semeraro, S.; Turco, G.; Gennaro, R.; Paoletti, S. Non-cytotoxic Silver Nanoparticle-Polysaccharide Nanocomposites with Antimicrobial Activity. *Biomacromolecules* **2009**, *10*, 1429–1435. [CrossRef] [PubMed]
15. GhavamiNejad, A.; Park, C.H.; Kim, C.S. In Situ Synthesis of Antimicrobial Silver Nanoparticles within Antifouling Zwitterionic Hydrogels by Catecholic Redox Chemistry for Wound Healing Application. *Biomacromolecules* **2016**, *17*, 1213–1223. [CrossRef] [PubMed]
16. Huang, J.-F.; Zhong, J.; Chen, G.-P.; Lin, Z.-T.; Deng, Y.; Liu, Y.-L.; Cao, P.-Y.; Wang, B.; Wei, Y.; Wu, T.; et al. A Hydrogel-Based Hybrid Theranostic Contact Lens for Fungal Keratitis. *ACS Nano* **2016**, *10*, 6464–6473. [CrossRef] [PubMed]
17. Fazly Bazzaz, B.S.; Khameneh, B.; Jalili-Behabadi, M.M.; Malaekeh-Nikouei, B.; Mohajeri, S.A. Preparation, characterization and antimicrobial study of a hydrogel (soft contact lens) material impregnated with silver nanoparticles. *Contact Lens Anterior Eye* **2014**, *37*, 149–152. [CrossRef] [PubMed]
18. Alarcon, E.I.; Udekwu, K.I.; Noel, C.W.; Gagnon, L.B.P.; Taylor, P.K.; Vulesevic, B.; Simpson, M.J.; Gkotzis, S.; Islam, M.M.; Lee, C.-J.; et al. Safety and efficacy of composite collagen-silver nanoparticle hydrogels as tissue engineering scaffolds. *Nanoscale* **2015**, *7*, 18789–18798. [CrossRef] [PubMed]
19. Marsich, E.; Bellomo, F.; Turco, G.; Travan, A.; Donati, I.; Paoletti, S. Nano-composite scaffolds for bone tissue engineering containing silver nanoparticles: Preparation, characterization and biological properties. *J. Mater. Sci. Mater. Med.* **2013**, *24*, 1799–1807. [CrossRef] [PubMed]
20. Alarcon, E.I.; Vulesevic, B.; Argawal, A.; Ross, A.; Bejjani, P.; Podrebarac, J.; Ravichandran, R.; Phopase, J.; Suuronen, E.J.; Griffith, M. Coloured cornea replacements with anti-infective properties: Expanding the safe use of silver nanoparticles in regenerative medicine. *Nanoscale* **2016**, *8*, 6484–6489. [CrossRef] [PubMed]
21. Sheeney-Haj-Ichia, L.; Sharabi, G.; Willner, I. Control of the electronic properties of thermosensitive poly(*N*-isopropylacrylamide) and Au-nanoparticle/poly(*N*-isopropylacrylamide) composite hydrogels upon phase transition. *Adv. Funct. Mater.* **2002**, *12*, 27–32. [CrossRef]
22. Wang, C.; Flynn, N.T.; Langer, R. Controlled structure and properties of thermoresponsive nanoparticle-hydrogel composites. *Adv. Mater.* **2004**, *16*, 1074–1079. [CrossRef]
23. Zhao, X.; Ding, X.; Deng, Z.; Zheng, Z.; Peng, Y.; Long, X. Thermoswitchable electronic properties of a gold nanoparticle/hydrogel composite. *Macromol. Rapid Comm.* **2005**, *26*, 1784–1787. [CrossRef]
24. Sershen, S.R.; Westcott, S.L.; Halas, N.J.; West, J.L. Temperature-sensitive polymer-nanoshell composites for photothermally modulated drug delivery. *J. Biomed. Mater. Res.* **2000**, *51*, 293–298. [CrossRef]
25. Uygun, M.; Kahveci, M.U.; Odaci, D.; Timur, S.; Yagci, Y. Antibacterial acrylamide hydrogels containing silver nanoparticles by simultaneous photoinduced free radical polymerization and electron transfer processes. *Macromol. Chem. Phys.* **2009**, *210*, 1867–1875. [CrossRef]
26. Saez, S.; Fasciani, C.; Stamplecoskie, K.G.; Gagnon, L.B.-P.; Mah, T.-F.; Marin, M.L.; Alarcon, E.I.; Scaiano, J.C. Photochemical synthesis of biocompatible and antibacterial silver nanoparticles embedded within polyurethane polymers. *Photochem. Photobiol. Sci.* **2015**, *14*, 661–664. [CrossRef] [PubMed]
27. Zhang, Z.; Ni, J.; Chen, L.; Yu, L.; Xu, J.; Ding, J. Biodegradable and thermoreversible PCLA–PEG–PCLA hydrogel as a barrier for prevention of post-operative adhesion. *Biomaterials* **2011**, *32*, 4725–4736. [CrossRef] [PubMed]
28. Soto-Quintero, A.; Meneses-Acosta, A.; Romo-Uribe, A. Tailoring the viscoelastic, swelling kinetics and antibacterial behavior of poly(ethylene glycol)-based hydrogels with polycaprolactone. *Eur. Polym. J.* **2015**, *70*, 1–17. [CrossRef]
29. Alvarado-Tenorio, B.; Romo-Uribe, A.; Mather, P.T. Microstructure and phase behavior of POSS/PCL shape memory nanocomposites. *Macromolecules* **2011**, *44*, 5682–5692. [CrossRef]
30. Kweon, H.; Yoo, M.K.; Park, I.K.; Kim, T.H.; Lee, H.C.; Lee, H.S.; Oh, J.S.; Akaike, T.; Cho, C.S. A novel degradable polycaprolactone networks for tissue engineering. *Biomaterials* **2003**, *24*, 801–808. [CrossRef]

31. París, R.; Marcos-Fernández, A.; Quijada-Garrido, I. Synthesis and characterization of poly(ethylene glycol)-based thermo-responsive polyurethane hydrogels for controlled drug release. *Polym. Adv. Technol.* **2013**, *24*, 1062–1067. [CrossRef]

32. Rydholm, A.E.; Bowman, C.N.; Anseth, K.S. Degradable thiol-acrylate photopolymers: polymerization and degradation behavior of an in situ forming biomaterial. *Biomaterials* **2005**, *26*, 4495–4506. [CrossRef] [PubMed]

33. Díez-Peña, E.; Quijada-Garrido, I.; Barrales-Rienda, J.M. Hydrogen-Bonding Effects on the Dynamic Swelling of P(N-iPAAm-co-MAA) Copolymers. A Case of Autocatalytic Swelling Kinetics. *Macromolecules* **2002**, *35*, 8882–8888. [CrossRef]

34. Grassie, N.; Perdomo Mendoza, G.A. Thermal degradation of polyether-urethanes: Part 1-Thermal degradation of poly(ethylene glycols) used in the preparation of polyurethanes. *Polym. Degrad. Stab.* **1984**, *9*, 155–165. [CrossRef]

35. Fu, H.; Gao, H.; Wu, G.; Wang, Y.; Fan, Y.; Ma, J. Preparation and tunable temperature sensitivity of biodegradable polyurethane nanoassemblies from diisocyanate and poly(ethylene glycol). *Soft Matter* **2011**, *7*, 3546–3552. [CrossRef]

36. Saeki, S.; Kuwahara, N.; Nakata, M.; Kaneko, M. Upper and lower critical solution temperatures in poly (ethylene glycol) solutions. *Polymer* **1976**, *17*, 685–689. [CrossRef]

37. Dormidontova, E.E. Influence of End Groups on Phase Behavior and Properties of PEO in Aqueous Solutions. *Macromolecules* **2004**, *37*, 7747–7761. [CrossRef]

38. Lee, K.J.; Jun, B.H.; Choi, J.; Lee, Y.I.; Joung, J.; Oh, Y.S. Environmentally friendly synthesis of organic-soluble silver nanoparticles for printed electronics. *Nanotechnology* **2007**, *18*, 335601. [CrossRef]

39. Shankar, R.; Groven, L.; Amert, A.; Whites, K.W.; Kellar, J.J. Non-aqueous synthesis of silver nanoparticles using tin acetate as a reducing agent for the conductive ink formulation in printed electronics. *J. Mater. Chem.* **2011**, *21*, 10871–10877. [CrossRef]

40. Karak, N.; Konwarh, R.; Voit, B. Catalytically Active Vegetable-Oil-Based Thermoplastic Hyperbranched Polyurethane/Silver Nanocomposites. *Macromol. Mater. Eng.* **2010**, *295*, 159–169. [CrossRef]

41. Van Hyning, D.L.; Klemperer, W.G.; Zukoski, C.F. Characterization of colloidal stability during precipitation reactions. *Langmuir* **2001**, *17*, 3120–3127. [CrossRef]

42. Graves, J.L.; Tajkarimi, M.; Cunningham, Q.; Campbell, A.; Nonga, H.; Harrison, S.H.; Barrick, J.E. Rapid evolution of silver nanoparticle resistance in *Escherichia coli*. *Front. Genet.* **2015**, *6*, 1–13. [CrossRef] [PubMed]

43. Raho, R.; Paladini, F.; Lombardi, F.A.; Boccarella, S.; Zunino, B.; Pollini, M. In-situ photo-assisted deposition of silver particles on hydrogel fibers for antibacterial applications. *Mater. Sci. Eng. C* **2015**, *55*, 42–49. [CrossRef] [PubMed]

44. Martínez-Castañón, G.A.; Niño-Martínez, N.; Martínez-Gutierrez, F.; Martínez-Mendoza, J.R.; Ruiz, F. Synthesis and antibacterial activity of silver nanoparticles with different sizes. *J. Nanopart. Res.* **2008**, *10*, 1343–1348. [CrossRef]

45. Ng, V.W.L.; Chan, J.M.W.; Sardon, H.; Ono, R.J.; García, J.M.; Yang, Y.Y.; Hedrick, J.L. Antimicrobial hydrogels: A new weapon in the arsenal against multidrug-resistant infections. *Adv. Drug Deliv. Rev.* **2014**, *78*, 46–62. [CrossRef] [PubMed]

46. Salick, D.A.; Pochan, D.J.; Schneider, J.P. Design of an injectable β-hairpin peptide hydrogel that kills methicillin-resistant staphylococcus aureus. *Adv. Mater.* **2009**, *21*, 4120–4123. [CrossRef]

nanomaterials

MDPI

Article

Halloysite Nanotubes: Controlled Access and Release by Smart Gates

Giuseppe Cavallaro [1], Anna A. Danilushkina [2], Vladimir G. Evtugyn [2], Giuseppe Lazzara [1,*], Stefana Milioto [1], Filippo Parisi [1], Elvira V. Rozhina [2] and Rawil F. Fakhrullin [2,*]

[1] Dipartimento di Fisica e Chimica, Università degli Studi di Palermo Viale delle Scienze, pad. 17, 90128 Palermo, Italy; giuseppe.cavallaro@unipa.it (G.C.); stefana.milioto@unipa.it (S.M.); filippo.parisi@unipa.it (F.P.)

[2] Institute of Fundamental Medicine and Biology, Kazan Federal University, Kreml uramı 18, Kazan, 420008 Republic of Tatarstan, Russia; anchutka124@gmail.com (A.A.D.); vevtugyn@gmail.com (V.G.E.); rozhinaelvira@gmail.com (E.V.R.)

* Correspondence: giuseppe.lazzara@unipa.it (G.L.); kazanbio@gmail.com (R.F.F.); Tel.: +39-091-2389-7962 (G.L.)

Received: 23 June 2017; Accepted: 26 July 2017; Published: 28 July 2017

Abstract: Hollow halloysite nanotubes have been used as nanocontainers for loading and for the triggered release of calcium hydroxide for paper preservation. A strategy for placing end-stoppers into the tubular nanocontainer is proposed and the sustained release from the cavity is reported. The incorporation of $Ca(OH)_2$ into the nanotube lumen, as demonstrated using transmission electron microscopy (TEM) imaging and Energy Dispersive X-ray (EDX) mapping, retards the carbonatation, delaying the reaction with CO_2 gas. This effect can be further controlled by placing the end-stoppers. The obtained material is tested for paper deacidification. We prove that adding halloysite filled with $Ca(OH)_2$ to paper can reduce the impact of acid exposure on both the mechanical performance and pH alteration. The end-stoppers have a double effect: they preserve the calcium hydroxide from carbonation, and they prevent from the formation of highly basic pH and trigger the response to acid exposure minimizing the pH drop-down. These features are promising for a composite nanoadditive in the smart protection of cellulose-based materials.

Keywords: halloysite; nanocomposite; cellulose; controlled release

1. Introduction

Halloysite clay (HNT) is a natural and abundantly available nanoparticle formed by rolled kaolin sheets. The main deposits of HNT are from Dragon Mine and Matauri Bay, which are in Utah (USA) and Northland (New Zealand), respectively. Due to its biocompatibility [1,2] HNT was recently studied for the development of innovative nanomaterials useful for biotechnological applications, such as the controlled release of drugs [3–6], tissue engineering [7–9], oil recovery [10], and eco-compatible packaging [11–13]. Furthermore, several studies proved that HNT is a suitable catalytic support [14,15], as well as an efficient removal agent [16], because of its geometrical and surface properties (large specific area, hollow tubular shape, and tunable surface chemistry). Both the sizes and polydispersity are influenced by the HNT geological deposit [17]. Typically, the HNTs lengths range between 0.1 and 3.0 μm, while their external and inner diameters are ca. 50–200 and 15–70 nm, respectively [18]. The HNT surfaces are oppositely charged within a large pH range (between 2 and 8) because of their different chemical compositions [19]. Particularly, the internal surface consists of gibbsite octahedral sheet (Al–OH) groups with a positive surface charge, whereas the outer surface is composed of siloxane groups (Si–O–Si) with a negative electrical potential. Accordingly, the selective HNT functionalization can be easily achieved through electrostatic interactions between the nanoparticle surfaces and ionic molecules, such as polymers [20], surfactants [21,22], enzymes [23], and proteins [24]. Inorganic hybrid nanoparticles are considered suitable building blocks for nanomaterials with smart properties [25–28]. The HNT cavity is an efficient nanocontainer for the loading of chemically- and biologically-active compounds allowing the fabrication of hybrid nanomaterials with functional properties (antibacterial, antioxidant, and anti-acid) [4,29–33]. Interestingly, the release of the encapsulated species can be controlled under specific external stimuli dependent of the environmental conditions, such as pH or temperature [5]. A Monte Carlo model was successfully used to describe the effect of environmental variables (pH and temperature) on the transport and release of dexamethasone molecules from HNT [34]. A recent review [35] highlighted that a typical release time of water-soluble active molecules from the nanotubes is 5–10 h. It should be noted that slower release kinetics are generally needed for composite materials with antioxidant, flame-retardant, and antimicrobial properties. A time-extended release can be achieved by the HNT coating with thin polymeric layers or through the formation of tube-end stoppers [36–39]. Using dextrin as a smart end-stopper endowed a targeted release of the payload within cancer cells [39].

The functionalized HNT can be employed as a filler for biopolymeric matrices in order to generate functional bionanocomposite films with long-term activity [40,41]. The paper consolidation with perfluorinated modified HNT induced a flame-retardant effect on the cellulose [42]. Similarly, pristine HNTs provided thermal stability and flame-retardant effects on poly(propylene) [43].

The mechanical resistance of cellulose-based materials is significantly influenced by the degree of hydrolytic and oxidative reactions. The material deterioration depends on the environmental conditions (temperature, presence of oxygen, humidity, etc.) and it might be retarded by adding nanoparticles with specific anti-acid [44] and antibacterial [45] properties. It was demonstrated that non-aqueous dispersions of alkaline nanoparticles, such as calcium and magnesium hydroxide, are efficient deacidifying treatments for cellulose-based works [44,46,47]. Due to their high reactivity, these nanoparticles provide a stable neutral environment by rapidly turning into slight alkaline species (carbonates). $Ca(OH)_2$ nanoparticles are typically stabilized in short-chain alcohol dispersions. A recent study proved that Ca-alkoxides are formed and they can hamper/delay the strengthening or consolidation effects of nanolimes [48]. In general, acid paper is a challenge and many approaches have been published and reviewed [49,50]. Industrial scale deacidification processes have been installed since the 1990s and the approach we propose in this study offers a benefit to the known technologies [49].

In this paper, we propose an innovative deacidification and consolidation treatment for paper based on HNT filled with calcium hydroxide and hydroxypropyl cellulose (HPC). The method represents an improvement of the consolidation obtained by HNT/HPC mixtures [42]. The selective loading of the alkaline molecule into the HNT cavity was investigated by using several microscopic techniques, while the kinetic release of calcium hydroxide was studied by pH and thermogravimetry measurements. The HNT modification with calcium salts (triphosphate) was explored as an original approach for the formation of tube end-stoppers, which can generate a time-extended release of the loaded calcium hydroxide and, consequently, a consolidation and deacidification for the treated paper. The acquired knowledge represents an advanced step for designing tubular alkaline nanoparticles with an extended deacidification activity towards cellulose-based materials.

2. Results and Discussion

2.1. Characterization of HNT/Ca(OH)$_2$ with and without End-Stoppers

The thermal behavior of loaded calcium hydroxide was determined by thermogravimetry. $Ca(OH)_2$ presents a mass loss from 400 to 600 °C due to the dehydration process and CaO formation (Figure S1). Halloysite nanotubes present ca. 20 wt % mass loss due to hydration water [11]. By comparing the thermoanalytical curves of pristine materials and the HNT/Ca(OH)$_2$ composite (Figure S1) it turned out that, in the composite material, the $Ca(OH)_2$ is likely present as an additional mass loss is observed. The $Ca(OH)_2$ loaded amount can be evaluated by considering the residual mass at 900 °C for pristine components and assuming the rule of mixtures. On this basis, one can calculate a value of $3.9 \pm 0.2\%$ w/w (corresponding to $4.5 \pm 0.3\%$ v/v) for the loading. Given that the full geometrical filling would provide ca. 10% v/v of loaded material [19], one may conclude that ca. half of the lumen is filled by the calcium hydroxide. The presence of $Ca(OH)_2$ in the HNTs-$Ca(OH)_2$ composite was confirmed by Fourier transform infrared spectroscopy (FTIR) spectra. As evidenced in Figure 1, the composite material presents the characteristic bands of both components, proving that during the loading procedure the $Ca(OH)_2$ is preserved and incorporated in the composite.

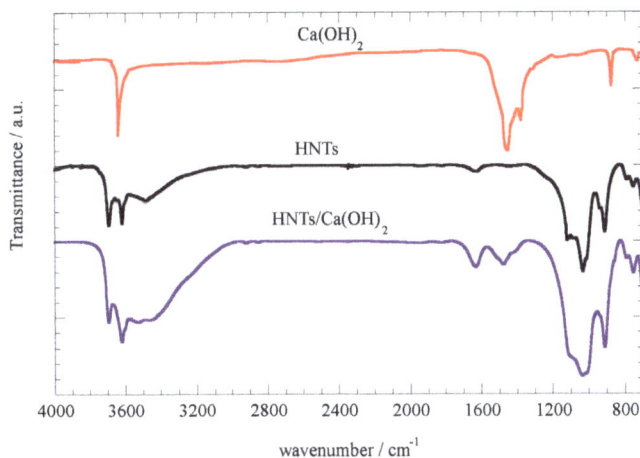

Figure 1. FTIR spectra for $Ca(OH)_2$, HNTs and HNTs/$Ca(OH)_2$.

Thermogravimetric analysis (TGA) data on HNTs-$Ca(OH)_2$ with calcium phosphate end-stoppers did not show any significant difference from the HNTs-$Ca(OH)_2$ sample as proof that the end-stopper treatment did not alter the general composition of the material to a large extent (Figure S1). To investigate the end-stopper formation, TEM experiments were carried out on HNTs-$Ca(OH)_2$ with calcium phosphate end-stoppers. Literature reports on TEM images for HNTs samples were able to identify the lumen filling especially for high electron density materials, such as metals [51,52]. The images for HNTs-$Ca(OH)_2$ based system show that the lumen of HNTs is filled (Figure 2, additional images are in Figure S2).

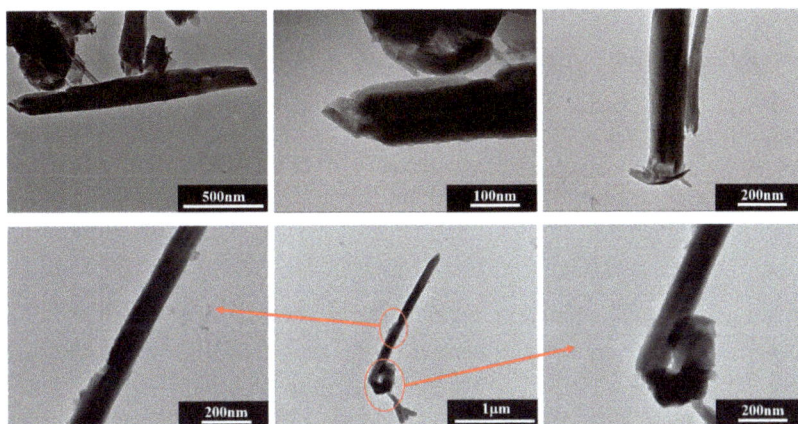

Figure 2. TEM images of HNTs/$Ca(OH)_2$ with calcium phosphate end-stoppers.

EDX mapping allowed us to make a proper identification of the filling; as Figure 3 shows, the Ca signals come from the same spots where tubular-like nanostructures are imaged. As a confirmation, this is also the case for Al and Si, which are HNT components, and the Ca signal is absent in pristine HNTs. Going further, a phosphorus signal was detected, proving that phosphate was, by some means, kept in the sample during the treatment. Its concentration is relatively small and far below that stoichiometrically expected for $Ca_3(PO_4)_2$. On the other hand, P is not phase separated within the observed sample. By a close look at the nanotube ends (Figure 1), it is revealed that they are closed by what appears to be a stopper, moreover the lumen cavity nearby the HNTs' termination appears empty. Such a morphological observation is in agreement with a mechanism of end stopper formation based on the reaction between partially-released $Ca(OH)_2$ and Na_3PO_4 in proximity of the nanotube ends forming $Ca_3(PO_4)_2$ due to a high local concentration. It should be noted that a flow of Na_3PO_4 aqueous solution is used and that a short solution-HNT/$Ca(OH)_2$ contact time is ensured by vacuum filtration in order to avoid a complete HNT unload. A schematic representation of end stoppers' formation is depicted in Figure 4.

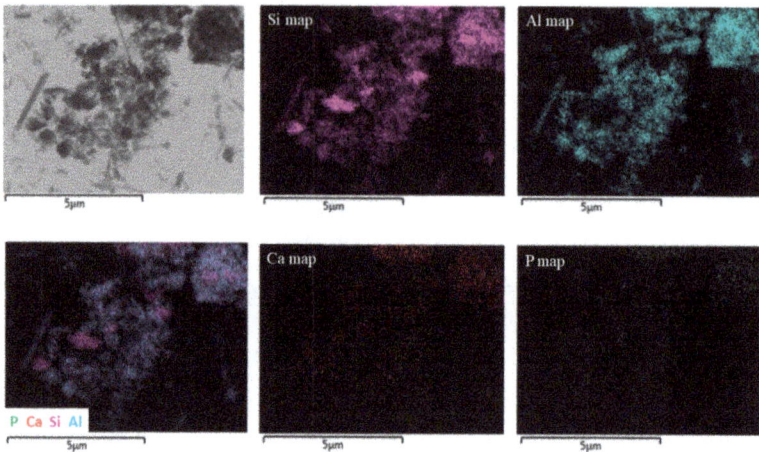

Figure 3. TEM image and EDX mapping of HNTs/$Ca(OH)_2$ with calcium phosphate end-stoppers.

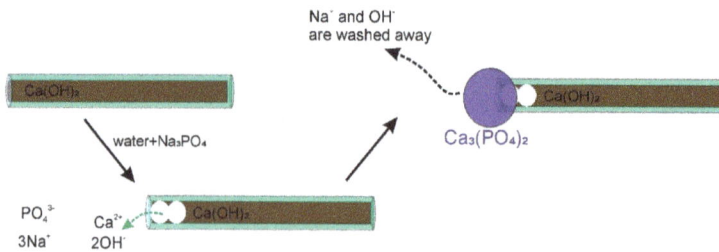

Figure 4. Sketch of the end-stopper formation.

Additional dark field optical images were taken from the aqueous dispersion HNTs-$Ca(OH)_2$ with calcium phosphate end-stoppers. Figure 5 shows that the nanotubes generate a uniform dispersion as they are not aggregated in water. Therefore, the preparation protocols avoid any clustering of nanoparticles. The literature reports that aggregation and dispersion behaviours of halloysite nanotubes (HNTs) can be influenced by pH [53]. In particular, it is reported that the pH variation could be used as a strategy for blocking and opening the halloysite cavity. In our system, based on the observed morphology by TEM and dark field (DF) microscopy, we can exclude a clustering of HNTs-Ca(OH)$_2$ nanoparticles and, therefore, the controlled access/release due to aggregative phenomena.

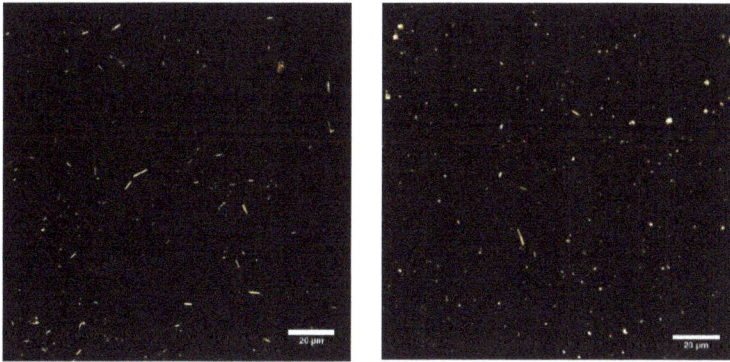

Figure 5. DF optical images of HNTs (**left**) and HNTs/Ca(OH)$_2$ (**right**) with calcium phosphate end-stoppers in water.

2.2. Kinetics Study on Carbonatation and Release of Ca(OH)$_2$ from HNT Lumen

In addition to the interesting molecular architecture, we investigated the functionality of the end stopper in playing any barrier role for gas or to control the release of Ca(OH)$_2$ from the lumen. Calcium hydroxide typically undergoes CO$_2$ capture with CaCO$_3$ formation. This process has been widely investigated due to the relevant applicative interest [54]. To explore the ability of HNT lumen in controlling such a process, we used thermogravimetric analysis under a CO$_2$ atmosphere. The degree of Ca(OH)$_2$ conversion to CaCO$_3$, based on measured mass gain and initial Ca(OH)$_2$ content in the measured sample is provided in Figure 6 as a function of time.

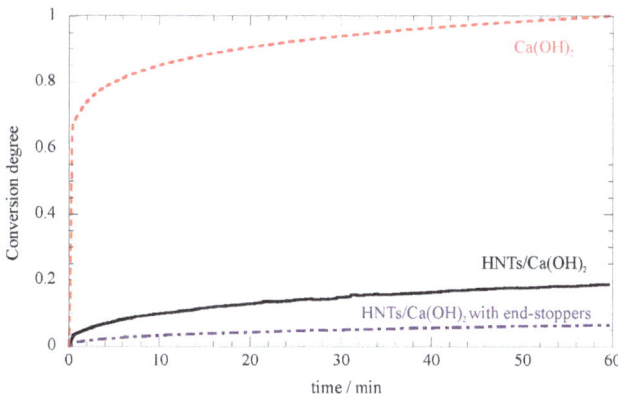

Figure 6. Degree of Ca(OH)$_2$ carbonation in a CO$_2$ atmosphere for Ca(OH)$_2$, HNT/Ca(OH)$_2$ and HNT/Ca(OH)$_2$ with phosphate end-stoppers.

It is worth noting that confining Ca(OH)$_2$ within the HNTs lumen cavity significantly retards the carbonation reaction. Furthermore, the end-stoppers prevent the CO$_2$ contact and less than 10% of the calcium hydroxide is converted to carbonate after 1 h under the experimental conditions. Although the time frame is relatively short (one hour), the experiment proves that encapsulated Ca(OH)$_2$ is still in its original form when bare Ca(OH)$_2$ undergoes complete conversion. This result is very promising for applications as it shows the possibility to keep $Ca(OH)_2$ preserved from carbonation during the treatment.

The release kinetics of $Ca(OH)_2$ in water were investigated by measuring the pH of the dispersion over time. To this aim, the aqueous dispersions of HNT and $HNT/Ca(OH)_2$ with and without end-stoppers (0.1 wt %) were left to equilibrate under static conditions while a glassy electrode was used to monitor the pH. A blank experiment reporting the kinetics for pure $Ca(OH)_2$, in the same amount as the loaded value in HNTs, revealed a quick dissolution of the hydroxide that occurs within 5 min. After that a constant pH value was approached, 0.15 cm^3 of HCl (0.1 M) was added to the dispersion, and the pH response was measured for 18 h. With respect to the release in water, the $HNT/Ca(OH)_2$ composite showed a sustained increase of pH (Figure 7). Even slower is the pH increase for the composites with the phosphate end-stoppers being the most efficient in retarding the hydroxide solubilization in water. It is reported that the dissolution kinetics of nanosized materials is influenced by the grain size due to high specific area and surface energy effects. It should be noted that even if the net $Ca(OH)_2$ amount was similar for all samples, a higher pH is approached at the plateau for the $HNT/Ca(OH)_2$ composite compared to the system with end-stoppers. This result reflects the ability of the end-stopping strategy to retain the hydroxide in the HNT lumen even in water media for a certain extent.

The HCl addition generates a sudden drop-down of the pH that slowly returns toward higher values due to a further release of the calcium hydroxide from the lumen. The pH increasing trend is significantly slowed by the end-stopper presence. Moreover, the ΔpH, due to the HCl addition and after equilibration, is 0.35 and 0.85 for $HNT/Ca(OH)_2$ with and without end-stoppers, respectively. From the stoichiometric calculation a pH change of 1.35 is expected if all of the calcium hydroxide would have been dissolved prior to HCl addition. Therefore, we might conclude that confining $Ca(OH)_2$ into the HNT lumen generates an alkaline reservoir which is released in response to acid addition.

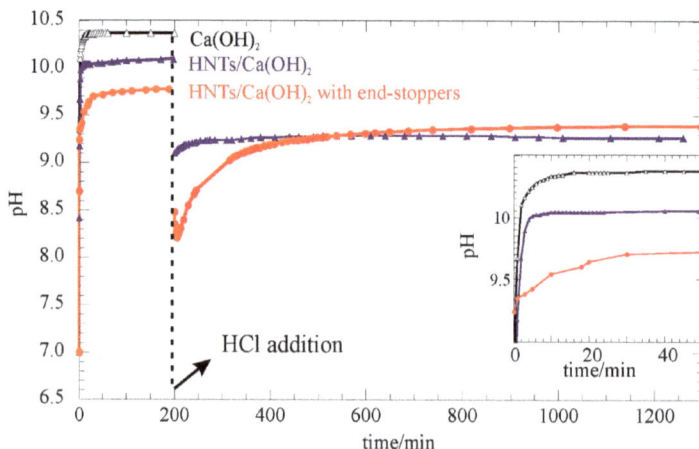

Figure 7. pH measurements in aqueous dispersion before and after HCl solution addition as functions of time. The inset reports an enlargement of the initial release.

2.3. Effect of $HNT/Ca(OH)_2$ on Paper Deacidification and Consolidation

The efficacy of the prepared nanomaterials on paper deacidification was monitored by cycling the aging protocols and controlling the paper conditions and its damage by pH measurements and tensile experiments.

The paper sample without a $Ca(OH)_2$ basic reservoir reaches acid pH values after the first aging cycle and it remains constant, not being able to compensate for the effect of acid gas presence (Table 1). The $HNTs/Ca(OH)_2$ system generated a paper alkaline pH which systematically decreases with aging approaching the value for the paper sample without the basic reservoir. Keeping in mind the strong acidic environment used in this investigation compared to the actual situations that might be experienced in the conventional conservation for books, the obtained results are already promising. On the other hand, the end-stopped system could be considered even more performant as the starting value for pH is only slightly basic and the pH change is kept within one unit even after two aging cycles when a still alkaline/neutral pH is maintained.

Tensile measurements provided information on the alteration of mechanical performance for paper samples after exposure to acidic gas. Stress at breaking point (σ_r) data showed that no treatments significantly altered the paper property, while only the samples with the alkaline reservoir were able to minimize the σ_r reduction upon aging (Table 1). The mechanical performance might also be described by the storage energy parameter (SE) that is obtained from the stress vs. strain curve integral and provides an idea on the maximum energy that can be adsorbed by the paper sample until it breaks down. Results in Figure 8 demonstrate that paper aging reduced the SE if the alkaline reservoir is not introduced within the paper. On the other hand, the end-stopped system is more efficient in strengthening the paper, maintaining a relatively high SE value even after the aging protocol.

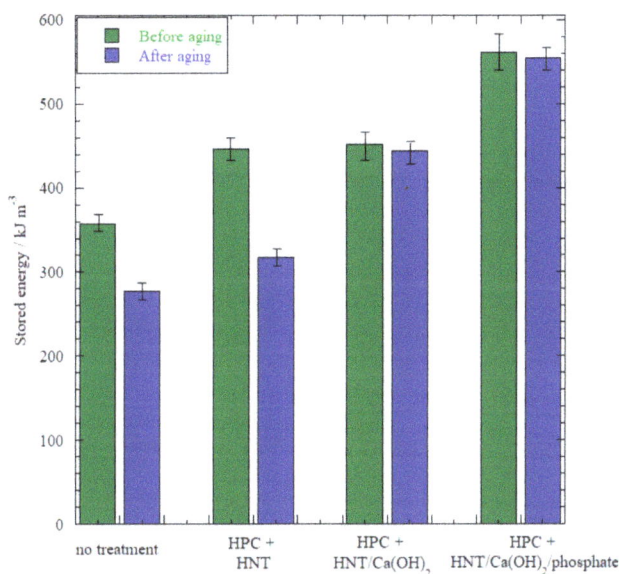

Figure 8. Stored energy up to sample breaking from tensile stress measurements. The error is based on the standard deviation from repeated experiments.

Table 1. Paper pH values and stress at the breaking point before and after aging under HNO_3 saturated vapours.

Sample	pH before Aging	pH after First Aging Cycle	pH after Second Aging Cycle	σ_r/Mpa before Aging	$\Delta\sigma_r$ [a]/MPa
Paper	6.7	6.3	6.2	24.3 ± 0.3	−8.6
Paper + HPC/HNTs	7.7	6.2	6.3	23.7 ± 0.2	−5.0
Paper + HPC/HNTs-Ca(OH)$_2$	10.4	8.5	6.2	22.8 ± 0.2	−3.3
Paper + HPC/HNTs-Ca(OH)$_2$ with phosphate end-stoppers	8.5	7.6	7.6	23.6 ± 0.2	−3.2

[a] $\Delta\sigma_r$ represents the reduction of the stress at breaking point induced by the aging cycle.

3. Materials and Methods

Materials: Halloysite nanotubes with a specific surface area of 65 $m^2 \cdot g^{-1}$ and a specific gravity of 2.53 $g \cdot cm^{-3}$ are from Sigma-Aldrich (Milan, Italy). $Ca(OH)_2$, $Na_3PO_4 \cdot 12H_2O$, HNO_3 60%, 2-hydroxypropylcellulose (HPC), and ethanol (96%) were purchased from Sigma-Aldrich (Milan, Italy) and used without further purification. The paper sample is cellulose based from Albet® (Milan, Italy) (70 $g \cdot m^{-2}$, thickness 0.138 mm and water capillary raise > 178 mm/h).

Thermogravimetry analysis (TGA): Experiments were done using the Q5000 IR (TA Instruments, Milan, Italy) under nitrogen flow (25 $cm^3 \cdot min^{-1}$) by heating the samples from room temperature to 900 °C. Each sample (ca. 5 mg) was placed in a platinum pan and heated under the temperature program of 10 °C·min^{-1}. Loading was calculated according to the procedure in the literature and errors were evaluated from standard deviations of three measurements [11]. The CO_2 capturing experiments were carried out by quickly heating the sample (200 °C·min^{-1}) to 600 °C in a N_2 flow (25 $cm^3 \cdot min^{-1}$). Afterwards, the gas flow was switched to CO_2 with 99.995% chemical purity (25 $cm^3 \cdot min^{-1}$). The mass gain was monitored for 60 min. The high temperature was chosen to accelerate the CO_2 capture based on literature reports [55]. Calibration was carried out as reported elsewhere [56].

Tensile Analysis: Tensile properties on paper samples were determined by means of a DMA Q800 instrument (TA Instruments, Milan, Italy). Tensile tests were performed on rectangular paper samples (10 mm × 4 mm) under a stress ramp of 1 MPa min^{-1} at 26.0 ± 0.5 °C. We determined the stress at which the material undergoes fractures (σ_r) and stored energy up to sample breaking by integrating the stress vs. strain curves. The reproducibility was checked by repeating the experiment three times.

pH measurements: The pH curves were obtained by using a PCD650 pH meter (Eutech Instruments, Landsmeer, The Netherlands) immersed in an aqueous dispersion of loaded nanoclay under stirring conditions. For all of the tested nanomaterials, dispersions were kept under a controlled environment, magnetic stirring, and measured continuously. Degassed water was used and the concentration of the dispersions was 0.1 wt %. The pH values of paper was measured by using a HI 1413B/50 portable pH meter with a flat-tip electrode (Hanna Instruments, Milan, Italy) in accordance with a non-destructive test that may be used to measure the hydrogen ion concentration (pH) on the surface of the paper in books and documents that constitute the collections of libraries and government archives (working procedure: TAPPI T 529 om-04).

TEM-EDX: For electron microscopy imaging and energy-dispersive X-ray analysis (EDX) a Hitachi HT7700 Exalens transmission electron microscope (Tokyo, Japan) was used. The samples were prepared by placing 10 μL of the suspension on a carbon-coated lace 3 mm copper grid, then dried at room temperature. TEM imaging was performed at a 100 kV accelerating voltage in TEM mode. EDX analysis was carried out in scanning transmission electron microscope (STEM) mode using an Oxford Instruments (High Wycomb, UK) X-Max™ 80T detector.

Enhanced dark-field imaging: During enhanced dark field microscopy experiments the images were obtained using a CytoViva® enhanced dark-field condenser attached to an Olympus BX51 upright microscope equipped with fluorite 100× objectives and a DAGE CCD camera. Extra-clean dust-free Nexterion® glass slides and coverslips (Schott, Mainz, Germany) were used for EDF microscopy imaging to minimise dust interference.

Loading of Ca(OH)₂ onto HNTs: Degassed aqueous solution of $Ca(OH)_2$ (1.5 $g \cdot dm^{-3}$) was mixed with halloysite powder (5 $g \cdot dm^{-3}$) and sonicated for 15 min. Then, the obtained suspension was stirred and kept under vacuum for 5 min resulting in light fizzling and the loaded compound condensated within the tube. This procedure was repeated three times to improve the loading efficiency. Successively, the nanotubes were separated from the aqueous phase by centrifugation and dried under vacuum at 70 °C overnight.

End-stopper formation: Aqueous phosphate solution was prepared by dissolving 40 g of trisodium phosphate dodecahydrate in 250 cm^3 of water. This solution was poured onto the $HNT/Ca(OH)_2$ powder placed in a Buechner funnel with filter paper placed on the perforated plate. Vacuum was applied during the pouring. The filtered material was dried by using a side-arm flask connected to a vacuum pump.

Paper treatments: For the paper treatment we prepared a 2 wt % HPC solution in ethanol. A certain amount of HNT (1 wt %) was added to the polymer solution and kept under stirring over night at 25 °C. The same procedure was followed by using HNT/Ca(OH)$_2$ with and without end-stoppers. It should be noted that ethanol was used as the solvent to avoid calcium hydroxide solubilization during the paper treatment. The paper samples were cut in a rectangular shape (40 mm × 8 mm) and they were deeply immersed into the well-dispersed aqueous mixtures for 24 h at 20 °C. The treated samples were dried at 35 °C.

Paper aging under acidic conditions: Paper specimens were placed in a closed desiccator. The vapours were saturated with HNO$_3$ by equilibrating the system with 30% acid solution. One aging cycle was three days. Before any characterization, the paper samples were re-equilibrated with air for 20 days.

4. Conclusions

We developed a novel strategy for sustained release and controlled access to the halloysite nanotubes lumen. Calcium hydroxide was loaded into the HNTs lumen and imaged by TEM and EDX mapping. End-stoppers were created when calcium hydroxide was partially released in the presence of phosphate anions. The obtained end-stoppers prevent CO$_2$ from entering the tube lumen and preserving the calcium hydroxide from carbonation. Moreover, they slow the release in water, minimizing the pH jumps if an acid is added to the dispersion. These features are very promising for paper preservation, as was demonstrated by aging experiments on treated and pristine cellulose paper samples. This composite nanomaterial would allow adding an alkaline reservoir to the paper and minimizing the pH changes, as well as the aging impact on the mechanical performance of the sample. The proposed strategy could be interesting in designing and building up nanocontainers with nanogates that are sensitive to external stimuli.

Supplementary Materials: The following are available online at http://www.mdpi.com/2079-4991/7/8/199/s1. Figure S1: TGA curves for HNT-based hybrid materials. Figure S2: Additional TEM figures.

Acknowledgments: The work was financially supported by the University of Palermo, FIRB 2012 (prot. RBFR12ETL5), PON-TECLA (PON03PE_00214_1), and RFBR grant 14-04-01474_a. The work was partially performed according to the Russian Government Program of Competitive Growth of Kazan Federal University. This work was partially funded by the subsidy allocated to Kazan Federal University for the state assignment in the sphere of scientific activities.

Author Contributions: Rawil F. Fakhrullin, Giuseppe Lazzara, and Stefana Milioto conceived and directed the project. Giuseppe Cavallaro and Filippo Parisi prepared the HNT hybrid materials and performed the TGA, release, and paper protection experiments. Anna A. Danilushkina, Vladimir G. Evtugyn, and Elvira V. Rozhina performed TEM, EDX, and enhanced dark-field imaging. All authors analysed the data, discussed their implications, wrote the paper, and revised the manuscript at all stages.

Conflicts of Interest: The authors declare no conflict of interest.

References

1. Fakhrullina, G.I.; Akhatova, F.S.; Lvov, Y.M.; Fakhrullin, R.F. Toxicity of halloysite clay nanotubes In Vivo: A Caenorhabditis elegans study. *Environ. Sci. Nano* **2015**, *2*, 54–59. [CrossRef]
2. Kryuchkova, M.; Danilushkina, A.; Lvov, Y.; Fakhrullin, R. Evaluation of toxicity of nanoclays and graphene oxide In Vivo: A Paramecium caudatum study. *Environ. Sci. Nano* **2016**, *3*, 442–452. [CrossRef]
3. Shutava, T.G.; Fakhrullin, R.F.; Lvov, Y.M. Spherical and tubule nanocarriers for sustained drug release. *Curr. Opin. Pharmacol.* **2014**, *18*, 141–148. [CrossRef] [PubMed]
4. Wei, W.; Minullina, R.; Abdullayev, E.; Fakhrullin, R.; Mills, D.; Lvov, Y. Enhanced efficiency of antiseptics with sustained release from clay nanotubes. *RSC Adv.* **2014**, *4*, 488–494. [CrossRef]
5. Cavallaro, G.; Lazzara, G.; Massaro, M.; Milioto, S.; Noto, R.; Parisi, F.; Riela, S. Biocompatible poly(*N*-isopropylacrylamide)-halloysite nanotubes for thermoresponsive curcumin release. *J. Phys. Chem. C* **2015**, *119*, 8944–8951. [CrossRef]

6. Lvov, Y.M.; DeVilliers, M.M.; Fakhrullin, R.F. The application of halloysite tubule nanoclay in drug delivery. *Expert Opin. Drug Deliv.* **2016**, *13*, 977–986. [CrossRef] [PubMed]

7. Bonifacio, M.A.; Gentile, P.; Ferreira, A.M.; Cometa, S.; De Giglio, E. Insight into halloysite nanotubes-loaded gellan gum hydrogels for soft tissue engineering applications. *Carbohydr. Polym.* **2017**, *163*, 280–291. [CrossRef] [PubMed]

8. Fakhrullin, R.F.; Lvov, Y.M. Halloysite clay nanotubes for tissue engineering. *Nanomedicine* **2016**, *11*, 2243–2246. [CrossRef] [PubMed]

9. Ji, L.; Qiao, W.; Zhang, Y.; Wu, H.; Miao, S.; Cheng, Z.; Gong, Q.; Liang, J.; Zhu, A. A gelatin composite scaffold strengthened by drug-loaded halloysite nanotubes. *Mater. Sci. Eng. C* **2017**, *78*, 362–369. [CrossRef] [PubMed]

10. Von Klitzing, R.; Stehl, D.; Pogrzeba, T.; Schomäcker, R.; Minullina, R.; Panchal, A.; Konnova, S.; Fakhrullin, R.; Koetz, J.; Möhwald, H.; et al. Halloysites stabilized emulsions for hydroformylation of long chain olefins. *Adv. Mater. Interfaces* **2017**, *4*, N1. [CrossRef]

11. Cavallaro, G.; Lazzara, G.; Milioto, S. Dispersions of nanoclays of different shapes into aqueous and solid biopolymeric matrices. Extended physicochemical study. *Langmuir* **2011**, *27*, 1158–1167. [CrossRef] [PubMed]

12. Cavallaro, G.; Lazzara, G.; Konnova, S.; Fakhrullin, R.; Lvov, Y. Composite films of natural clay nanotubes with cellulose and chitosan. *Green Mater.* **2014**, *2*, 232–242. [CrossRef]

13. Gorrasi, G.; Pantani, R.; Murariu, M.; Dubois, P. PLA/Halloysite nanocomposite films: Water vapor barrier properties and specific key characteristics. *Macromol. Mater. Eng.* **2014**, *299*, 104–115. [CrossRef]

14. Machado, G.S.; de Freitas Castro, K.A.D.; Wypych, F.; Nakagaki, S. Immobilization of metalloporphyrins into nanotubes of natural halloysite toward selective catalysts for oxidation reactions. *J. Mol. Catal. Chem.* **2008**, *283*, 99–107. [CrossRef]

15. Wang, R.; Jiang, G.; Ding, Y.; Wang, Y.; Sun, X.; Wang, X.; Chen, W. Photocatalytic activity of heterostructures based on TiO_2 and halloysite nanotubes. *ACS Appl. Mater. Interfaces* **2011**, *3*, 4154–4158. [CrossRef] [PubMed]

16. Luo, P.; Zhao, Y.; Zhang, B.; Liu, J.; Yang, Y.; Liu, J. Study on the adsorption of Neutral Red from aqueous solution onto halloysite nanotubes. *Water Res.* **2010**, *44*, 1489–1497. [CrossRef] [PubMed]

17. Pasbakhsh, P.; Churchman, G.J.; Keeling, J.L. Characterisation of properties of various halloysites relevant to their use as nanotubes and microfibre fillers. *Appl. Clay Sci.* **2013**, *74*, 47–57. [CrossRef]

18. Joussein, E.; Petit, S.; Churchman, G.J.; Theng, B.; Righi, D.; Delvaux, B. Halloysite clay minerals—A review. *Clay Miner.* **2005**, *40*, 383–426. [CrossRef]

19. Lvov, Y.M.; Shchukin, D.G.; Mohwald, H.; Price, R.R. Halloysite clay nanotubes for controlled release of protective agents. *ACS Nano* **2008**, *2*, 814–820. [CrossRef] [PubMed]

20. Cavallaro, G.; Lazzara, G.; Milioto, S.; Parisi, F. Steric stabilization of modified nanoclays triggered by temperature. *J. Colloid Interface Sci.* **2016**, *461*, 346–351. [CrossRef] [PubMed]

21. Cavallaro, G.; Lazzara, G.; Milioto, S.; Parisi, F.; Sanzillo, V. Modified halloysite nanotubes: Nanoarchitectures for enhancing the capture of oils from vapor and liquid phases. *ACS Appl. Mater. Interfaces* **2014**, *6*, 606–612. [CrossRef] [PubMed]

22. Cavallaro, G.; Lazzara, G.; Milioto, S. Exploiting the colloidal stability and solubilization ability of clay nanotubes/ionic surfactant hybrid nanomaterials. *J. Phys. Chem. C* **2012**, *116*, 21932–21938. [CrossRef]

23. Tully, J.; Yendluri, R.; Lvov, Y. Halloysite clay nanotubes for enzyme immobilization. *Biomacromolecules* **2016**, *17*, 615–621. [CrossRef] [PubMed]

24. Della Porta, V.; Bramanti, E.; Campanella, B.; Tine, M.R.; Duce, C. Conformational analysis of bovine serum albumin adsorbed on halloysite nanotubes and kaolinite: A Fourier transform infrared spectroscopy study. *RSC Adv.* **2016**, *6*, 72386–72398. [CrossRef]

25. Ruiz-Hitzky, E.; Aranda, P.; Darder, M.; Ogawa, M. Hybrid and biohybrid silicate based materials: Molecular vs. block-assembling bottom-up processes. *Chem. Soc. Rev.* **2011**, *40*, 801–828. [CrossRef] [PubMed]

26. Nagy, D.; Firkala, T.; Drotar, E.; Szegedi, A.; Laszlo, K.; Szilagyi, I.M. Photocatalytic WO_3/TiO_2 nanowires: WO_3 polymorphs influencing the atomic layer deposition of TiO_2. *RSC Adv.* **2016**, *6*, 95369–95377. [CrossRef]

27. Szilágyi, I.M.; Santala, E.; Heikkilä, M.; Pore, V.; Kemell, M.; Nikitin, T.; Teucher, G.; Firkala, T.; Khriachtchev, L.; Räsänen, M.; et al. Photocatalytic properties of WO_3/TiO_2 Core/Shell Nanofibers prepared by Electrospinning and Atomic Layer Deposition. *Chem. Vap. Depos.* **2013**, *19*, 149–155. [CrossRef]

28. Andres, C.M.; Larraza, I.; Corrales, T.; Kotov, N.A. Nanocomposite microcontainers. *Adv. Mater.* **2012**, *24*, 4597–4600. [CrossRef] [PubMed]

29. Lvov, Y.; Abdullayev, E. Functional polymer-clay nanotube composites with sustained release of chemical agents. *Prog. Polym. Sci.* **2013**, *38*, 1690–1719. [CrossRef]

30. Abdullayev, E.; Price, R.; Shchukin, D.; Lvov, Y. Halloysite tubes as nanocontainers for anticorrosion coating with Benzotriazole. *ACS Appl. Mater. Interfaces* **2009**, *1*, 1437–1443. [CrossRef] [PubMed]

31. Viseras, M.T.; Aguzzi, C.; Cerezo, P.; Viseras, C.; Valenzuela, C. Equilibrium and kinetics of 5-aminosalicylic acid adsorption by halloysite. *Microporous Mesoporous Mater.* **2008**, *108*, 112–116. [CrossRef]

32. Fix, D.; Andreeva, D.V.; Lvov, Y.M.; Shchukin, D.G.; Möhwald, H. Application of inhibitor-loaded halloysite nanotubes in active anti-corrosive coatings. *Adv. Funct. Mater.* **2009**, *19*, 1720–1727. [CrossRef]

33. Shchukin, D.G.; Möhwald, H. Surface-engineered nanocontainers for entrapment of corrosion inhibitors. *Adv. Funct. Mater.* **2007**, *17*, 1451–1458. [CrossRef]

34. Elumalai, D.N.; Tully, J.; Lvov, Y.; Derosa, P.A. Simulation of stimuli-triggered release of molecular species from halloysite nanotubes. *J. Appl. Phys.* **2016**, *120*, 134311. [CrossRef]

35. Lvov, Y.; Wang, W.; Zhang, L.; Fakhrullin, R. Halloysite clay nanotubes for loading and sustained release of functional compounds. *Adv. Mater.* **2016**, *28*, 1227–1250. [CrossRef] [PubMed]

36. Du, M.; Guo, B.; Jia, D. Newly emerging applications of halloysite nanotubes: A review. *Polym. Int.* **2010**, *59*, 574–582. [CrossRef]

37. Joshi, A.; Abdullayev, E.; Vasiliev, A.; Volkova, O.; Lvov, Y. Interfacial modification of clay nanotubes for the sustained release of corrosion inhibitors. *Langmuir* **2013**, *29*, 7439–7448. [CrossRef] [PubMed]

38. Abdullayev, E.; Lvov, Y. Clay nanotubes for corrosion inhibitor encapsulation: Release control with end stoppers. *J. Mater. Chem.* **2010**, *20*, 6681–6687. [CrossRef]

39. Dzamukova, M.R.; Naumenko, E.A.; Lvov, Y.M.; Fakhrullin, R.F. Enzyme-activated intracellular drug delivery with tubule clay nanoformulation. *Sci. Rep.* **2015**, *5*, 10560. [CrossRef] [PubMed]

40. Biddeci, G.; Cavallaro, G.; Di Blasi, F.; Lazzara, G.; Massaro, M.; Milioto, S.; Parisi, F.; Riela, S.; Spinelli, G. Halloysite nanotubes loaded with peppermint essential oil as filler for functional biopolymer film. *Carbohydr. Polym.* **2016**, *152*, 548–557. [CrossRef] [PubMed]

41. Gorrasi, G. Dispersion of halloysite loaded with natural antimicrobials into pectins: Characterization and controlled release analysis. *Carbohydr. Polym.* **2015**, *127*, 47–53. [CrossRef] [PubMed]

42. Cavallaro, G.; Lazzara, G.; Milioto, S.; Parisi, F. Halloysite nanotubes with fluorinated cavity: An innovative consolidant for paper treatment. *Clay Miner.* **2016**, *51*, 445–455. [CrossRef]

43. Du, M.; Guo, B.; Jia, D. Thermal stability and flame retardant effects of halloysite nanotubes on poly(propylene). *Eur. Polym. J.* **2006**, *42*, 1362–1369. [CrossRef]

44. Poggi, G.; Giorgi, R.; Toccafondi, N.; Katzur, V.; Baglioni, P. Hydroxide nanoparticles for deacidification and concomitant inhibition of iron-gall ink corrosion of paper. *Langmuir* **2010**, *26*, 19084–19090. [CrossRef] [PubMed]

45. Dong, C.; Cairney, J.; Sun, Q.; Maddan, O.L.; He, G.; Deng, Y. Investigation of $Mg(OH)_2$ nanoparticles as an antibacterial agent. *J. Nanopart. Res.* **2010**, *12*, 2101–2109. [CrossRef]

46. Giorgi, R.; Dei, L.; Ceccato, M.; Schettino, C.; Baglioni, P. Nanotechnologies for conservation of cultural heritage: Paper and canvas deacidification. *Langmuir* **2002**, *18*, 8198–8203. [CrossRef]

47. Poggi, G.; Toccafondi, N.; Melita, L.N.; Knowles, J.C.; Bozec, L.; Giorgi, R.; Baglioni, P. Calcium hydroxide nanoparticles for the conservation of cultural heritage: New formulations for the deacidification of cellulose-based artifacts. *Appl. Phys. A* **2014**, *114*, 685–693. [CrossRef]

48. Rodriguez-Navarro, C.; Vettori, I.; Ruiz-Agudo, E. Kinetics and mechanism of calcium hydroxide conversion into calcium alkoxides: Implications in heritage conservation using nanolimes. *Langmuir* **2016**, *32*, 5183–5194. [CrossRef] [PubMed]

49. Cheradame, H.; Ipert, S.; Rousset, E. Mass deacidification of paper and books. I: Study of the limitations of the gas phase processes. *Restorator* **2008**, *24*, 227–239. [CrossRef]

50. Baty, J.W.; Maitland, C.L.; Minter, W.; Hubbe, M.A.; Jordan-Mowery, S.K. Deacidification for the conservation and preservation of paper-based works: A review. *BioResources* **2010**, *5*, 1955–2023. [CrossRef]

51. Abdullayev, E.; Sakakibara, K.; Okamoto, K.; Wei, W.; Ariga, K.; Lvov, Y. Natural tubule clay template synthesis of silver nanorods for antibacterial composite coating. *ACS Appl. Mater. Interfaces* **2011**, *3*, 4040–4046. [CrossRef] [PubMed]

52. Cavallaro, G.; Lazzara, G.; Milioto, S.; Parisi, F. Hydrophobically modified halloysite nanotubes as reverse micelles for water-in-oil emulsion. *Langmuir* **2015**, *31*, 7472–7478. [CrossRef] [PubMed]

53. Joo, Y.; Sim, J.H.; Jeon, Y.; Lee, S.U.; Sohn, D. Opening and blocking the inner-pores of halloysite. *Chem. Commun.* **2013**, *49*, 4519–4521. [CrossRef] [PubMed]
54. Materic, V.; Hyland, M.; Jones, M.I.; Northover, B. High temperature carbonation of $Ca(OH)_2$: The effect of particle surface area and pore volume. *Ind. Eng. Chem. Res.* **2014**, *53*, 2994–3000. [CrossRef]
55. Materic, V.; Smedley, S.I. High temperature carbonation of $Ca(OH)_2$. *Ind. Eng. Chem. Res.* **2011**, *50*, 5927–5932. [CrossRef]
56. Blanco, I.; Abate, L.; Bottino, F.A.; Bottino, P. Thermal behaviour of a series of novel aliphatic bridged polyhedral oligomeric silsesquioxanes (POSSs)/polystyrene (PS) nanocomposites: The influence of the bridge length on the resistance to thermal degradation. *Polym. Degrad. Stab.* **2014**, *102*, 132–137. [CrossRef]

nanomaterials

MDPI

Article

Highly Efficient Near Infrared Photothermal Conversion Properties of Reduced Tungsten Oxide/Polyurethane Nanocomposites

Tolesa Fita Chala [1], Chang-Mou Wu [1,*], Min-Hui Chou [1], Molla Bahiru Gebeyehu [1] and Kuo-Bing Cheng [2]

[1]   Department of Materials Science and Engineering, National Taiwan University of Science and Technology, Taipei 10607, Taiwan, R.O.C; tolesafita@gmail.com (T.F.C.); bear200718@gmail.com (M.-H.C.); ycom5647@gmail.com (M.B.G.)

[2]   Department of Fiber and Composite Materials, Feng Chia University, Taichung 40724, Taiwan, R.O.C; kbcheng@fcu.edu

*   Correspondence: cmwu@mail.ntust.edu.tw; Tel.: +886-22-737-6530

Received: 21 June 2017; Accepted: 13 July 2017; Published: 22 July 2017

Abstract: In this work, novel WO_{3-x}/polyurethane (PU) nanocomposites were prepared by ball milling followed by stirring using a planetary mixer/de-aerator. The effects of phase transformation ($WO_3 \rightarrow WO_{2.8} \rightarrow WO_{2.72}$) and different weight fractions of tungsten oxide on the optical performance, photothermal conversion, and thermal properties of the prepared nanocomposites were examined. It was found that the nanocomposites exhibited strong photoabsorption in the entire near-infrared (NIR) region of 780–2500 nm and excellent photothermal conversion properties. This is because the particle size of WO_{3-x} was greatly reduced by ball milling and they were well-dispersed in the polyurethane matrix. The higher concentration of oxygen vacancies in WO_{3-x} contribute to the efficient absorption of NIR light and its conversion into thermal energy. In particular, $WO_{2.72}$/PU nanocomposites showed strong NIR light absorption of ca. 92%, high photothermal conversion, and better thermal conductivity and absorptivity than other WO_3/PU nanocomposites. Furthermore, when the nanocomposite with 7 wt % concentration of $WO_{2.72}$ nanoparticles was irradiated with infrared light, the temperature of the nanocomposite increased rapidly and stabilized at 120 °C after 5 min. This temperature is 52 °C higher than that achieved by pure PU. These nanocomposites are suitable functional materials for solar collectors, smart coatings, and energy-saving applications.

Keywords: nanocomposites; tungsten trioxide; photothermal conversion; polyurethane; near infrared ray

1. Introduction

The development of nanomaterials capable of effectively absorbing near-infrared (NIR) radiation with a broad working waveband has increasingly attracted attention from the viewpoint of energy economization for applications in photothermal therapy, solar collectors, smart windows, and optical filters [1–3]. Solar energy is the most promising sustainable energy source among the existing sources of renewable energy and it can be converted to thermal energy by using solar collectors. The efficiency of conversion of solar energy into heat energy is mainly determined by the optical properties of the surface of the absorber, which should show strong absorption of solar radiation [4–7]. Photothermal materials convert light to heat and are widely used as absorbers. The process of converting light to heat involves absorption of photon energy of a specific wavelength by the photothermal material followed by conversion into thermal energy under optical illumination [8,9]. In this regard, plasmonic nanoparticles have received considerable attention in the past decade because of their novel and tunable localized surface plasmon resonance in the visible and NIR regions [10]. Several plasmonic nanomaterials with NIR photothermal conversion properties have been studied in the field of biological medicine, especially in photothermal therapy. Typical examples include carbon-based materials such as carbon nanotubes [11], graphene, and reduced graphene oxide [12,13],which exhibit relatively low absorption coefficients in the NIR region, and noble metal nanostructures including Pd-based nanosheets [14], gold nanorods [11,15], gold nanoshells [16,17], and gold nanocages [18]. However, all of the above mentioned materials can only absorb certain frequencies of NIR radiation and are not effective in the entire NIR wavelength range [19]. In addition, gold is an expensive noble metal and the preparation of its nanostructures with NIR photothermal conversion properties usually requires accurate synthesis conditions or depositions process which is relatively expensive and limit its further application [20]. Thus, there is a need to develop a low-cost and simple method for the preparation of NIR-absorbing nanomaterials for photothermal applications.

Tungsten trioxide (WO_3) has been recognized as one of the most promising semiconductor materials for gas sensors, photovoltaic organic solar cells, and electrochromic and photocatalytic applications owing to its suitable band gap (2.62 eV) and environmental benignity [21–23]. It has been reported that transition metal oxides are interesting candidates for photothermal applications as they exhibit localized surface plasmon resonance (LSPR) [24]. Particularly, non-stoichiometric tungsten oxide (WO_{3-x}) nanocrystals are of significant interest because of their strong LSPR effect, which gives rise to strong photoabsorption peaks in the NIR region [25,26]. The strong NIR absorption properties of WO_{3-x} can be obtained by either reducing the oxygen content or adding ternary alkali metals [27–29]. Takeda and Adachi have reported the optical properties of reduced tungsten oxide under the H_2/N_2 gas atmosphere [28]. Recently, tungsten oxide-like monoclinic $W_{18}O_{49}$ ($WO_{2.72}$) has attracted considerable attention for various applications, such as transparent smart windows, photocatalysts, and imaging guided photothermal therapy [30–33] because of its unusual defect structure and intense NIR photoabsorption. These properties motivated us to develop novel WO_{3-x}/polyurethane (PU) nanocomposites for NIR photothermal conversion applications. PU is an attractive material and one of the most actively investigated polymers because of its outstanding properties, such as thermal and chemical stabilities, high impact strength, and easy processing [34,35]. Due to these advancements, PU has been widely applied in many fields, such as breathable waterproof textiles, functional coatings, paints, adhesives and foams, etc. [36–39]. The easy fabrication and low cost of PU composites is highly desirable for practical applications. Therefore, PU was selected as the matrix and incorporated with WO_{3-x} nanoparticles to provide more functions. It is believed that the good dispersion states of nanoparticles in polyurethane matrix using ball milling significantly affects the NIR absorption and photothermal conversion properties of the nanocomposites.

In this work, reduced tungsten oxide (WO_{3-x}) nanoparticles were first prepared from pure WO_3 by reduction in a tube furnace under a carbon monoxide atmosphere [40]. The WO_{3-x} nanoparticles were mixed with PU in a dimethylformamide (DMF) solution and then stirred by ball milling, followed by continuous stirring using a planetary mixer/de-aerator. Polyurethane was used as the matrix for WO_{3-x} nanoparticles because good dispersion states of the particles were achieved by ball milling them together. The dispersion state of a nanocomposite has a significant effect on its NIR absorption and photothermal conversion properties. Thermal properties of the resulting nanocomposites such as conductivity, absorptivity, and resistivity were investigated. In addition, the effects of phase transformations and weight fractions of reduced tungsten oxide on the photothermal performance and the thermal properties of the nanocomposites were studied.

2. Results and Discussion

2.1. Characterization of WO$_{3-x}$ Nanoparticles

Reduced tungsten oxide was prepared from pure tungsten trioxide via reduction under an atmosphere of carbon monoxide in a tube furnace. The reduction of mechanisms of tungsten oxides by CO could be expected to proceed as follows:

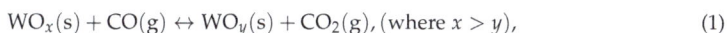

$$WO_x(s) + CO(g) \leftrightarrow WO_y(s) + CO_2(g), \text{(where } x > y\text{)}, \tag{1}$$

The overall procedure for the preparation of WO$_{3-x}$ and its nanocomposites (WO$_{3-x}$/PU) is schematically depicted in Figure 1.

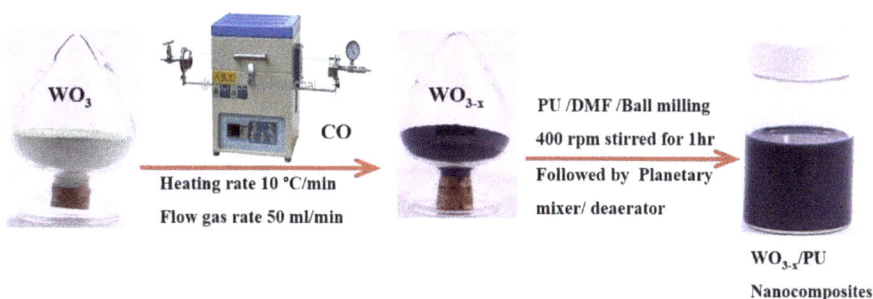

Figure 1. Schematic illustrations of the preparation of WO$_{3-x}$ and WO$_{3-x}$/PU nanocomposites.

Figure 2 shows the typical X-ray diffraction (XRD) patterns of pure WO$_3$ before reduction and its sub-oxides after reduction. The WO$_{3-x}$ phase undergoes phase transformations during reduction, which affects its stoichiometry. As shown in Figure 2, multiple peaks were observed in the XRD pattern of pure WO$_3$. At 550 °C thermal treatment, WO$_3$ exhibited an orthorhombic crystal structure (JCPDS-05-0364). However, increasing reduction temperature the multiple diffraction peaks of pure WO$_3$ at 23.3° became sharper and narrower, indicating the phase transformation of WO$_3$ nanoparticles. The intermediate phases WO$_{2.8}$ (JCPDS-05-0386), WO$_{2.72}$ (JCPDS-05-0392), and WO$_2$ (JCPDS-02-0414) were obtained at reduction temperatures of 600, 650, and 700 °C, respectively. A further increase in the reduction temperature to 1000 °C resulted in the formation of WC particles (JCPDS-02-1055). These results match well with those described in previous reports wherein the final product of reduction that was formed under an atmosphere of hydrogen and carbon was tungsten (W) [41,42]. In addition, based on the experimental data, the diffraction peaks of reduced tungsten oxide at 650 °C indicated the formation of a monoclinic phase. The interplanar spacing of 0.37 nm determined from the XRD pattern corresponded to that of the (010) plane of the monoclinic crystal structure of WO$_{2.72}$ (W$_{18}$O$_{49}$). Figure 3 shows the variation in the color of the WO$_{3-x}$ powder obtained after reduction at different temperatures under CO atmosphere. The color changed from yellow for WO$_3$ to dark blue for WO$_{3-x}$ and black for WC.

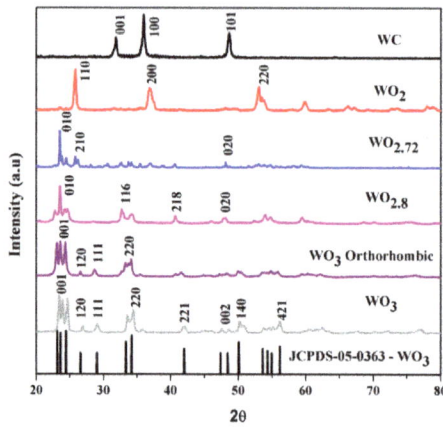

Figure 2. X-ray diffraction (XRD) patterns of pure tungsten trioxide before reduction and its sub-oxides after reduction under CO atmosphere.

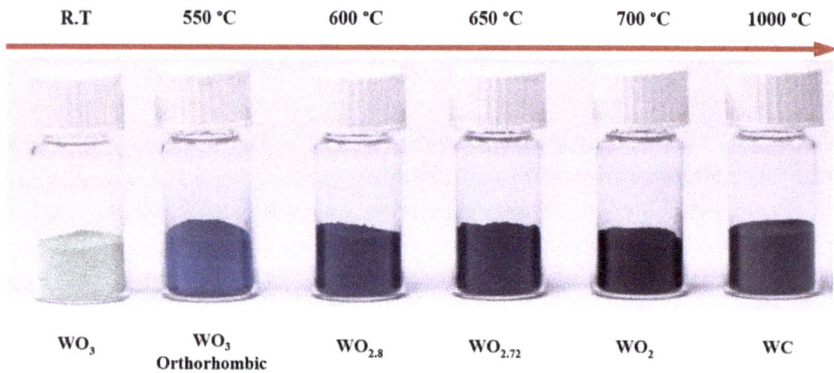

Figure 3. Variation in the color of reduced tungsten oxide powder prepared by reduction at different temperatures under CO atmosphere.

The chemical composition and the valence states of the prepared nanoparticles were examined by X-ray photoelectron spectroscopy (XPS). A complex energy distribution of W4f (where W is tungsten atoms, 4 is principal quantum number and f is core or inner atomic orbital) photoelectrons was obtained, as shown in Figure 4. The W4f core-level spectrum was fitted to three spin-orbit doublets corresponding to the three different oxidation states of W atoms. The $W4f_{5/2}$ and $W4f_{7/2}$ peaks at 37.86 and 35.77 eV, respectively, can be attributed to the +6 oxidation state of the W atoms. The second doublet at lower binding energy values of 34.8 and 36.9 eV arises due to the emissions from $W4f_{7/2}$ and $W4f_{5/2}$ core levels, respectively, corresponding to the +5 oxidation state of W. The third doublet observed at 33.8 and 35.75 eV corresponds to the tungsten +4 oxidation state. These three oxidation states are typically found in $WO_{2.72}$ nanomaterials [43,44].

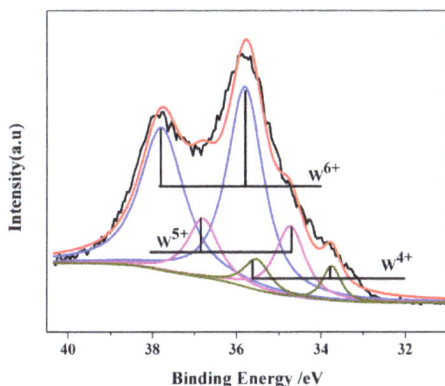

Figure 4. W4f X-ray photoelectron spectroscopy (XPS) spectra of $WO_{2.72}$ prepared by reduction at 650 °C under CO atmosphere.

The Field Emission Scanning Electron Microscopy (FESEM) image of the as-prepared $WO_{2.72}$ powder is shown in Figure 5. It is evident from Figure 5a that the sample consists of spherical particles with a relatively uniform size ranging from 57 to 106 nm. The morphologies and microstructures of the powders were further investigated by Transmission electron microscopy (TEM) and High-Resolution Transmission Electron Microscopy (HRTEM) analyses. The TEM analysis confirmed the formation of nanoparticles with an average particle size of 78 nm (Figure 5b), which was consistent with the particle size determined from the SEM analysis. The energy dispersive X-ray analysis (EDX) spectrum shown in Figure 5c confirmed the presence of W and O elements in the sample; the peaks corresponding to Cu originate from the copper grid substrate which was used for the TEM measurements. The spacing between adjacent lattice planes was found to be 0.37 nm from the HRTEM image (Figure 5d). This spacing corresponds to the (010) plane of monoclinic $WO_{2.72}$ phase, which is consistent with the results of XRD analysis.

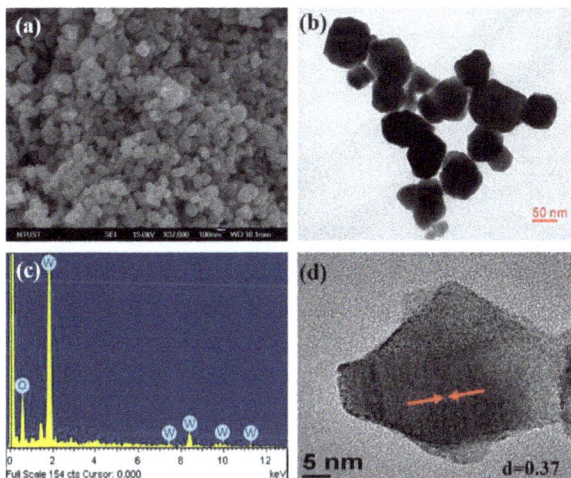

Figure 5. (a) Field Emission Scanning Electron Microscopy (FESEM); (b) Transmission electron microscopy (TEM); (c) Energy dispersive X-ray analysis (EDX) spectrum; and (d) High-Resolution Transmission Electron Microscopy (HRTEM) images of $WO_{2.72}$ powder.

2.2. Optical Properties and Morphologies of WO$_{3-x}$/PU Nanocomposites

The optical properties of the prepared nanocomposites were evaluated by using a UV-Vis-NIR spectrophotometer in the range of 300–2500 nm. The homogeneous sample solution prepared by the stirred ball milling method was spin-coated on quartz glass substrates at 800–1500 rpm. Figure 6a shows the UV-Vis-NIR transmittance spectra of the nanocomposites prepared with the same weight fraction (7 wt %) of reduced tungsten oxide. The transmittance values of WO$_{2.8}$/PU and WO$_{2.72}$/PU nanocomposites in the visible region (400–780 nm) were ca. 85.6 and 75%, respectively. The transmittance of the WO$_{2.72}$/PU nanocomposites was very low (8%) in the range of 780–2500 nm, which suggested that the WO$_{2.72}$/PU nanocomposites exhibit stronger NIR absorption (ca. 92%) compared to other nanocomposites. This is because the absorption of NIR radiation is closely related to the presence of free electrons or oxygen-deficiency-induced small polarons formed during the reduction process [45]. For comparison, the transmittance spectra of pure PU and WO$_3$/PU nanocomposites were also recorded. It was found that these nanocomposites exhibited high transmittance (>85%) in the entire UV-Vis-NIR region (300–2500 nm), indicating negligible photoabsorption in the NIR region. The WO$_2$/PU nanocomposites exhibited very low transmittance in the visible region in comparison with the WO$_{2.8}$/PU nanocomposites and a lower absorption in the NIR region compared to the WO$_{2.72}$/PU nanocomposites. This behavior may be attributed to the excessive reduction of WO$_2$ and consequently the lower number of free electrons or polarons that are formed [28]. The strong NIR photoabsorption of the WO$_{3-x}$/PU nanocomposites is attributed to the presence of WO$_{3-x}$ nanoparticles that efficiently absorb NIR radiation and convert it to thermal energy via the strong localized surface plasma resonance effect [25,46]. Thus, it is necessary to evaluate the effect of different weight fractions of WO$_{2.72}$ on the optical properties of WO$_{2.72}$/PU nanocomposites.

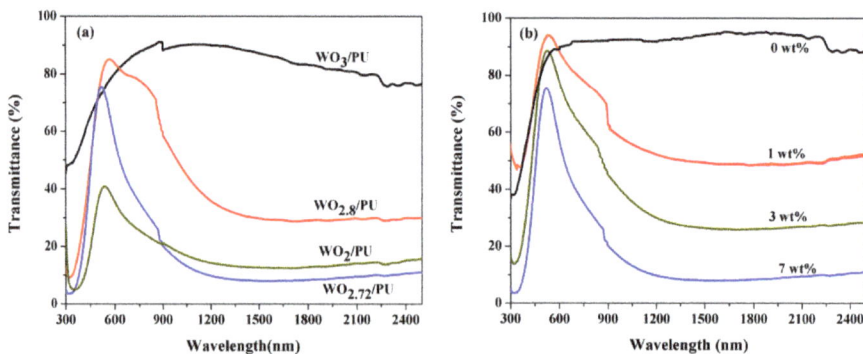

Figure 6. UV-Vis-NIR transmittance spectra of (**a**) WO$_3$/PU, WO$_{2.8}$/PU, and WO$_{2.72}$/PU nanocomposites prepared with tungsten oxide weight fraction of 7 wt %; and (**b**) WO$_{2.72}$/PU nanocomposites prepared with different weight fractions of WO$_{2.72}$ (0–7 wt %).

Figure 6b shows the UV-Vis-NIR-transmittance spectra of WO$_{2.72}$/PU nanocomposites prepared with different weight fractions of WO$_{2.72}$. According to the experimental results, as the weight fraction of WO$_{2.72}$ was increased from 0 to 7 wt %, the NIR transmittance of the nanocomposites decreased, indicating that the absorption of NIR radiation had significantly increased. This implies that the amount of nano-sized WO$_{2.72}$/PU required for efficient absorption of NIR radiation increases with increase in the content of reduced tungsten oxide (WO$_{2.72}$). The strong absorption of WO$_{3-x}$/PU nanocomposites in the NIR region motivated us to further study their morphologies and photothermal conversion properties.

The physical properties and efficiency of inorganic-organic composites can generally be enhanced by dispersing the inorganic filler in a polymer matrix [47]. The morphologies and dispersion states of reduced tungsten oxide nanoparticles in the PU matrix were investigated by FESEM and TEM techniques. Figure 7a,b show the typical FESEM images of pure PU and $WO_{2.72}$/PU nanocomposites prepared with 7 wt % of $WO_{2.72}$. While the pure PU sample exhibited a porous and rough surface, whereas the surface of $WO_{2.72}$/PU nanocomposites showed the presence of white spots owing to the incorporation of $WO_{2.72}$ nanoparticles in PU. In addition, the surface of the nanocomposite samples was uniform and smooth, and no cracks were observed. The TEM images also revealed that the $WO_{2.72}$ nanoparticles were well-dispersed in the PU matrix, and nanocomposites with particle sizes ranging from 20 to 40 nm (Figure 7c) were formed. The particle sizes of $WO_{2.72}$ in the PU matrix decreased significantly as a result of the grinding process. Ball-milling ground the materials to powder with shear stress and reduced the average particle sizes effectively, thus increasing the performance of these nanocomposites. However, the detail study of about the effects size, morphology of $WO_{2.72}$ before and after ball milling on NIR absorption properties of nanocomposites will be our next work.

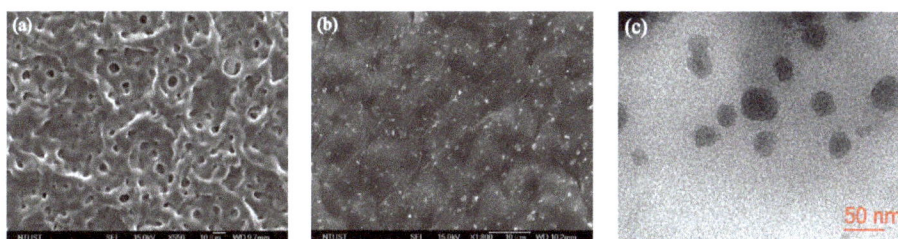

Figure 7. FESEM images of (**a**) pure polyurethane (PU); (**b**) $WO_{2.72}$/PU prepared with 7 wt % of $WO_{2.72}$; and (**c**) TEM image of $WO_{2.72}$/PU nanocomposites prepared with 7 wt % of $WO_{2.72}$.

2.3. NIR Photothermal Conversion and Thermal Properties of Nanocomposites

Figure 8a shows the effect of different weight fractions of $WO_{2.72}$ nanoparticles on the photothermal conversion characteristics of the corresponding nanocomposites. The results showed that the temperature of the nanocomposites increases rapidly with increase in the weight fraction of $WO_{2.72}$ from 0 to 7 wt %. The temperature changes (ΔT) were 44.8, 77, 87.5, and 96.5 °C for $WO_{2.72}$/PU nanocomposites prepared with 0, 1, 3, and 7 wt % $WO_{2.72}$, respectively, after light irradiation for 300 s. It is worth noting that the temperature for the 7 wt % sample increased rapidly and stabilized at 120 °C, which is 52 °C higher than the temperature attained by pure PU. For comparison, the NIR photothermal conversion properties of $WO_{2.8}$/PU and WO_3/PU nanocomposites were also examined at the same weight fraction of 7 wt % under identical conditions. These results are shown in Figure 8b. The temperature of the $WO_{2.72}$/PU nanocomposites increased rapidly to reach $\Delta T = 32.5$ °C after 10 s and $\Delta T = 58.9$ °C after 30 s, and gradually stabilized after 300 s. However, the ΔT for $WO_{2.8}$/PU, WO_3/PU, and pure PU were 41.9, 30.9, and 9.9 °C after 30 s, and stabilized at 86.6, 75.9, and 44.8 °C after 300 s, respectively. These results suggest that the $WO_{2.72}$/PU nanocomposites exhibit faster photothermal conversion rate than $WO_{2.8}$/PU, WO_3/PU, and pure PU. After an irradiation time of 30 s, the photothermal conversion rates were determined to be 108.20, 57.04, 44.43, and 19.94 °C min^{-1} for $WO_{2.72}$/PU, $WO_{2.8}$/PU, WO_3/PU, and pure PU, respectively. These results indicate that the photothermal conversion characteristics improve significantly when the oxygen content of the nanocomposites is reduced. The reduced oxygen content is responsible for the introduction of free electrons into the crystal structure and the resultant strong NIR absorption. Generally, during the irradiation process, the temperature of the nanocomposites initially increases sharply and then shows a gradual increase with increasing irradiation time. The photothermal conversion rate becomes lower with the further increase in the temperature owing to faster heat loss at higher temperatures [45,48–51]. The uniform dispersion and decrease in particle size of $WO_{2.72}$ powders in polyurethane matrix after grinding resulted in much higher photothermal conversion properties (56.5 °C after 10 s) under IR irradiation compared to those reported for $WO_{2.72}$ (36.5 °C after 10 s) [30,52]. In addition $WO_{2.72}$/PU nanocomposites also shows higher photothermal conversion performance than gold nanostars coated with polydopamine and graphene oxide modified PLA microcapsules containing gold nanoparticles, which the temperature increment was only 35–50 °C after 300 s [53,54]. The developed $WO_{2.72}$/PU nanocomposites exhibit extremely high photothermal conversions and the temperature reaches 120 °C after 5 min. To the best of our knowledge, this is the highest temperature that has been reported in the literature to date. Water can be efficient evaporated at such high temperature and, thus, the $WO_{2.72}$/PU nanocomposites show great potential applications in solar energy collectors, such as vapor power steam generators and others functional foams and coatings, such as warm/heat coatings, etc.

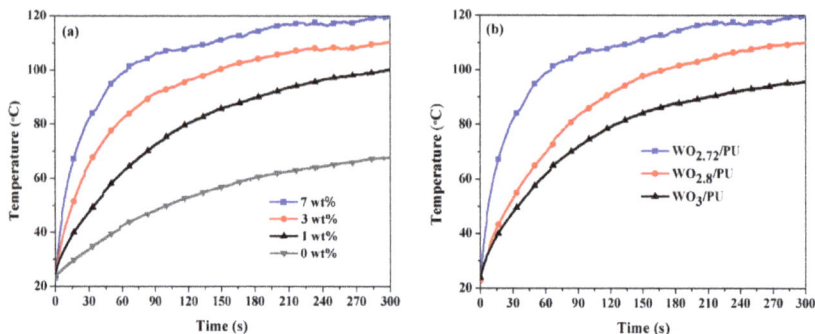

Figure 8. Temperature distribution of (**a**) $WO_{2.72}$/PU nanocomposites at different weight fractions (0–7 wt %) of $WO_{2.72}$ and (**b**) WO_3/PU, $WO_{2.8}$/PU, and $WO_{2.72}$/PU with 7 wt % as a function of time under infrared light irradiation.

Table 1 shows the thermal properties of the nanocomposites prepared with different contents of $WO_{2.72}$. The thermal conductivity and absorptivity of the nanocomposites was found to increase with increasing weight fraction of $WO_{2.72}$. Thus, the highest thermal conductivity and thermal absorptivity of 97.10 $mWm^{-1}K^{-1}$ and 496.80 $Ws^{1/2}m^{-2}K^{-1}$, respectively, was observed for the nanocomposite with 7 wt % of $WO_{2.72}$. However, the thermal resistance of the nanocomposites decreased with increasing weight fraction of $WO_{2.72}$. This is because thermal resistance (resistance to heat flow) is inversely proportional to thermal conductivity [55]. For comparison, the thermal properties of the WO_3/PU and the $WO_{2.8}$/PU nanocomposites were also studied under similar conditions at the same weight fraction of 7 wt %. These results are summarized in Table 2. It was found that the $WO_{2.72}$/PU nanocomposites showed the highest values of thermal conductivity and thermal absorptivity when compared to those of WO_3/PU nanocomposites owing to the presence of unusual oxygen defect structures.

Table 1. Thermal properties of $WO_{2.72}$/PU nanocomposites prepared with different weight fractions of $WO_{2.72}$.

Weight Fractions of $WO_{2.72}$ (wt %)	0	1	3	7
Thermal Conductivity ($mWm^{-1}K^{-1}$)	34.40	76.80	87.70	97.10
Thermal Absorption ($Ws^{1/2}m^{-2}K^{-1}$)	211.37	446.80	467.43	496.80
Thermal Resistance (m^2mkW^{-1})	11.40	9.50	7.75	7.2

Table 2. Thermal properties of WO$_3$/PU, WO$_{2.8}$/PU, and WO$_{2.72}$/PU nanocomposites prepared with 7 wt % tungsten oxide.

Parameter	WO$_3$/PU	WO$_{2.8}$/PU	WO$_{2.72}$/PU
Thermal Conductivity (mWm^{-1}K^{-1})	37.20	68.40	97.10
Thermal Absorption (Ws$^{1/2}$m^{-2}K^{-1})	153.60	384.83	496.80
Thermal Resistance (m^{2}mkW^{-1})	9.92	8.10	7.20

3. Materials and Methods

3.1. Materials

The tungsten trioxide slurry in water dispersion that was prepared by the ball milling process and had its particle size measured and confirmed as being typically 60–110 nm was obtained from Advanced Ceramics Nanotech Co. Ltd., Taipei, Taiwan. PU/DMF solution with 30 wt % solid content was purchased from Gabriel Advanced Materials Co. Ltd., Taipei, Taiwan.

3.2. Preparation of WO$_{3-x}$ Nanoparticles

The homogeneous yellow dispersion of WO$_3$ was separated by centrifugation and dried in an oven at 60 °C. Subsequently, the as-obtained WO$_3$ powder was reduced in a temperature-programmed tubular furnace under a carbon monoxide atmosphere at a heating rate of 10 °C min^{-1} and a carrier gas flow rate of 50 mL min^{-1}. The reduction of time of WO$_3$ was hold for 30 min and conducted under non-isothermal conditions in the temperature range of 550–1000 °C.

3.3. Preparation of WO$_{3-x}$/PU Nanocomposites

The WO$_{3-x}$/PU nanocomposites were prepared by a stirred ball milling process. The ball mill used was a high-performance batch-type stirred bead mill, Pulverisette classic line (Utek International Co. Ltd., Idar-oberstein, Germany). Yttrium-stabilized zirconia (95% ZrO$_2$, 5% Y$_2$O$_3$) stirred beads with a diameter of 5 mm were used. For the typical stirred bead milling process, various amounts (wt %) of the WO$_{3-x}$ powder were added to the PU-containing DMF solution and then dispersed in the stirred ball mill at an agitation speed of 400 rpm for 1 h. After ball milling, the solution was continuously stirred using a planetary mixer/deaerator (Mazerustar KK-250S Satellite Motion Mixer, Osaka, Japan) for 6 min to enable uniform dispersion and avoid the formation of air bubbles. Finally, a well-dispersed solution of WO$_{3-x}$/PU was obtained.

For the preparation of nanocomposite films, the resultant homogenized solution was casted onto a cleaned slide glass and dried at 60 °C in a vacuum oven to remove the solvent. The as-prepared nanocomposites with different contents of WO$_{2.72}$ (0, 1, 3, and 7 wt %) could be easily peeled off from the glass slides. For comparison, nanocomposites WO$_{2.8}$/PU and WO$_3$/PU with 7 wt % tungsten oxide were also prepared under identical conditions.

3.4. Characterization

X-ray diffraction (XRD) measurements were recorded with a BrukerD2 phaser diffractometer (Karlsruhe, Germany) using a Cu Kα radiation source in the scan range of 20–80° (2θ) at a scan rate of 2° min^{-1} and step size of 0.02°. The morphologies and sizes of the prepared powder samples and nanocomposites were studied by field emission scanning electron microscopy (FESEM, JSM6500F, JEOL, Tokyo, Japan) and transmission electron microscopy (TEM, JEOLJEM-2010, Tokyo, Japan). The surface compositions of the samples and the binding energies of the W4f core levels were determined by X-ray photoelectron spectroscopy (XPS, Perkin-Elmer PHI 5600, Waltham, MA, USA). The optical response of the coating was measured by using a spectrophotometer (JASCO V-670, Keith Link Technology, Jasco Analytical Instruments, Easton, MD, USA), which provided the transmittance in the UV, visible, and infrared ranges (300–2500 nm). In order to evaluate the photothermal conversion properties of the nanocomposites, the samples were irradiated with an infrared lamp at a power of 150 W and the temperature distribution was recorded by using a thermal imaging camera (FLIR P384A3-20, CTCT, Co. Ltd., Taipei, Taiwan). The thermal properties of the nanocomposites such as conductivity, absorptivity, and resistivity were measured by using the Alambeta instrument (Sensora Instruments, Thurmansbang, Germany).

4. Conclusions

In this work, WO_{3-x} nanoparticles were prepared from pure WO_3 via thermal reduction and, subsequently, novel WO_{3-x}/PU nanocomposites were prepared using the WO_{3-x} nanoparticles and PU by a simple stirred ball milling method. The particle size of the as-prepared nanocomposites was significantly reduced after ball milling. In addition, the FESEM, TEM, and UV-Vis-NIR absorption spectral analyses of the nanocomposites confirmed that the WO_{3-x} nanoparticles were well-dispersed in the PU matrix. The WO_{3-x} nanoparticles showed strong absorption of NIR light and rapid NIR photothermal conversion characteristics in the PU matrix. Among the different reduced tungsten oxide nanocomposites prepared in this work, $WO_{2.72}$/PU with 7 wt % $WO_{2.72}$ exhibited strong NIR light absorption, high thermal conductivity, high thermal absorptivity, and the highest photothermal conversion characteristics upon infrared light irradiation, owing to its unusual oxygen defect structure. The temperature change (ΔT) of the $WO_{2.72}$/PU nanocomposites increased rapidly and reached 32.5 °C after 10 s and 58.9 °C after 30 s, before gradually stabilizing at 96.5 °C after 300 s under infrared light irradiation. In addition, the photothermal conversion rate of the $WO_{2.72}$/PU nanocomposites was 108.20 °C min^{-1}, which is very fast when compared to that of $WO_{2.8}$/PU, WO_3/PU, and pure PU after an irradiation time of 30 s. These results indicate a quick conversion of the absorbed NIR light energy to local heat energy on the $WO_{2.72}$/PU nanocomposites.

Acknowledgments: The Ministry of Science and Technology of Taiwan, ROC, financially supported part of this work, under contract numbers: MOST 105-2218-E-035-006.

Author Contributions: Chang-Mou Wu supervised the experiments, and reviewed and revised the manuscript. Tolesa Fita Chala performed the experiments, analyzed the results, and wrote the manuscript. Min-Hui Chou provided assistance in the experimental work. Molla Bahiru Gebeyehu characterized the samples using SEM. Kuo-Bing Cheng provided the ball-milled WO_3 and PU materials for this work. All authors have read and approved the final manuscript.

Conflicts of Interest: The authors declare no conflict of interest.

References

1. Choi, J.; Moon, K.; Kang, I.; Kim, S.; Yoo, P.J.; Oh, K.W.; Park, J. Preparation of quaternary tungsten bronze nanoparticles by a thermal decomposition of ammonium metatungstate with oleylamine. *Chem. Eng. J.* **2015**, *281*, 236–242. [CrossRef]
2. Yan, M.; Gu, H.; Liu, Z.; Guo, C.; Liu, S. Effective near-infrared absorbent: Ammonium tungsten bronze nanocubes. *RSC Adv.* **2015**, *5*, 967–973. [CrossRef]
3. Guo, C.; Yin, S.; Dong, Q.; Sato, T. Near-infrared absorption properties of Rb_xWO_3 nanoparticles. *CrystEngComm* **2012**, *14*, 7727–7732. [CrossRef]
4. Guo, C.; Yin, S.; Huang, L.; Yang, L.; Sato, T. Discovery of an excellent ir absorbent with a broad working waveband: Cs_xWO_3 nanorods. *Chem. Commun.* **2011**, *47*, 8853–8855. [CrossRef] [PubMed]
5. Uhuegbu, C.C. Photo-thermal solar energy conversion device. *JETEAS* **2011**, *2*, 96–101.
6. Lizama-Tzec, F.; Macias, J.; Estrella-Gutiérrez, M.; Cahue-López, A.; Arés, O.; de Coss, R.; Alvarado-Gil, J.; Oskam, G. Electrodeposition and characterization of nanostructured black nickel selective absorber coatings for solar-thermal energy conversion. *J. Mater. Sci. Mater. Electron.* **2015**, *26*, 5553–5561. [CrossRef]
7. Li, B.; Nie, S.; Hao, Y.; Liu, T.; Zhu, J.; Yan, S. Stearic-acid/carbon-nanotube composites with tailored shape-stabilized phase transitions and light-heat conversion for thermal energy storage. *Energy Convers. Manag.* **2015**, *98*, 314–321. [CrossRef]

8. Hua, Z.; Li, B.; Li, L.; Yin, X.; Chen, K.; Wang, W. Designing a novel photothermal material of hierarchical microstructured copper phosphate for solar evaporation enhancement. *J. Phys. Chem. C* **2017**, *121*, 60–69. [CrossRef]

9. Chen, C.-J.; Chen, D.-H. Preparation and near-infrared photothermal conversion property of cesium tungsten oxide nanoparticles. *Nanoscale Res. Lett.* **2013**, *8*, 57. [CrossRef] [PubMed]

10. Chen, C.-J.; Chen, D.-H. Preparation of LaB$_6$ nanoparticles as a novel and effective near-infrared photothermal conversion material. *Chem. Eng. J.* **2012**, *180*, 337–342. [CrossRef]

11. Huang, X.; El-Sayed, M.A. Gold nanoparticles: Optical properties and implementations in cancer diagnosis and photothermal therapy. *J. Adv. Res.* **2010**, *1*, 13–28. [CrossRef]

12. Yang, K.; Zhang, S.; Zhang, G.; Sun, X.; Lee, S.-T.; Liu, Z. Graphene in mice: Ultrahigh in vivo tumor uptake and efficient photothermal therapy. *Nano Lett.* **2010**, *10*, 3318–3323. [CrossRef] [PubMed]

13. Yang, K.; Feng, L.; Shi, X.; Liu, Z. Nano-graphene in biomedicine: Theranostic applications. *Chem. Soc. Rev.* **2013**, *42*, 530–547. [CrossRef] [PubMed]

14. Huang, X.; Tang, S.; Liu, B.; Ren, B.; Zheng, N. Enhancing the photothermal stability of plasmonic metal nanoplates by a core-shell architecture. *Adv. Mater.* **2011**, *23*, 3420–3425. [CrossRef] [PubMed]

15. Horiguchi, Y.; Honda, K.; Kato, Y.; Nakashima, N.; Niidome, Y. Photothermal reshaping of gold nanorods depends on the passivating layers of the nanorod surfaces. *Langmuir* **2008**, *24*, 12026–12031. [CrossRef] [PubMed]

16. Liu, Z.; Song, H.; Yu, L.; Yang, L. Fabrication and near-infrared photothermal conversion characteristics of au nanoshells. *Appl. Phys. Lett.* **2005**, *86*, 113109. [CrossRef]

17. Ke, H.; Wang, J.; Dai, Z.; Jin, Y.; Qu, E.; Xing, Z.; Guo, C.; Yue, X.; Liu, J. Gold-nanoshelled microcapsules: A theranostic agent for ultrasound contrast imaging and photothermal therapy. *Angew. Chem.* **2011**, *123*, 3073–3077. [CrossRef]

18. Chen, J.; Wang, D.; Xi, J.; Au, L.; Siekkinen, A.; Warsen, A.; Li, Z.-Y.; Zhang, H.; Xia, Y.; Li, X. Immuno gold nanocages with tailored optical properties for targeted photothermal destruction of cancer cells. *Nano Lett.* **2007**, *7*, 1318. [CrossRef] [PubMed]

19. Li, G.; Zhang, S.; Guo, C.; Liu, S. Absorption and electrochromic modulation of near-infrared light: Realized by tungsten suboxide. *Nanoscale* **2016**, *8*, 9861–9868. [CrossRef] [PubMed]

20. An, Y.; Li, X.-F.; Zhang, Y.-J.; Tao, F.-J.; Zuo, M.-C.; Dong, L.-H.; Yin, Y.-S. Fabrication and application of a new-type photothermal conversion nano composite coating. *J. Nanosci. Nanotechnol.* **2015**, *15*, 3151–3156. [CrossRef] [PubMed]

21. Yang, C.; Chen, J.-F.; Zeng, X.; Cheng, D.; Cao, D. Design of the alkali-metal-doped WO$_3$ as a near-infrared shielding material for smart window. *Ind. Eng. Chem. Res.* **2014**, *53*, 17981–17988. [CrossRef]

22. Bai, H.; Su, N.; Li, W.; Zhang, X.; Yan, Y.; Li, P.; Ouyang, S.; Ye, J.; Xi, G. W$_{18}$O$_{49}$ nanowire networks for catalyzed dehydration of isopropyl alcohol to propylene under visible light. *J. Mater. Chem. A* **2013**, *1*, 6125. [CrossRef]

23. Zhou, J.; Ding, Y.; Deng, S.Z.; Gong, L.; Xu, N.S.; Wang, Z.L. Three-dimensional tungsten oxide nanowire networks. *Adv. Mater.* **2005**, *17*, 2107–2110. [CrossRef]

24. Wen, L.; Chen, L.; Zheng, S.; Zeng, J.; Duan, G.; Wang, Y.; Wang, G.; Chai, Z.; Li, Z.; Gao, M. Ultrasmall biocompatible WO$_{3-x}$ nanodots for multi-modality imaging and combined therapy of cancers. *Adv. Mater.* **2016**, *28*, 5072–5079. [CrossRef] [PubMed]

25. Manthiram, K.; Alivisatos, A.P. Tunable localized surface plasmon resonances in tungsten oxide nanocrystals. *J. Am. Chem. Soc.* **2012**, *134*, 3995–3998. [CrossRef] [PubMed]

26. Fang, Z.; Jiao, S.; Kang, Y.; Pang, G.; Feng, S. Photothermal conversion of W$_{18}$O$_{49}$ with a tunable oxidation state. *Chem. Open* **2017**. [CrossRef]

27. Lee, W.H.; Hwang, H.; Moon, K.; Shin, K.; Han, J.H.; Um, S.H.; Park, J.; Cho, J.H. Increased environmental stability of a tungsten bronze NIR-absorbing window. *Fibers Polym.* **2013**, *14*, 2077–2082. [CrossRef]

28. Takeda, H.; Adachi, K. Near infrared absorption of tungsten oxide nanoparticle dispersions. *J. Am. Ceram. Soc.* **2007**, *90*, 4059–4061. [CrossRef]

29. Wang, T.; Xiong, Y.; Li, R.; Cai, H. Dependence of infrared absorption properties on the Mo doping contents in M$_x$WO$_3$ with various alkali metals. *New J. Chem.* **2016**, *40*, 7476–7481. [CrossRef]

30. Guo, C.; Yin, S.; Yan, M.; Kobayashi, M.; Kakihana, M.; Sato, T. Morphology-controlled synthesis of $W_{18}O_{49}$ nanostructures and their near-infrared absorption properties. *Inorg. Chem.* **2012**, *51*, 4763–4771. [CrossRef] [PubMed]

31. Li, B.; Shao, X.; Liu, T.; Shao, L.; Zhang, B. Construction of metal/$WO_{2.72}$/rGO ternary nanocomposites with optimized adsorption, photocatalytic and photoelectrochemical properties. *Appl. Catal. B Environ.* **2016**, *198*, 325–333. [CrossRef]

32. Huo, D.; He, J.; Li, H.; Huang, A.J.; Zhao, H.Y.; Ding, Y.; Zhou, Z.Y.; Hu, Y. X-ray CT guided fault-free photothermal ablation of metastatic lymph nodes with ultrafine HER-2 targeting $W_{18}O_{49}$ nanoparticles. *Biomaterials* **2014**, *35*, 9155–9166. [CrossRef] [PubMed]

33. Zhang, Y.; Li, B.; Cao, Y.; Qin, J.; Peng, Z.; Xiao, Z.; Huang, X.; Zou, R.; Hu, J. $Na_{0.3}WO_3$ nanorods: A multifunctional agent for in vivo dual-model imaging and photothermal therapy of cancer cells. *Dalton Trans.* **2015**, *44*, 2771–2779. [CrossRef] [PubMed]

34. Chen, J.; Zhou, Y.; Nan, Q.; Ye, X.; Sun, Y.; Zhang, F.; Wang, Z. Preparation and properties of optically active polyurethane/TiO_2 nanocomposites derived from optically pure 1,1′-binaphthyl. *Eur. Polym. J.* **2007**, *43*, 4151–4159. [CrossRef]

35. Mirabedini, S.; Sabzi, M.; Zohuriaan-Mehr, J.; Atai, M.; Behzadnasab, M. Weathering performance of the polyurethane nanocomposite coatings containing silane treated TiO_2 nanoparticles. *Appl. Surf. Sci.* **2011**, *257*, 4196–4203. [CrossRef]

36. Chen, J.; Zhou, Y.; Nan, Q.; Sun, Y.; Ye, X.; Wang, Z. Synthesis, characterization and infrared emissivity study of polyurethane/TiO_2 nanocomposites. *Appl. Surf. Sci.* **2007**, *253*, 9154–9158. [CrossRef]

37. Chen, X.D.; Wang, Z.; Liao, Z.F.; Mai, Y.L.; Zhang, M.Q. Roles of anatase and rutile TiO_2 nanoparticles in photooxidation of polyurethane. *Polym Test.* **2007**, *26*, 202–208. [CrossRef]

38. Nikje, M.M.A.; Moghaddam, S.T.; Noruzian, M. Preparation of novel magnetic polyurethane foam nanocomposites by using core-shell nanoparticles. *Polímeros* **2016**. [CrossRef]

39. Meng, Q.B.; Lee, S.-I.; Nah, C.; Lee, Y.-S. Preparation of waterborne polyurethanes using an amphiphilic diol for breathable waterproof textile coatings. *Prog. Org. Coat.* **2009**, *66*, 382–386. [CrossRef]

40. Venables, D.S.; Brown, M.E. Reduction of tungsten oxides with carbon monoxide. *Thermochim. Acta* **1997**, *291*, 131–140. [CrossRef]

41. Venables, D.S.; Brown, M.E. Reduction of tungsten oxides with hydrogen and with hydrogen and carbon. *Thermochim. Acta* **1996**, *285*, 361–382. [CrossRef]

42. Fouad, N.; Attyia, K.; Zaki, M. Thermogravimetry of WO_3 reduction in hydrogen: Kinetic characterization of autocatalytic effects. *Powder Technol.* **1993**, *74*, 31–37. [CrossRef]

43. Leftheriotis, G.; Papaefthimiou, S.; Yianoulis, P.; Siokou, A. Effect of the tungsten oxidation states in the thermal coloration and bleaching of amorphous WO_3 films. *Thin Solid Films* **2001**, *384*, 298–306. [CrossRef]

44. Jeon, S.; Yong, K. Direct synthesis of $W_{18}O_{49}$ nanorods from W_2N film by thermal annealing. *Nanotechnology* **2007**, *18*, 245602. [CrossRef]

45. Xu, W.; Tian, Q.; Chen, Z.; Xia, M.; Macharia, D.K.; Sun, B.; Tian, L.; Wang, Y.; Zhu, M. Optimization of photothermal performance of hydrophilic $W_{18}O_{49}$ nanowires for the ablation of cancer cells in vivo. *J. Mater. Chem. B* **2014**, *2*, 5594–5601. [CrossRef]

46. Chen, Z.; Wang, Q.; Wang, H.; Zhang, L.; Song, G.; Song, L.; Hu, J.; Wang, H.; Liu, J.; Zhu, M. Ultrathin pegylated $W_{18}O_{49}$ nanowires as a new 980 nm-laser-driven photothermal agent for efficient ablation of cancer cells in vivo. *Adv. Mater.* **2013**, *25*, 2095–2100. [CrossRef] [PubMed]

47. Kim, D.; Jang, M.; Seo, J.; Nam, K.-H.; Han, H.; Khan, S.B. UV-cured poly (urethane acrylate) composite films containing surface-modified tetrapod ZnO whiskers. *Compos. Sci. Technol.* **2013**, *75*, 84–92. [CrossRef]

48. Xu, W.; Meng, Z.; Yu, N.; Chen, Z.; Sun, B.; Jiang, X.; Zhu, M. Pegylated Cs_xWO_3 nanorods as an efficient and stable 915 nm-laser-driven photothermal agent against cancer cells. *RSC Adv.* **2015**, *5*, 7074–7082. [CrossRef]

49. Jia, G.Z.; Lou, W.K.; Cheng, F.; Wang, X.L.; Yao, J.H.; Dai, N.; Lin, H.Q.; Chang, K. Excellent photothermal conversion of core/shell CdSe/Bi_2Se_3 quantum dots. *Nano Res.* **2015**, *8*, 1443–1453. [CrossRef]

50. Mebrouk, K.; Debnath, S.; Fourmigué, M.; Camerel, F. Photothermal control of the gelation properties of nickel bis(dithiolene) metallogelators under near-infrared irradiation. *Langmuir* **2014**, *30*, 8592–8597. [CrossRef] [PubMed]

51. Zhong, W.; Yu, N.; Zhang, L.; Liu, Z.; Wang, Z.; Hu, J.; Chen, Z. Synthesis of cus nanoplate-containing pdms film with excellent near-infrared shielding properties. *RSC Adv.* **2016**, *6*, 18881–18890. [CrossRef]

52. Guo, C.; Yin, S.; Huang, Y.; Dong, Q.; Sato, T. Synthesis of $W_{18}O_{49}$ nanorod via ammonium tungsten oxide and its interesting optical properties. *Langmuir* **2011**, *27*, 12172–12178. [CrossRef] [PubMed]

53. Li, D.; Zhang, Y.; Wen, S.; Song, Y.; Tang, Y.; Zhu, X.; Shen, M.; Mignani, S.; Majoral, J.-P.; Zhao, Q. Construction of polydopamine-coated gold nanostars for CT imaging and enhanced photothermal therapy of tumors: An innovative theranostic strategy. *J. Mater. Chem. B* **2016**, *4*, 4216–4226. [CrossRef]

54. Jin, Y.; Wang, J.; Ke, H.; Wang, S.; Dai, Z. Graphene oxide modified PLA microcapsules containing gold nanoparticles for ultrasonic/CT bimodal imaging guided photothermal tumor therapy. *Biomaterials* **2013**, *34*, 4794–4802. [CrossRef] [PubMed]

55. Matusiak, M. Investigation of the thermal insulation properties of multilayer textiles. *Fibres Text. East. Eur.* **2006**, *14*, 98–102.

nanomaterials

MDPI

Article

Biodegradable FeMnSi Sputter-Coated Macroporous Polypropylene Membranes for the Sustained Release of Drugs

Jordina Fornell [1], Jorge Soriano [2], Miguel Guerrero [1,*], Juan de Dios Sirvent [1], Marta Ferran-Marqués [1], Elena Ibáñez [2], Leonardo Barrios [2], Maria Dolors Baró [1], Santiago Suriñach [1], Carme Nogués [2,*], Jordi Sort [1,3,*] and Eva Pellicer [1]

[1] Departament de Física, Universitat Autònoma de Barcelona, E-08193 Bellaterra, Cerdanyola del Vallès, Spain; jordina.fornell@uab.cat (J.F.); juandesirvent@gmail.com (J.d.D.S); marta.ferranmarques@gmail.com (M.F.-M.); dolors.baro@uab.es (M.D.B.); Santiago.Surinyach@uab.cat (S.S.); eva.pellicer@uab.cat (E.P.)

[2] Departament de Biologia Cel·lular, Fisiologia i Immunologia, Universitat Autònoma de Barcelona, E-08193 Bellaterra, Cerdanyola del Vallès, Spain; jorge.soriano@uab.cat (J.S.); elena.ibanez@uab.cat (E.I.); lleonard.barrios@uab.cat (L.B.)

[3] Institució Catalana de Recerca i Estudis Avançats (ICREA), Pg. Lluís Companys 23, E-08010 Barcelona, Spain

*   Correspondence: Miguel.Guerrero@uab.cat (M.G.); Carme.Nogues@uab.cat (C.N.); Jordi.Sort@uab.cat (J.S.); Tel.: 0034-935813247 (M.G.)

Received: 6 June 2017; Accepted: 22 June 2017; Published: 24 June 2017

Abstract: Pure Fe and FeMnSi thin films were sputtered on macroporous polypropylene (PP) membranes with the aim to obtain biocompatible, biodegradable and, eventually, magnetically-steerable platforms. Room-temperature ferromagnetic response was observed in both Fe- and FeMnSi-coated membranes. Good cell viability was observed in both cases by means of cytotoxicity studies, though the FeMnSi-coated membranes showed higher biodegradability than the Fe-coated ones. Various strategies to functionalize the porous platforms with transferrin-Alexa Fluor 488 (Tf-AF488) molecules were tested to determine an optimal balance between the functionalization yield and the cargo release. The distribution of Tf-AF488 within the FeMnSi-coated PP membranes, as well as its release and uptake by cells, was studied by confocal laser scanning microscopy. A homogeneous distribution of the drug within the membrane skeleton and its sustained release was achieved after three consecutive impregnations followed by the addition of a layer made of gelatin and maltodextrin, which prevented exceedingly fast release. The here-prepared organic-inorganic macroporous membranes could find applications as fixed or magnetically-steerable drug delivery platforms.

Keywords: biodegradable material; porous membrane; drug delivery; hybrid material

1. Introduction

The most widely used forms of drug delivery are oral ingested tablets, injection, transdermal patches, and implantable fixtures. Once administered, controlling the rate at which the drug is released in the body and its transport to the desired location becomes crucial. One very important requirement is that the drug must display some preferential selectivity to the target tissue or organ in order to avoid side effects due to systemic exposure of healthy organ systems to the drug. This is of utmost importance in cancer treatment and has led to the discovery of many innovative approaches for drug delivery, frequently taking advantage of concepts from nanotechnology [1–3].

Polymeric materials are particularly amenable for medical applications because they offer an interesting combination of properties, such as diffusivity, permeability, biocompatibility, and solubility.

In addition, they can respond to pH or temperature changes. The diffusion, dissolution, permeation, and swelling characteristics of polymeric materials have been exploited to obtain a constant release of entrapped molecules [4]. Pore-containing polymer systems have emerged as a new class of efficient drug delivery platforms [5–7]. By controlling the size and shape of the material, the number of "reservoirs" and their volume, the wall thickness, and the surface chemistry, permeability, and resorption rate, porous polymers enable better control of the vectorisation and the release kinetics of the drug.

Traditional, degradation-resistant polymers for implantable medical devices, such as polytetrafluoroethylene (PTFE), polyether ether ketone (PEEK), poly(methyl methacrylate) (PMMA), polyethylene, and silicones have been frequently associated with clinical problems, including microbial biofilm formation and medical device related infection, encrustation, poor biocompatibility, and low lubricity [8–10]. For this reason, bioresorbable polymer implants, employed in tissue engineering, absorbable sutures, arterial stent coatings and implants for controlled drug delivery have been devised.

For the therapeutic delivery of molecules, the polymer can act as a passive entity that hosts the drug and ensures its sustained release (e.g., subcutaneous patch) or can be envisioned as a micro-robotic platform that enables both transport of the molecule to the target site and its subsequent release. Guidance of micro-robotic platforms under the action of magnetic fields is a promising approach because they are capable of penetrating most materials with minimal interaction, and are almost harmless to human beings even at relatively high field strengths (provided no high-frequency AC magnetic fields are used). Magnetic fields have been successfully used to wirelessly manipulate microdevices of various sizes and shapes [11,12]. Polymers have been already combined with magnetic materials such as magnetic nanoparticles (MNPs) to obtain carriers consisting of an inorganic core, composed by the MNP, and a biocompatible polymer coating, in which functional ligands (e.g., targeting agents, therapeutic agents, permeation enhancers) can be integrated. Magnetite (Fe_3O_4) and maghemite (γ-Fe_2O_3) MNPs have been mostly used as cores. Magnetic layer-by-layer (LBL) capsules consisting of multilayer polymeric capsules containing MNPs are another example of polymer-magnetic material hybrids [13].

Here, we propose the combination of a porous polymeric backbone (polypropylene, PP) and a sputtered ferromagnetic Fe-based metallic layer for biomedical applications. There are a few examples of polymeric membranes sputtered with single metals like Pt and Ni for different purposes [14,15]. In this work, a ternary alloy (FeMnSi) is sputtered onto the biodegradable, macroporous PP membrane [16] for the first time. Among ferromagnetic metallic materials, Fe-based ones are very convenient because of their cytocompatibility and magnetic response. During the last few years, Fe has been typically co-alloyed with other elements, such as Mn, Pd, Si, C, or P, to increase the degradation rate of Fe-structures without compromising their mechanical integrity and their cyto- and hemocompatibility [17–20]. It is, therefore, the aim of this work to produce porous, fully-biodegradable, magnetic organic-inorganic polymer-metal hybrids that could be used either as implantable patches or as steerable micro-robotic platforms if appropriately miniaturized. With this purpose, commercial macroporous PP membranes (pore size ~450 nm) were coated with FeMnSi to confer magnetic properties to the membranes whilst avoiding clogging of the pores. The microstructure, morphology, biodegradability, and cytocompatibility of the FeMnSi-coated membranes were studied and compared with those of Fe-coated membranes (taken as a reference). These magnetic organic-inorganic hybrids were loaded with Transferrin-Alexa Fluor 488 (Tf-AF488) to investigate its distribution within the porous skeleton, its kinetics of release, and its cellular uptake.

2. Materials and Methods

2.1. Preparation and Characterization of the Macroporous Fe- and FeMnSi-Coated PP Membranes

2.1.1. Sputtering of Fe and FeMnSi on Macroporous PP Membranes

Pure Fe and FeMnSi thin films were grown by magnetron sputtering using a ATC Orion 5 from AJA International equipment (N. Scituate, MA, USA). The films were sputtered on PP membranes of 47 mm in diameter and a pore size ~450 nm (purchased from Sigma-Aldrich, St. Louis, MO, USA). The sputtering process to coat the PP membranes with Fe was carried out at 100 W in DC using a Fe target (99.9% purity) for 15 min. To coat the PP membranes with a targeted composition of Fe-14Mn-4Si (wt %), simultaneous sputtering from a Fe target at 200 W in RF and a Fe-30Mn-6Si-1Pd (99.95% purity) target at 100 W in DC was carried out for 5 min. To determine the thickness of the resulting coatings, Fe and FeMnSi thin films were sputtered on flat Si substrates under the same conditions as for the membranes. The thickness was assessed with a 3D Optical Surface Metrology System (DCM 3D) from Leica which combines confocal and interferometry technology. The deposition time was adjusted to obtain films with similar thickness and magnetization response.

2.1.2. Structural and Magnetic Characterization of the Fe- and FeMnSi-coated PP Membranes

The morphology and chemical composition of the materials were investigated by field emission scanning electron microscopy (FE-SEM, Zeiss, Oberkochen, Germany) using a Merlin Zeiss microscope equipped with an energy dispersive X-ray (EDX) detector. Transmission Electron Microscopy (TEM) characterization was carried out on a JEOL JEM-2011 (Tokyo, Japan) microscope operated at 200 kV. For TEM observations, cross-sections were prepared by embedding the FeMnSi-coated PP membrane in an epoxy resin followed by cutting thin slices with an ultramicrotome (Leica EM UC6, Leica Microsystems Ltd., Milton Keynes, UK) using a 35° diamond knife at room temperature (RT).

X-ray diffraction (XRD) patterns were recorded on a Philips X'Pert diffractometer using Cu Kα radiation, in the 40°–60° 2θ range. For XRD measurements, the targeted compositions were sputtered following the same conditions on flat Si/Ti/Cu substrates for 15 min to increase the amount of material on the diffraction focal plane. Hysteresis loops were acquired at RT using a vibrating sample magnetometer (VSM) from Micro-Sense (from LOT-Quantum Design, Darmstadt, Germany), varying the magnetic field from −20 to 20 kOe.

2.1.3. Biodegradability Studies

In order to evaluate the release of metal cations from the samples, coated PP membranes with an exposed geometrical area of 0.5 cm^2 were introduced in sterilized plastic containers filled with Hank's balanced salt solution (HBSS, Sigma-Aldrich, Nuaillé, France). A blank PP membrane (without coating) was also immersed in HBSS for comparison purposes. The containers were then sealed and placed in a thermostatic bath at 37 °C. After 15 days the solution was pipetted off and placed in tube tests for inductively coupled plasma-mass spectroscopy (ICP-MS) measurements, which were carried in an Agilent 7500 ce apparatus (Midland, ON, Canada). The ratio between the HBSS volume to geometrical sample surface was 25 mL cm^{-2}, in accordance with the ASTM-G31-72 standard [21]. Notice that this is an upper threshold since the porosity (i.e., the real surface area in contact with the solution) is not taken into account.

2.1.4. Cell Culture

Cellular studies were conducted using a tumoral human mammary epithelial cell line, SKBR-3 American Type Culture Collection (ATCC, Manassas, VA, USA). The cells were cultured in McCoy's 5A modified medium (Gibco, Paisley, UK) supplemented with 10% fetal bovine serum (FBS, Gibco, Paisley, UK). The cell line was maintained at 37 °C and 5% CO$_2$ (standard conditions). The culture medium was refreshed every 72 h.

2.1.5. MTT Assay

To evaluate a possible cytotoxic effect of the Fe- and FeMnSi-coated PP membranes, cell viability (metabolic activity) was tested by MTT (3-(4,5-dimethylthiazol-2-yl)-2,5-diphenyltetrazolium bromide) assay (Sigma-Aldrich, Nuaillé, France).

A total of 10,000 cells were seeded in each well of a 24-well dish in order to carry out the experiment. At 24 h after seeding, pieces of 3 mm × 3 mm (9 mm^2) were cut from the Fe and FeMnSi-coated PP membranes, cleaned with ethanol, and introduced individually in Millicell$^®$ 8 μm cell culture plate inserts (Millipore, Bedford, MA, USA), which were then placed inside the wells of the 24-well dish. This approach was used to prevent the membranes to crush the cells. After one week, the MTT assay was performed and the absorbance was recorded at 540 nm using a Victor 3 Multilabel Plate Reader (PerkinElmer, Waltham, MA, USA). For each treatment, viability was compared with that of cells incubated without the membranes (control). Three independent experiments were performed for each condition.

2.2. Functionalization of the Macroporous FeMnSi-Coated PP Membranes

In order to functionalize the FeMnSi porous membranes, 9 mm^2 cuts were loaded with Transferrin-Alexa Fluor 488 (Tf-AF488, 1 mg mL^{-1}, Molecular Probes, Eugene, OR, USA), maltodextrin (25 mg L^{-1}) and gelatin from bovine skin (0.25 mg mL^{-1}) purchased from Sigma-Aldrich (Nuaillé, France). Transferrin was used in order to target SKBR-3 cells as the transferrin receptor has shown to be overexpressed in SKBR-3 cells [22]. The maltodextrin and gelatin were used as co-gelling agents. Fifteen μL of the previous solution was added dropwise onto the surface of the coated membranes. The wetted FeMnSi-PP porous membranes were placed in a vacuum desiccator (0.5 mbar) for 30 min at RT to force the gel to penetrate into the network. This process was done once or repeated three consecutive times. After each impregnation, the membranes were rinsed with HBSS. In some cases, a drop of maltodextrin and gelatin was placed on top of the loaded membranes to partially block and hence delay the release of Tf-AF488. Table 1 lists the steps followed for each sample. Sample 1 was loaded with Tf-AF488 once and dried in vacuum. Sample 2 was also loaded once, rinsed in HBSS, and dried in vacuum. Subsequently, a stopper made of gelatine and maltodextrin was added on top of the membrane to prevent a fast release. Sample 3 was loaded three times with Tf-AF488, washed with HBSS after the last load, dried in vacuum and blocked twice with the stopper.

Table 1. Summary of the different procedures (in chronological order) followed to incorporate Tf-AF488 in the FeMnSi-coated PP membranes for three different cases (denoted as Samples 1, 2 and 3).

Steps (# of Times)	Sample 1	Sample 2	Sample 3
Loading with Tf-AF488	1	1	3
Cleaning with HBSS	0	1	1
Drying under vacuum	1	1	1
Stopper	0	1	2
Drying under vacuum	0	1	1

2.3. Characterization and Performance of the Tf-AF488-Loaded Membranes

2.3.1. Distribution of Tf-AF488

To evaluate the distribution of the Tf-AF488 inside the membranes, the photoluminescence response of the materials was studied by confocal scanning laser microscopy (CSLM). The samples were mounted on Ibidi culture dishes (Ibidi GmbH, Martinsried, Germany). For 3D reconstruction, the samples were visualized with a TCS-SP5 CSLM microscope (Leica Microsystems CMS GmbH, Mannheim, Germany) using a Plan Apochromat (Leica, Mannheim, Germany) 20.0×/0.70 (dry) objective. The membranes were excited with a green diode laser (488 nm) and the fluorescence of the Tf-AF488 was detected in the 500–560 nm range. The PP membranes (non-fluorescent) were imaged in the reflected light mode from an argon laser (488 nm) and subsequently detected in the 480–495 nm range. Stacks were collected at every 0.5 µm along the material's thickness. The three-dimensional images were processed by using Surpass Module in Imaris ×64 v. 6.4.0 software (Bitplane; Zurich, Switzerland). A set of 15 regions of interest (ROIs) of 400 µm² each was used to analyze the mean fluorescence intensity of the samples in relation to the emission wavelength.

2.3.2. Release of Tf-AF488

The kinetics of Tf-AF488 release from the loaded membranes was studied by monitoring the fluorescence signal of Tf-AF488 using a fluorimeter (Varian Cary Eclipse, Palo Alto, CA, USA). To carry out the fluorescence measurements, 1 mL of HBSS and one single Tf-AF488-loaded membrane piece (9 mm²) were placed inside a cuvette. A stirring bar was added to homogenize the solution. Samples were excited at 488 nm and emission was measured every 5 min at 517 nm, as this was the point of maximum emission of Tf-AF488. Two replicates for each condition were prepared to evaluate the reproducibility of the absorption and release of Tf-AF488. After fluorometric measurements, the membranes were imaged by confocal microscopy to confirm the total release of Tf-AF488.

2.3.3. Tf-AF488 Uptake by Cells

A total of 50,000 cells were seeded in each well of four-well dishes. After 24 h, the membranes previously charged with Tf-AF488 were added in a Millicell® 8 µm cell culture plate insert and images of the cells were taken at different times of incubation using a fluorescence microscope (Olympus IX70, Olympus, Hamburg, Germany).

3. Results and Discussion

3.1. Characterization of the Fe- and FeMnSi-Coated Membranes

3.1.1. Structural and Morphological Characterization

To verify that the porosity of the resulting platforms was not compromised by the sputtering process, SEM imaging was carried out on the Fe- and FeMnSi-coated membranes. Pores were clearly visible after Fe sputtering on the PP membrane at 100 W for 15 min (see Figure 1a). A representative SEM image of the PP membrane co-sputtered from Fe and FeMnSiPd targets for 5 min is presented in Figure 1b. In this case, the co-sputtering process (i.e., applied voltage), was adjusted to render a nominal composition of Fe-14Mn-4Si (in wt %). Additionally, the deposition time was adjusted to obtain nanocoatings of similar thickness (~70 nm). Although the co-sputtering resulted in a decrease of the porosity degree (i.e., slightly smaller pores and thicker pore walls), the coated membrane is still rather porous (see the inset of Figure 1b). It has been reported that the addition of Mn within the solubility limit of Fe reduces the standard electrode potential of Fe, thereby making it more susceptible to corrosion [23–25]. The composition of the co-sputtered film coating the PP membrane measured by EDX.

The influence of the alloying elements (Mn and Si) on the lattice structure of the resulting nanocoatings was investigated by X-ray diffraction (Figure S1). Note that Fe and FeMnSi were sputtered onto a non-porous Cu-coated Si substrate instead of the porous PP membrane with the aim to increase the amount of material on the diffraction focal plane. For the FeMnSi layer, the slight shift of the (110) diffraction peak characteristic of the body-centered cubic (bcc) phase of Fe towards lower diffraction angles is indicative of the substitutional solid solution. No additional peaks belonging to Mn, Si, or oxides could be observed, corroborating the formation of a single-phase nanocoating.

To gain further information about the structure and morphology of the coated membranes, TEM imaging was carried out on the FeMnSi-coated membrane. In the low magnification TEM image presented in Figure 2a, a discontinuous Fe-based layer can be observed at the outer side of the PP membrane. Furthermore, it can also be observed that the FeMnSi penetrates inside the PP membrane up to ~2 µm. A higher magnification image (Figure 2b) reveals that, besides the thicker FeMnSi layer grown on top of the membrane, nanometer-sized crystals surrounding the pores can be observed within the first few hundred nanometers from the surface. From Figure 2c an average thickness of 40 nm of the top layer can be measured. The selected area electron diffraction (SAED) image (inset of Figure 2c) confirms the crystalline nature of the outer Fe-based layer. The interplanar distances' values are slightly larger than the theoretical bcc-Fe interplanar distances due to the presence of Mn atoms dissolved in the Fe lattice. The spots/rings of the SAED pattern belong to (110), (200), (211), and (310) planes.

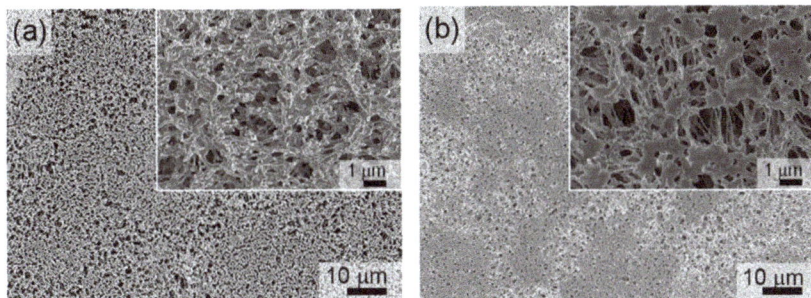

Figure 1. SEM image of PP membranes coated with (**a**) Fe and (**b**) Fe-14Mn-4Si alloy. The insets are higher magnification images.

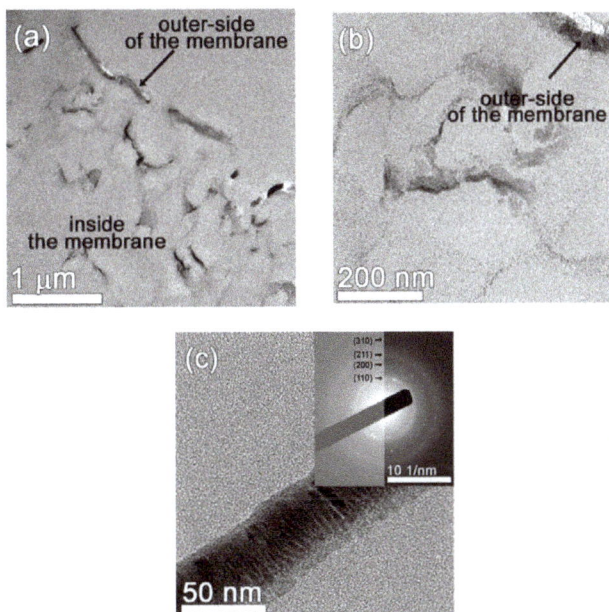

Figure 2. Cross-section TEM image of the FeMnSi-coated PP membrane (**a**) showing the outer and inner FeMnSi coating (**b**) higher magnification image taken inside the membrane; and (**c**) the FeMnSi layer grown on top of the membrane of ~40 nm in thickness. The inset belongs to the corresponding SAED pattern.

3.1.2. Magnetic Characterization

The hysteresis loops of the PP membranes coated with Fe and FeMnSi (Figure 3) exhibit similar magnetization saturation (M_S) values (~2 emu g^{-1}). Note that the overall mass of the hybrid material (membrane plus coating) was taken into account for normalization; thus, resulting in a much smaller M_S value than expected ($M_{S,Fe}$ = 217 emu g^{-1}). However, the coercivity (H_c) of the Fe-14Mn-4Si alloy (~90 Oe) is lower than that of pure Fe (~200 Oe). According to these values, they can be classified as soft-magnetic materials. The soft magnetic behavior of these coatings (in particular, their relatively high magnetic susceptibility), ensures that they might eventually be manipulated (magnetically-guided) using moderate external magnetic fields (easy to be applied using conventional electromagnets, for example).

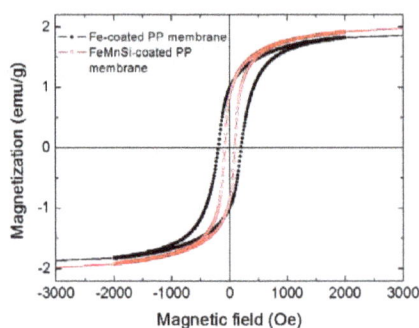

Figure 3. Room-temperature hysteresis loops of Fe- and FeMnSi-coated PP membranes.

3.1.3. Biodegradability

The nature and quantity of metallic ions released from the Fe- and FeMnSi-coated membranes after incubation in HBSS solution for 15 days at 37 °C (Table 2) were evaluated by ICP analysis. For comparison purposes, a PP membrane (without coating) was also immersed in HBSS solution. The concentration of Fe ions released from the Fe-coated PP membrane and from the pristine PP membrane was always below the detection limit of the technique. Meanwhile, both Mn and Fe (in much larger amount) were detected in the HBSS solution in contact with the FeMnSi-coated PP membranes. These results indicated that the FeMnSi nanocoating degraded faster than pure Fe, probably as a result of the Mn addition. Figure S2 displays SEM images of the Fe- and FeMnSi- coated PP membranes after immersion in HBSS for 15 days. The morphology of the Fe-coated PP membrane remained almost the same as before the immersion (Figure 1), keeping the porous structure unaltered. On the contrary, the morphology of the FeMnSi-coated PP membranes changed after incubation, showing a spongy appearance.

Table 2. Concentration of Fe and Mn released from the Fe- and FeMnSi-coated PP membranes after immersion in HBSS solution for 15 days at 37 °C.

Membrane Type	[Fe] (µg L^{-1})	[Mn] (µg L^{-1})
PP membrane	<10	<10
Fe	<10	<10
Fe-14Mn-4Si	19 ± 6	17 ± 2

3.1.4. Cytotoxicity Results

To assess cell viability after one week of culture in the presence of the Fe- and FeMnSi-coated PP membranes, the MTT cytotoxicity assay was carried out. Since the values obtained were slightly higher for cell cultures in presence of Fe- and FeMnSi-coated PP membranes than for control cells, it can be concluded that neither Fe nor FeMnSi nanocoatings are cytotoxic (Figure 4). This increase could be attributed to Fe released during the experiment, which could increase the metabolic activity [26].

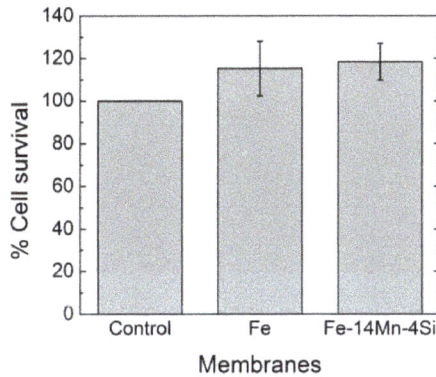

Figure 4. Cell viability measured with MTT cell proliferation assay protocol for the control, Fe- and FeMnSi-coated membranes.

3.2. Transferrin-Alexa Fluor 488 Functionalization of the FeMnSi-Coated Membranes

Taking into account the previous results (morphology, structural properties, biodegradability, and cell viability results), transferrin-Alexa Fluor 488 functionalization studies were carried out on the FeMnSi-coated membrane as this composition showed enhanced biodegradability and was proven not to be toxic. A schematic picture of the whole process is shown in Figure 5.

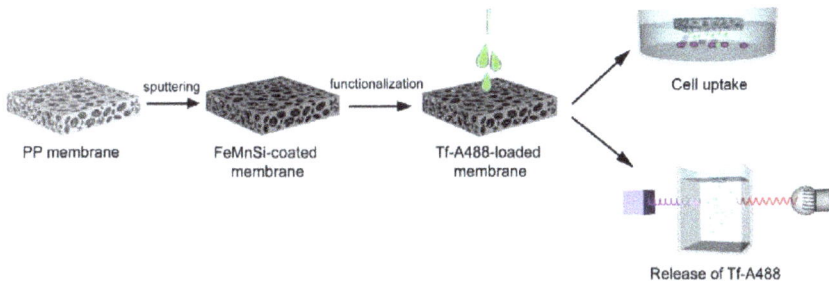

Figure 5. Schematic picture of the functionalization of the PP membranes process.

Information about the distribution of the Tf-AF488 inside the porous membranes was provided by CSLM. Fluorescence and reflectance mode confocal images of Sample 1 (Figure 6a) point toward an inhomogeneous distribution of the Tf-AF488 over the surface. Tf-AF488 cumulative release kinetics (Figure 6b) showed that the Tf-AF488 was immediately released after placing the loaded membrane in the media (when the highest fluorescence intensity was observed). After the peak, fluorescence decayed during the next 15 min until it stabilized. It is worth mentioning that when the membrane was placed inside the fluorimeter cuvette, it remained on top of the media. As the fluorescence reader was located just below, a high concentration of Tf-AF488 was detected in this zone immediately after placing the membrane, as a result of the extremely fast release. Afterwards, the concentration of TF-AF488 began to balance along the cuvette until it reached an equilibrium. This can explain the fluorescence decay observed at the beginning of the test. Accordingly, uptake tests demonstrated that green fluorescence, belonging to Tf-AF488, was present inside of the cells only after 4 h of incubation (Figure S3d). After 24 h of incubation (Figure 6c and Figure S3f), fluorescence had increased and nearly all the cells showed a green spotted fluorescence pattern that corresponds to the Tf-AF488 accumulated in the endosomes and lysosomes [27].

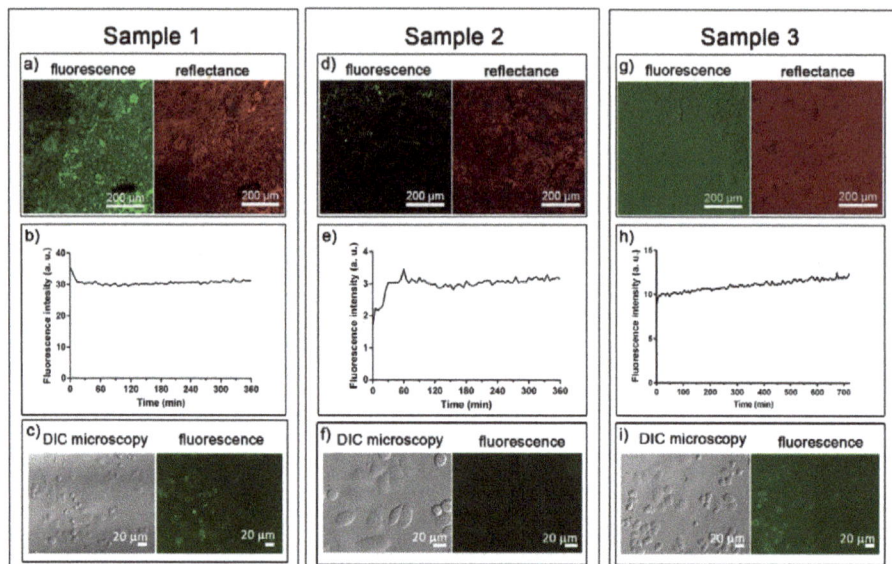

Figure 6. Fluorescence and reflectance images of (**a**) Sample 1; (**d**) Sample 2; and (**g**) Sample 3. Transferrin kinetic cumulative release of (**b**) Sample 1; (**e**) Sample 2 and (**h**) Sample 3. SKBR-3 cells observed under differential interference contrast (DIC) microscopy and fluorescence microscopy at 24 h of incubation for (**c**) Sample 1; (**f**) Sample 2, and (**i**) Sample 3.

With the aim to slow down the kinetics of release, Sample 2 was washed with HBSS after loading to eliminate the excess of Tf-AF488 accumulated onto the surface of the membrane, and a stopper of maltodextrin and gelatin was added on top of the membrane to delay the Tf-AF488 release (Table 1). Regarding Tf-AF488 distribution, confocal images (Figure 6d) indicated also a non-homogeneous distribution of Tf-AF488. However, the cumulative release kinetics (Figure 6e) showed sustained release during the first hour. Nonetheless, the cargo yield was low, as corroborated with the uptake studies in which only a diffuse fluorescence could be appreciated after 24 h of incubation (Figure 6f).

To increase the cargo yield while keeping the progressive release achieved in Sample 2, Sample 3 was consecutively loaded three times to allow more Tf-AF488 to penetrate inside the membrane. Confocal imaging in fluorescence mode revealed that Tf-AF488 homogeneously covered the entire surface of the platform (Figure 6g). 3D image reconstruction, together with confocal microscopy, offer great advantages for studying complex nanocomposites by providing valuable 3D information. The isosurface module of Imaris 3D photoluminescence (3D-PL, green) and the structural (reflection mode, red) image of the porous Fe-14Mn-4Si membrane (Figure 7) showed a homogeneously distributed Tf-AF488 within the membrane. Regarding cellular uptake, no fluorescence could be detected at 4 h of incubation but, at 24 h of incubation (Figure 6i), Tf-AF488 fluorescence could be observed in many cells, similarly to the results observed in Sample 1 after 4 h of incubation.

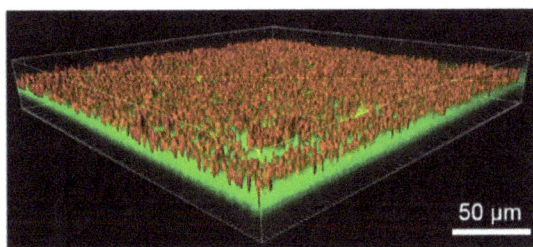

Figure 7. 3D reconstruction obtained from CSLM images showing the distribution of Tf-AF488, in green, and the structural image of the PP membrane, in red (Sample 3).

Among the two replicates produced for each approach, differences could be observed for the cargo yield, especially for Samples 2 and 3, but all the replicates exhibited the same release pattern, evidencing the reproducibility of the loading conditions and Tf-AF488 release. Moreover, fluorescence was barely detectable on the membranes at the end of the fluorometric test, confirming the complete release of Tf-AF488 (Figure S4).

For fixed bio-applications, the presence of the FeMnSi covering the PP membrane might seem unclear. However, it has been shown that the release of certain molecules can be modulated by the action of an external magnetic stimuli. Concerning the use of these materials in micro-robotic applications, the issue of the size needs to be considered. Given that erythrocytes take on average 60 s to complete one cycle of circulation [28], the FeMnSi-coated PP membranes, once introduced into the circulatory system, could reach the target tissue before all of the cargo is released. Once there, they could be retained by an externally applied magnetic field, enhancing the specific accumulation of the cargo in the target tissue. In order to be able to circulate through the blood vessels, the membranes should be, in any case, miniaturized to a micrometric scale. We are currently developing a top-down approach to obtain micrometric FeMnSi-coated PP membranes able to circulate through arteries and veins.

4. Conclusions

The results presented in this study demonstrate that macroporous FeMnSi-coated PP membranes are interesting organic-inorganic materials for biomedical purposes. By using a physical deposition method (sputtering), macroporous PP membranes are coated with a ternary alloy. Of note, both components, the FeMnSi alloy nanocoating and the PP polymer, are biodegradable. Indeed, FeMnSi shows higher degradability in HBSS than pure Fe and is not cytotoxic. The macroporous FeMnSi-coated PP membranes have been functionalized with Tf-AF488 molecule using various protocols, which provide a different balance between cargo yield and release. Thus, their functionalization can be adapted to host molecules that require different concentration or release kinetics to obtain an optimal response. Interestingly, after the cargo is released, both the FeMnSi alloy and the polymer would degrade within a reasonable timescale without the need of a post-extraction surgery.

Supplementary Materials: The following are available online at www.mdpi.com/2079-4991/7/7/155/s1, Figure S1: X-ray diffraction patterns of Fe (black line) and Fe-14Mn-4Si (red line) sputtered on flat Si/Ti/Cu substrates. The first diffraction peak (*) belongs to the Cu substrate and the second one (+) to Fe (110) BCC Im$\bar{3}$m, Figure S2: SEM images of (a) Fe- and (b) FeMnSi-coated PP membranes after incubation in HBSS for 15 days, Figure S3: SKBR-3 cells observed under differential interference contrast (DIC) microscopy and fluorescence microscopy after different times of incubation in the presence of Sample 1: (a,b) control cells, (c,d) cells incubated for 4 h, (e,f) cells incubated for 24 h. Scale Bar, 20 μm, Figure S4: Fluorescence image of Sample 3 after the fluorimetric assay indicating total absence of Tf-AF488.

Acknowledgments: This work has been partially funded by the 2014-SGR-1015 and the 2014-SGR-524 projects from the Generalitat de Catalunya, and the MAT2014-57960-C3-1-R (co-financed by the *Fondo Europeo de Desarrollo Regional*, FEDER), and the MAT2014-57960-C3-3-R projects from the Spanish *Ministerio de Economía y Competitividad* (MINECO). Eva Pellicer is grateful to MINECO for the "Ramon y Cajal" contract (RYC-2012-10839). Jordina Fornell acknowledges the Juan de la Cierva Fellowship from MINECO (IJCI-2015-27030).

Author Contributions: Jordina Fornell, Miguel Guerrero, Jordi Sort, and Eva Pellicer developed the overall concept and designed the experiments for the synthesis of the samples, their structural characterization, and the assessment of their magnetic properties. Juan de Dios Sirvent and Marta Ferran-Marqués optimized the synthetic procedures (sputtering) and acquired the scanning electron microscopy images. Elena Ibáñez, Leonardo Barrios, and Carme Nogués were in charge of the biological part of the work. Jorge Soriano performed the cell adhesion and proliferation experiments. Miguel Guerrero optimized the drug delivery protocols. Santiago Suriñach and Maria Dolors Barógave technical support and conceptual advice. Jordina Fornell and Eva Pellicer wrote the manuscript, and all authors thoroughly reviewed it.

Conflicts of Interest: The authors declare no conflict of interest.

References

1. De Jong, W.H.; Borm, P.J. Drug delivery and nanoparticles: Applications and hazards. *Int. J. Nanomed.* **2008**, *3*, 133–149. [CrossRef]
2. Nelson, B.J.; Kaliakatsos, I.K.; Abbott, J.J. Microrobots for minimally invasive medicine. *Annu. Rev. Biomed. Eng.* **2010**, *12*, 55–85. [CrossRef] [PubMed]
3. Li, S.D.; Chen, Y.C.; Hackett, M.J.; Huang, L. Tumor-targeted delivery of siRNA by self-assembled nanoparticles. *Mol. Ther.* **2007**, *16*, 163–169. [CrossRef] [PubMed]
4. Ferrari, E. BioMEMS and Biomedical Technology. In *Therapeutic Micro/Nanotechnology*; Desai, T., Bhatia, S., Eds.; Springer: New York, NY, USA, 2006; Volume III.
5. Caka, M.; Türkcan, C.; Uygun, D.A.; Uygun, M.; Akgöl, S.; Denizli, A. Controlled release of curcumin from poly(HEMA-MAPA) membrane. *Artif. Cells Nanomed. Biotechnol.* **2017**, *45*, 426–431. [CrossRef] [PubMed]
6. Salazar, H.; Lima, A.C.; Lopes, A.C.; Botelho, G.; Lanceros-Mendez, S. Poly(vinylidene fluoride-trifluoroethylene)/NAY zeolite hybrid membranes as a drug release platform applied to ibuprofen release. *Colloid. Sur. A* **2015**, *469*, 93–99. [CrossRef]
7. Temtem, M.; Pompeu, D.; Jaraquemada, G.; Cabrita, E.J.; Casimiro, T.; Aguiar-Ricardo, A. Development of PMMA membranes functionalized with hydroxypropyl-β-cyclodextrins for controlled drug delivery using a supercritical CO_2-assisted technology. *Int. J. Pharm.* **2009**, *376*, 110–115. [CrossRef] [PubMed]
8. Minelli, E.B.; Bora, T.D.; Benini, A. Different microbial biofilm formation on polymethylmethacrylate [PMMA] bone cement loaded with gentamicin and vancomycin. *Anaerobe* **2011**, *17*, 380–383. [CrossRef] [PubMed]
9. Rochford, E.T.; Richards, R.G.; Moriarty, T.F. Influence of material on the development of device-associated infections. *Clin. Microbiol. Infect.* **2012**, *18*, 1162–1167. [CrossRef] [PubMed]
10. Gristina, A.G.; Costerton, J.W. Bacterial adherence to biomaterials and tissue. The significance of its role in clinical sepsis. *J. Bone Jt. Surg. Am.* **1985**, *67*, 264–273. [CrossRef]
11. Kummer, M.P.; Abbott, J.J.; Kratochvil, B.E.; Borer, R.; Sengul, A.; Nelson, B.J. OctoMag: An electromagnetic system for 5-DOF wireless micromanipulation. *IEEE Trans. Robot.* **2010**, *26*, 1006–1017. [CrossRef]
12. Zhang, L.; Peyer, K.E.; Nelson, B.J. Artificial bacterial fagella for micromanipulation. *Lab Chip* **2010**, *10*, 2203–2215. [CrossRef] [PubMed]
13. Lopes, J.R.; Santos, G.; Barata, P.; Oliveira, R.; Lopes, C.M. Physical and chemical stimuli-responsive drug delivery systems: Targeted delivery and main routes of administration. *Curr. Pharm. Des.* **2013**, *19*, 7169–7184. [CrossRef] [PubMed]
14. Singh, D.; Rezac, M.E.; Pfromm, P.H. Partial hydrogenation of soybean oil with minimal trans fat production using a Pt-decorated polymeric membrane reactor. *J. Am. Oil Chem. Soc.* **2009**, *86*, 93–101. [CrossRef]
15. Himy, A.; Wagner, O.C. Stabilized Nickel-Zinc Battery. U.S. Patent 4327157A, 27 April 1982.
16. Iakovlev, V.; Guelcher, S.; Bendavid, R. Degradation of polypropylene in vivo: A microscopic analysis of meshes explanted from patients. *J. Biomed. Mater. Res. B* **2017**, *105*, 237–248. [CrossRef] [PubMed]
17. Liu, B.; Zheng, Y.F. Effects of alloying elements [Mn, Co, Al, W, Sn, B, C and S] on biodegradability and in vitro biocompatibility of pure iron. *Acta Biomater.* **2011**, *7*, 1407–1420. [CrossRef] [PubMed]

18. Schinhammer, M.; Steiger, P.; Moszner, F.; Löffler, J.F.; Uggowitzer, P.J. Degradation performance of biodegradable Fe Mn C (Pd) alloys. *Mat. Sci. Eng. C* **2013**, *33*, 1882–1893. [CrossRef] [PubMed]

19. Feng, Y.P.; Blanquer, A.; Fornell, J.; Zhang, H.; Solsona, P.; Baró, M.D.; Suriñach, S.; Ibáñez, E.; García-Lecina, E.; Wei, X.; et al. Novel Fe–Mn–Si–Pd alloys: Insights into mechanical, magnetic, corrosion resistance and biocompatibility performances. *J. Mat. Chem. B* **2016**, *4*, 6402–6412. [CrossRef]

20. Zheng, Y.F.; Gu, X.N.; Witte, F. Biodegradable metals. *Mat. Sci. Eng. R* **2014**, *77*, 1–34. [CrossRef]

21. ASTM G31-72. *Standard Practice for Laboratory Immersion Corrosion Testing of Metals*; ASTM International: West Conshohocken, PA, USA, 2004.

22. Kawamoto, M.; Horibe, T.; Kohno, M.; Kawakami, K. A novel transferrin receptor-targeted hybrid peptide disintegrates cancer cell membrane to induce rapid killing of cancer cells. *BMC Cancer* **2011**, *11*, 359. [CrossRef] [PubMed]

23. Schinhammer, M.; Hänzi, A.C.; Löffler, J.F.; Uggowitzer, P.J. Design strategy for biodegradable Fe-based alloys for medical applications. *Acta Biomater.* **2010**, *6*, 1705–1713. [CrossRef] [PubMed]

24. Hermawan, H.; Purnama, A.; Dube, D.; Couet, J.; Mantovani, D. Fe–Mn alloys for metallic biodegradable stents: Degradation and cell viability studies. *Acta Biomater.* **2010**, *6*, 1852–1860. [CrossRef] [PubMed]

25. Hermawan, H.; Alamadari, H.; Mantovani, D.; Dubé, D. Iron-manganese: New class of metallic degradable biomaterials prepared by powder metallurgy. *Powder Metall.* **2008**, *51*, 38–45. [CrossRef]

26. Moravej, M.; Purnama, A.; Fiset, M.; Couet, J.; Mantovani, D. Electroformed pure iron as a new biomaterial for degradable stents: In vitro degradation and preliminary cell viability studies. *Acta Biomater.* **2010**, *6*, 1843–1851. [CrossRef] [PubMed]

27. Baravalle, G.; Schober, D.; Huber, M.; Bayer, N.; Murphy, R.F.; Fuchs, R. Transferrin recycling and dextran transport to lysosomes is differentially affected by bafilomycin, nocodazole, and low temperature. *Cell Tissue Res.* **2005**, *320*, 99–113. [CrossRef] [PubMed]

28. Bloom, J.A. *Monitoring of Respiration and Circulation*; CRC Press: Boca Raton, FL, USA, 2003.

nanomaterials

MDPI

Article

Heteromer Nanostars by Spontaneous Self-Assembly

Caitlin Brocker [1], Hannah Kim [2], Daniel Smith [1] and Sutapa Barua [1,*

[1]   Department of Chemical and Biochemical Engineering, Missouri University of Science and Technology, 110 Bertelsmeyer Hall, 1101 N. State Street, Rolla, MO 65409, USA; cebfk6@mst.edu (C.B.); dlsrt8@mst.edu (D.S.)

[2]   Department of Biological Sciences, Missouri University of Science and Technology, 143 Schrenk Hall, 400 W. 11th St., Rolla, MO 65409, USA; hhk4hf@mst.edu

*   Correspondence: baruas@mst.edu; Tel.: +1-(573)-341-7551

Academic Editor: Giuliana Gorrasi
Received: 23 April 2017; Accepted: 23 May 2017; Published: 31 May 2017

Abstract: Heteromer star-shaped nanoparticles have the potential to carry out therapeutic agents, improve intracellular uptake, and safely release drugs after prolonged periods of residence at the diseased site. A one-step seed mediation process was employed using polylactide-*co*-glycolic acid (PLGA), polyvinyl alcohol (PVA), silver nitrate, and tetrakis(hydroxymethyl)phosphonium chloride (THPC). Mixing these reagents followed by UV irradiation successfully produced heteromer nanostars containing a number of arm chains attached to a single core with a high yield. The release of THPC from heteromer nanostars was tested for its potential use for breast cancer treatment. The nanostars present a unique geometrical design exhibiting a significant intracellular uptake by breast cancer cells but low cytotoxicity that potentiates its efficacy as drug carriers.

Keywords: nanostar; polylactide-*co*-glycolic acid (PLGA); star shape; tetrakis (hydroxylmethyl) phosphonium chloride (THPC)

1. Introduction

Synthesis of self-assembly systems is of significant interest to optimize drug delivery efficacy. Polymeric assembly may offer variation in composition, chemical functionality, size, and shape including disks, rods, spheres, cubes, and filaments to name just a few [1–9]. The choice of non-spherical nanoparticles exhibits improved blood circulation time [7], specific receptor binding [10], cellular internalization [9,10], and low phagocytosis compared to its spherical counterparts [11,12]. Non-spherical lipomers composed of poly(methylvinylether-*co*-maleic anhydride) and lipids have shown higher splenic accumulation in rats, rabbits, and dogs and are more effective at evading non-specific uptake by macrophages than spherical lipomers, suggesting a potential for the carrier to be used for drug delivery to the spleen [13]. Star-shaped poly(L-lactide) [14], poly(ethylene-*co*-propylene) [15], and polybutadienes [16] have been synthesized with one end chemically linked to a hydrophilic core, while the other end is functionalized creating a hydrophobic corona [14]. Multistep polymerization reactions have been involved in synthesizing star-shaped particles such as dendrimers and micelles [8,17–20]. These macromolecules require polymers of various lengths and particle generations. The interest in such systems originates from, in addition to their biocompatible properties, their applications in slow release drug delivery systems, bioresorbable surgical sutures, and surgical implants. To this end, heteromer star-shaped nanoparticles are synthesized with independent sets of arm chains attached to a single polymer core. Poly(lactide)-*co*-glycolic acid (PLGA) polymer is chosen for its biodegradable characteristics by hydrolytic cleavage of ester groups in the physiological microenvironment, forming non-toxic (biodegradable and biocompatible) lactic acid and glycolic acid groups [21]. While PLGA scaffold has been widely used for developing drug delivery systems [21–30], the preparation of heteromer-shaped PLGA particles is unknown. Herein, a combinatorial method is described where a phosphonium salt—tetrakis (hydroxymethyl) phosphonium chloride (THPC)—is adsorbed on PLGA in the presence of silver nitrate seed. THPC has been used as a crosslinker of hydrogels [31] and a reducing agent for metal nanoparticles [32,33]. It is hypothesized that the chemical reduction of silver in the presence of THPC impregnated in PLGA nanoparticles causes the intraparticle morphology to vary. In addition, due to its chemical structure, THPC is available in aqueous solution and cytocompatible for pharmaceutical applications. The focus of this study is the synthesis of heteromer star-shaped nanoparticles as an anticancer drug delivery platform, combining a self-assembled THPC reducing agent with PLGA and the resulting nanostars for intracellular uptake by breast cancer cells.

2. Materials and Methods

2.1. Synthesis of PLGA-THPC Heteromer Nanostars

Most reagents were purchased from Sigma-Aldrich (St. Louis, MO, USA) unless specified otherwise. THPC was kindly donated by Kattesh Katti from the University of Missouri Columbia (Columbia, MO, USA). Nanostars were developed using a silver core surrounded by THPC particles coated with PLGA polymer [34]. Polyvinyl alcohol (PVA; MW 33,000–70,000) of 5 mg was dissolved in 10 mL of reverse osmosis water at 80 °C. Once this solution was cooled to room temperature, 1 mL of 1 mg/mL THPC was pipetted into the PVA solution. Fifty milligrams of PLGA was dissolved in 2 mL of acetone and added dropwise to the PVA solution under sonication. Prepared PLGA particles were mixed with 450 µL of 0.1 N silver nitrate ($AgNO_3$) solution (Acros Organics, Fisher Scientific, Waltham, MA, USA). The mixture was gently mixed, transferred into round, shallow Petri dishes with a thickness of ~3 mm, and exposed to 8 W of UV light (254 nm) for 40 min. In a separate beaker, trisodium citrate (TSC; Alfa Aesar, Haver Hill, MA, USA) of 0.7 mmol was added to a mixture of 10 mg/mL PVA. The silver-nucleated PLGA solution was transferred into the TSC-PVA mixture and stirred for 5 min. Finally, 100 µL of ascorbic acid was added to the mixture and stirred vigorously at room temperature for 5 min. PVA was removed by repeated washing using water and centrifugation.

2.2. Measurement of THPC in Nanostars

A standard curve was generated by plotting absorbance values versus various concentrations of THPC for measuring THPC concentrations in nanostars. Given the chemical structure of THPC, the chloride ion on THPC is highly reactive with the silver ion in silver nitrate ($(HOCH_2)_4PCl + AgNO_3 \rightarrow AgCl \downarrow$). This reaction was utilized to generate the THPC standard curve for measuring its soluble concentrations. Briefly, equal parts of THPC and silver nitrate ($AgNO_3$) were mixed in a 96-well plate (Corning, Corning, NY, USA) at different concentrations. The formation of silver chloride (AgCl) precipitates was detected by measuring the absorbance at 395 nm using a microplate reader (BioTek Synergy 2, BioTek, Winooski, VT, USA). Water was used as a blank.

2.3. Characterization of Nanostars

The morphology and size of PLGA-THPC nanostars were examined using a transmission electron microscope (TEM; Tecnai F20; FEI company, Hillsboro, OR, USA) at an accelerating voltage of 120 kV. A drop of 10 µL of a previously prepared PLGA-THPC nanostar suspension was pipetted onto carbon-coated copper grids (Ted Pella, Redding, CA, USA) and air-dried. The diameter of nanostars was measured using ImageJ (version 1.45S, NIH, Bethesda, MD, USA) for at least 20 particles. The size and surface charges of nanostars were measured by dynamic light scattering using a Nanoseries Zetasizer ZS 90 (Malvern Instruments Ltd., Malvern, Worcestershire, UK), with backscattering detection at 90°.

2.4. Quantification of THPC Release

The PLGA-THPC nanostars were suspended in water at both pH 7.4 and 6.2. Each sample of 5 mL was placed in a 37 °C water bath to mimic the body temperature. At t = 0, 0.5, 2, 4, 8, 24, 36, and 72 h, a 500 μL solution was withdrawn from each sample. The solution was analyzed for THPC concentration using the THPC standard curve. Phosphate buffer saline (PBS) could not be used in this study, as it has been used in other drug release studies [9,35,36], due to the interference of sodium chloride with silver chloride during the absorbance measurement assay.

2.5. Intracellular Uptake

MDA-MB-231 breast cancer cell monolayers were seeded at a density of 50,000 cells/mL in 8-well chamber plates and grown overnight. The medium was replaced with fresh medium including nanostars for 2 h of incubation with THPC alone as a control. The nuclei were stained by 4′,6-diamidino-2-phenylindole (DAPI). The cells were washed using PBS. The intracellular accumulation of nanostars was visualized using a fluorescence microscope (Zeiss, Oberkochen, Germany) equipped with a transmitted light illuminator (Zeiss, Oberkochen, Germany), fluorescence filter set of 390/450 ex/em, 63× water immersion objective, an Axiocam camera (Zeiss, Oberkochen, Germany), and ZEN2 Pro software (Zeiss, Oberkochen, Germany).

2.6. In Vitro Cytotoxicity of Heteromer Nanostars in MDA-MB-231 Breast Cancer Cells

MDA-MB-231 cells (ATCC, Manassas, VA, USA) were cultured in RPMI 1640 medium supplemented with 10% FBS (Corning) and 1% (100 units/mL) penicillin-streptomycin (Gibco, Gaithersberg, MD, USA) in a 5% CO_2 and 37 °C incubator. The cells were plated in 96-well tissue plates (Corning) at a density of 10,000 cells/well in 200 μL of medium. PLGA-THPC nanostars were added to the cells at 0, 0.1, 0.5, 1, 5, 10, 15, 20, and 25 μg/mL of THPC. PBS was used as a negative control (100% live cells) along with Triton X-100 as a positive control (100% dead cells). After 3 h of incubation, the medium was replaced; cells were incubated for another 72 h. Live cells were measured using the live/dead assay (Life Technologies, Carlsbad, CA, USA). Briefly, the medium was removed followed by an addition of 2 μM calcein in PBS to stain the live cells in each well. Cells were incubated for 30 min at room temperature. The fluorescence intensity of calcein AM was measured at an excitation/emission of 485/528 using a plate reader (BioTek Synergy 2, BioTek, Winooski, VT, USA). The percent inhibition of cell growth was calculated using the following equation:

$$\% \ inhibition \ of \ cell \ growth = \frac{F.I._{PBS \ treated \ cells} - F.I._{samples}}{F.I._{PBS \ treated \ cells}} \times 100.$$

2.7. Statistical Analysis

Each experiment was carried out with three independent experiments of at least triplicate measurements. The mean differences and standard deviations were evaluated.

3. Results and Discussion

3.1. Synthesis of Heteromer Nanostars

A star-like growth is a seed-mediated process in the presence of the PLGA core, THPC molecules, and a silver ion [34]. A silver-polymer composite star synthesis protocol was employed to encapsulate THPC in PLGA nanoparticles (Figure 1a). Three steps are required for this synthesis: phase separation, PLGA-THPC nanostar synthesis, and photoreduction. PLGA-THPC nanostars were formed employing oil in a water emulsion technique, also known as phase separation. By dissolving PLGA polymers in acetone to create an oil phase and gently pipetting the oil phase dropwise to the water phase under sonication, PLGA-THPC nanoparticles were formed. When the nanoparticles were placed in a Petri dish and irradiated with a UV light, THPC served as a reducing agent for Ag^+ in solution, as shown mechanistically in Figure 1b. THPC is a strong reducing agent for synthesizing nanoparticles [32,37]. Photoreduction, along with adding a strong reducing agent, by reducing silver nitrate, allowed for a change in its morphology to a star shape. The TEM image describes the formation of a uniform shape of PLGA–Ag nanoparticles and star-shaped morphology of a THPC-encapsulated PLGA silver nanoparticle. The average number of arms (sharp corners) was calculated at 10 ± 2. The average hydrodynamic radius was ~315 ± 140 nm (Figure 2a) that reflects their average ensemble size when dispersed in PBS. The three intensity distribution curves (blue, red, and green) in Figure 2a represent light scattering data of nanoparticles from three different batches. All samples showed the highest peak at around 315 nm in size, indicating the reproducibility of the presented nanostar synthesis method. The differences in percent intensity on the *y*-axis are due to variation in particle concentrations. A small peak at around 40 nm is seen in the green curve and may be due to impurities in the system or PLGA fragments from biodegradation in the storing solution. However, the strong peak at 315 nm shows that the majority of particles are in this range. The surface charge of the nanoparticles was -39 ± 5.5 mV (Figure 2b) in water (pH 6.8), indicating a stable nanoparticle suspension in the aqueous medium. The encapsulation of THPC was investigated from silver chloride precipitation and absorbance curve. A calibration curve was constructed for THPC by measuring absorbance at 395 nm versus different concentrations of THPC (Figure 3). A linear curve ($y = mx$) was generated to find out the slope and regression coefficient ($R^2 > 0.99$) using the least squares regression method. The percentage encapsulation efficiency was calculated ~$60\% \pm 5\%$ based on the initial loading of the THPC.

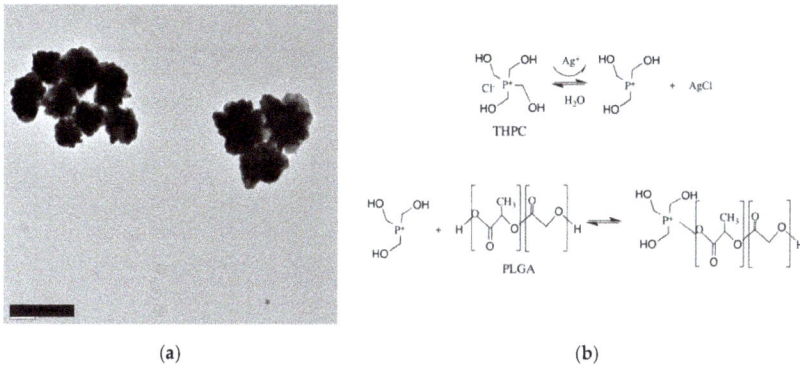

(a)   (b)

Figure 1. (a) TEM image of heteromer PLGA-THPC star-shaped nanoparticles (scale bar = 500 nm); (b) A hypothesized mechanism of PLGA-THPC nanostar assembly.

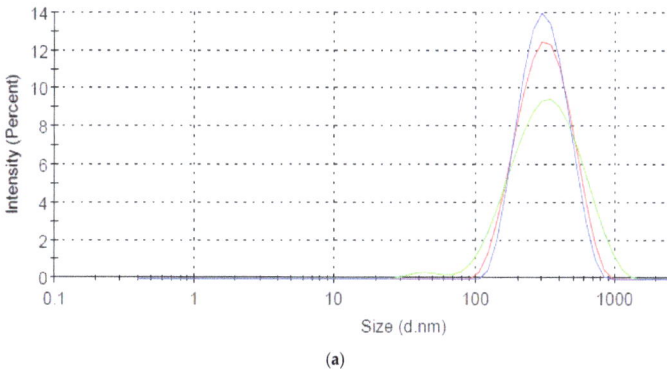

(a)

Figure 2. *Cont.*

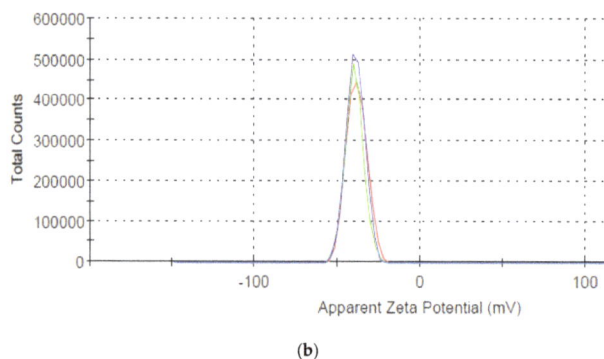

(b)

Figure 2. (a) Size distribution of PLGA-THPC nanostars; (b) surface charges by dynamic light scattering in water (pH 6.8). The three colored (blue, red, and green) curves represent the reproducibility in nanostar size and surface charges, as synthesized from three separate batches.

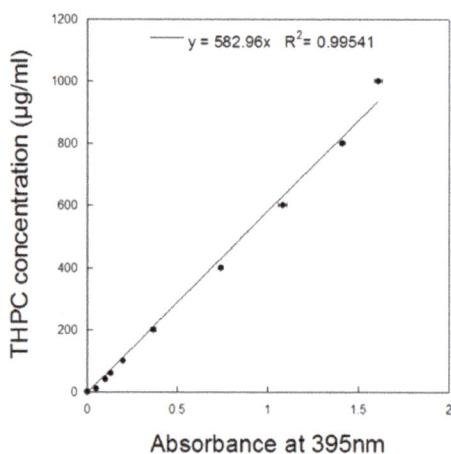

Figure 3. Standard curve demonstrating the linear relationship between absorbance of reacted silver chloride at 395 nm and THPC concentration in μg/mL.

3.2. THPC Release

The percentage of THPC release at different time intervals and different pH values are shown in Figure 4. A similar amount (~5%) of THPC was released from PLGA-THPC nanostars up to 5 min in PBS buffer at both pH 6.2 (solid point) and pH 7.4 (open point). The incorporated THPC release occurred due to the hydrolysis of PLGA in PBS through cleavage of ester linkages in its backbone [21]. The biodegradation is faster in the slightly acidic medium. Faster and higher release of THPC was observed when nanostars were exposed to a pH 6.2 environment compared to a pH 7.4 environment. At pH 7.4 (open point), it took nearly 72 h to reach ~40% release, while ~75% of the release rate was reached at pH 6.2 (solid point). It is known that the pH of an inflammatory disease site such as a tumor and a cytoplasm is lower than that of blood plasma [38]. The THPC release data shows a higher release at lower pH than the physiological blood pH.

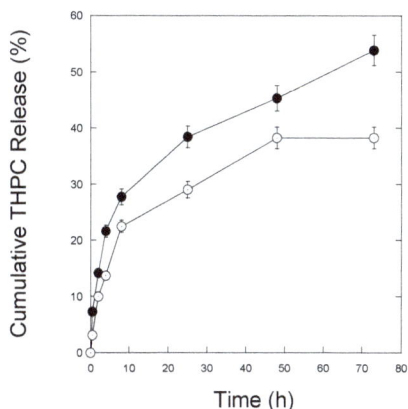

Figure 4. The cumulative release amount of THPC from PLGA-THPC nanostars at pH 6.2 (solid filled circles) and pH 7.4 (open circles) as determined by the silver chloride precipitation and absorbance (A_{395}) assay. The complete release profile assay was conducted in a 37 °C water bath. The release rate is faster in breast cancer cells mimicking pH 6.2 than the body pH at 7.4.

3.3. Cellular Uptake and Cytotoxicity in Breast Cancer Cells

The therapeutic effects of drug nanoparticles depend on their internalization and sustained retention inside diseased cells [20]. In order to evaluate the intracellular localization and in vitro cytotoxicity of THPC, breast cancer cell MDA-MB-231 was chosen. Phase microscopy was used to visualize the internalization and distribution of PLGA-THPC nanostars in MDA-MB-231 cells (Figure 5). The nanostars accumulated inside endosomes after adsorptive endocytosis through fluid phase pinocytosis and clathrin-coated endocytosis process [39,40]. PLGA nanoparticles have been shown to degrade in acidic endolysosomal compartments [40,41], where it may release THPC in a sustained manner. The phosphonium cation of THPC facilitates the accumulation from endosomes into the mitochondria, which is promising for mitochondria-targeted cancer therapy [42,43]. The nanostars accumulated inside the endosomes after intracellular internalization. The cytotoxic activity of PLGA-THPC star-shaped nanoparticles was evaluated in MDA-MB-231 cells (Figure 6; filled circles) at varying concentrations. The *x*-axis represents the concentrations of THPC in the PLGA-THPC star-shaped nanoparticles. The corresponding effects by THPC solution alone are shown in Figure 6 (open circles). The PLGA-THPC nanostars did not show much cytotoxicity against MDA-MB-231 cells, with up to 25% MDA-MB-231 cell death at 25 μg/mL of THPC, which followed a trend similar to a solution of THPC alone. Surprisingly, THPC nanostars yielded a conspicuously higher cell growth inhibition than did pure THPC, without showing statistically significant differences. Although the underlying mechanism of such behavior is unknown, a possible explanation could be related to the enhancement in intracellular uptake and the reduction in drug efflux by P-glycoprotein (P-gp) membrane transporter proteins [44–46] for the nanostar-mediated THPC delivery. The drug efflux pump, P-gp, is overexpressed in MDA-MB-231 breast cancer cells [47]. Nanoparticles have been developed to increase the intracellular concentration of drugs in cancer cells by circumventing the P-gp exerted resistance [46,48]. The mechanisms of carriers to overcome P-gp have been reported for various drug delivery systems, including PLGA [49], *N*-(2-hydroxypropyl)methacrylamide (HPMA) drug conjugates [50], polymeric micelles [51,52], hybrid lipid nanoparticles [53], lipid-based nanoparticles [54], and liposomes [55]. Doxorubicin-loaded PLGA/lipid nanoassemblies increased drug uptake and enhanced cytotoxicity in MCF-7 human breast cancer cells bypassing drug resistance [56]. Doxorubicin-loaded nanoparticles increased intracellular drug concentration and efficiently suppressed P-gp expression in multidrug resistant osteosarcoma cell lines [57]. Nanoparticles of doxorubicin and curcumin overcome multidrug resistance in multiple in vivo models, including multiple myeloma, acute leukemia, and prostate and ovarian cancers [58]. The low cytotoxicity effects by nanostars indicate a new possibility of using this geometrical shape as a potential drug carrier for cancer treatment.

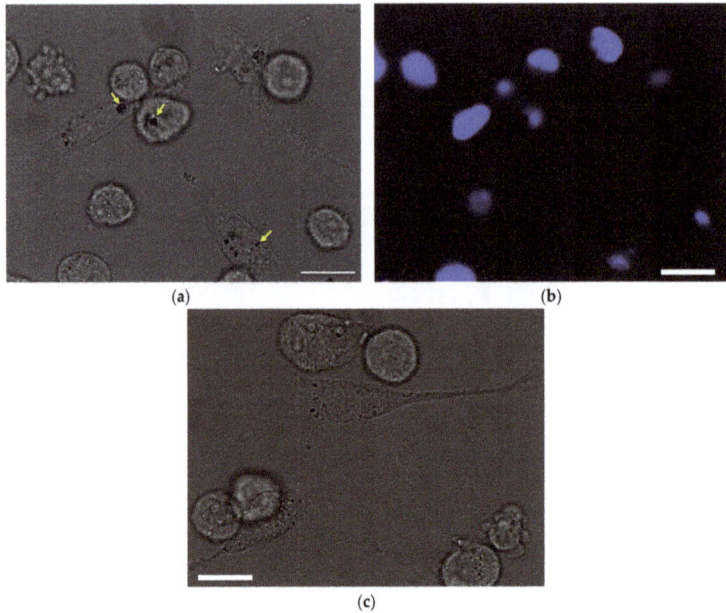

Figure 5. (**a**) Phase contrast image showing the intracellular uptake of PLGA-THPC nanostars by MDA-MB-231 cells. The black aggregates (as indicated by yellow arrows) indicate the spatial distribution of the nanostars; (**b**) Fluorescent nuclei of MDA-MB-231 breast cancer cells after 72 h incubation with nanostars; (**c**) Control MDA-MB-231 breast cancer cells were incubated with similar amount of THPC solution without any nanoparticles. No black spots in this control confirm the intracellular uptake of nanostars in (**a**). Scale bar = 20 µm.

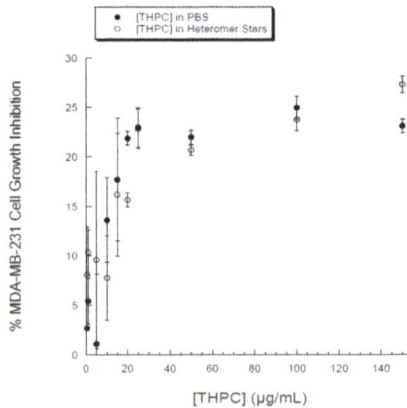

Figure 6. Cytotoxicity of PLGA-THPC nanostars (filled circles) as measured by its dose-dependent effects on MDA-MB-231 breast cancer cell growth inhibition. THPC solution (open circles) was used as a control.

4. Conclusions

Heteromer nanostars containing ten arm species with PLGA and THPC were prepared. Simple mixing of biodegradable polymers and subsequent linking reaction with THPC and silver ions successfully yielded heteromer nanostars. The particles are highly stable as predicted from their high zeta potential value. Approximately 75% THPC was released in pH mimicking buffer conditions. In particular, breast cancer cells took up the nanostars inside the endosomes, where pH was slightly more acidic (pH 6) compared to the blood pH at 7.4 [39,40]. Furthermore, the nanostars were evaluated to induce cytotoxicity in breast cancer cells. Although the cytotoxic effect by the nanostars alone is low, the features of heteromer chains were clearly demonstrated in terms of their potential drug delivery carriers in breast cancer treatment.

Acknowledgments: This work was funded by Missouri S&T's OURE program, and the PI's start-up. The authors would like to thank Jesicca Terbush for assistance using TEM and acknowledge the facilities available for use at the Environmental Resource Center.

Author Contributions: Caitlin Brocker and Hannah Kim performed experiments, wrote the manuscript and reviewed it. Daniel Smith conducted particle size distribution experiments, interpreted the data and revised the manuscript. Sutapa Barua supervised experiments, wrote, reviewed and revised the manuscript.

Conflicts of Interest: The authors declare no conflict of interest.

References

1. Gratton, S.E.A.; Ropp, P.A.; Pohlhaus, P.D.; Luft, J.C.; Madden, V.J.; Napier, M.E.; DeSimone, J.M. The effect of particle design on cellular internalization pathways. *Proc. Natl. Acad. Sci. USA* **2008**, *105*, 11613–11618. [CrossRef] [PubMed]
2. Perry, J.L.; Herlihy, K.P.; Napier, M.E.; DeSimone, J.M. PRINT: A Novel Platform Toward Shape and Size Specific Nanoparticle Theranostics. *Acc. Chem. Res.* **2011**, *44*, 990–998. [CrossRef] [PubMed]
3. Champion, J.A.; Katare, Y.K.; Mitragotri, S. Making polymeric micro- and nanoparticles of complex shapes. *Proc. Natl. Acad. Sci. USA* **2007**, *104*, 11901–11904. [CrossRef] [PubMed]
4. Mitragotri, S.; Lahann, J. Physical approaches to biomaterial design. *Nat. Mater.* **2009**, *8*, 15–23. [CrossRef] [PubMed]
5. Kolhar, P.; Doshi, N.; Mitragotri, S. Polymer Nanoneedle-Mediated Intracellular Drug Delivery. *Small* **2011**, *7*, 2094–2100. [CrossRef] [PubMed]
6. Bhaskar, S.; Pollock, K.M.; Yoshida, M.; Lahann, J. Towards designer microparticles: Simultaneous control of anisotropy, shape and size. *Small* **2010**, *6*, 404–411. [CrossRef] [PubMed]
7. Geng, Y.; Dalhaimer, P.; Cai, S.; Tsai, R.; Tewari, M.; Minko, T.; Discher, D.E. Shape effects of filaments versus spherical particles in flow and drug delivery. *Nat. Nanotechnol.* **2007**, *2*, 249–255. [CrossRef] [PubMed]
8. Geng, Y.; Discher, D.E. Hydrolytic degradation of poly(ethylene oxide)-block-polycaprolactone worm micelles. *J. Am. Chem. Soc.* **2005**, *127*, 12780–12781. [CrossRef] [PubMed]
9. Laemthong, T.; Kim, H.H.; Dunlap, K.; Brocker, C.; Barua, D.; Forciniti, D.; Huang, Y.-W.; Barua, S. Bioresponsive polymer coated drug nanorods for breast cancer treatment. *Nanotechnology* **2017**, *28*, 045601. [CrossRef] [PubMed]
10. Barua, S.; Yoo, J.-W.; Kolhar, P.; Wakankar, A.; Gokarn, Y.R.; Mitragotri, S. Particle shape enhances specificity of antibody-displaying nanoparticles. *Proc. Natl. Acad. Sci. USA* **2013**, *110*, 3270–3275. [CrossRef] [PubMed]
11. Champion, J.; Mitragotri, S. Shape Induced Inhibition of Phagocytosis of Polymer Particles. *Pharm. Res.* **2009**, *26*, 244–249. [CrossRef] [PubMed]
12. Champion, J.A.; Mitragotri, S. Role of target geometry in phagocytosis. *Proc. Natl. Acad. Sci. USA* **2006**, *103*, 4930–4934. [CrossRef] [PubMed]
13. Devarajan, P.V.; Jindal, A.B.; Patil, R.R.; Mulla, F.; Gaikwad, R.V.; Samad, A. Particle shape: A new design parameter for passive targeting in splenotropic drug delivery. *J. Pharm.Sci.* **2010**, *99*, 2576–2581. [CrossRef] [PubMed]
14. Danko, M.; Libiszowski, J.; Biela, T.; Wolszczak, M.; Duda, A. Molecular dynamics of star-shaped poly(L-lactide)s in Tetrahydrofuran as Solvent Monitored by Fluorescence Spectroscopy. *J. Polym. Sci. Part A Polym. Chem.* **2005**, *43*, 4586–4599. [CrossRef]

15. Elkins, C.L.; Viswanathan, K.; Long, T.E. Synthesis and characterization of star-shaped poly(ethylene-*co*-propylene) polymers bearing terminal self-complementary multiple hydrogen-bonding sites. *Macromolecules* **2006**, *39*, 3132–3139. [CrossRef]
16. Pitsikalis, M.; Hadjichristidis, N. Model mono-, di-, and tri-ω-functionalized three-arm star polybutadienes. Synthesis and association in dilute solutions by membrane osmometry and static light scattering. *Macromolecules* **1995**, *28*, 3904–3910. [CrossRef]
17. Nanjwade, B.K.; Bechra, H.M.; Derkar, G.K.; Manvi, F.V.; Nanjwade, V.K. Dendrimers: Emerging polymers for drug-delivery systems. *Eur. J. Pharm. Sci.* **2009**, *38*, 185–196. [CrossRef] [PubMed]
18. Torchilin, V.P. Micellar Nanocarriers: Pharmaceutical Perspectives. *Pharm. Res.* **2006**, *24*, 1. [CrossRef] [PubMed]
19. Grayson, S.M.; Fréchet, J.M.J. Convergent dendrons and dendrimers: From synthesis to applications. *Chem. Rev.* **2001**, *101*, 3819–3867. [CrossRef] [PubMed]
20. Li, Y.-L.; van Cuong, N.; Hsieh, M.-F. Endocytosis Pathways of the Folate Tethered Star-Shaped PEG-PCL Micelles in Cancer Cell Lines. *Polymers* **2014**, *6*, 634–650. [CrossRef]
21. Makadia, H.K.; Siegel, S.J. Poly Lactic-*co*-Glycolic Acid (PLGA) as Biodegradable Controlled Drug Delivery Carrier. *Polymers* **2011**, *3*, 1377–1397. [CrossRef] [PubMed]
22. Chung, Y.I.; Kim, J.C.; Kim, Y.H.; Tae, G.; Lee, S.Y.; Kim, K.; Kwon, I.C. The effect of surface functionalization of PLGA nanoparticles by heparin- or chitosan-conjugated Pluronic on tumor targeting. *J. Control. Release* **2010**, *143*, 374–382. [CrossRef] [PubMed]
23. Thomasin, C.; Nam-Trân, H.; Merkle, H.P.; Gander, B. Drug microencapsulation by PLA/PLGA coacervation in the light of thermodynamics. 1. Overview and theoretical considerations. *J. Pharm. Sci.* **1998**, *87*, 259–268. [CrossRef] [PubMed]
24. Murakami, H.; Kobayashi, M.; Takeuchi, H.; Kawashima, Y. Preparation of poly(DL-lactide-*co*-glycolide) nanoparticles by modified spontaneous emulsification solvent diffusion method. *Int. J. Pharm.* **1999**, *187*, 143–152. [CrossRef]
25. Yang, Y.-Y.; Chung, T.-S.; Ng, N.P. Morphology, drug distribution, and in vitro release profiles of biodegradable polymeric microspheres containing protein fabricated by double-emulsion solvent extraction/evaporation method. *Biomaterials* **2001**, *22*, 231–241. [CrossRef]
26. Betancourt, T.; Brown, B.; Brannon-Peppas, L. Doxorubicin-loaded PLGA nanoparticles by nanoprecipitation: Preparation, characterization and in vitro evaluation. *Nanomedicine* **2007**, *2*, 219–232. [CrossRef] [PubMed]
27. Kocbek, P.; Obermajer, N.; Cegnar, M.; Kos, J.; Kristl, J. Targeting cancer cells using PLGA nanoparticles surface modified with monoclonal antibody. *J. Control. Release* **2007**, *120*, 18–26. [CrossRef] [PubMed]
28. Bhardwaj, V.; Ankola, D.D.; Gupta, S.C.; Schneider, M.; Lehr, C.M.; Kumar, M.R. PLGA Nanoparticles Stabilized with Cationic Surfactant: Safety Studies and Application in Oral Delivery of Paclitaxel to Treat Chemical-Induced Breast Cancer in Rat. *Pharm. Res.* **2009**, *26*, 2495–2503. [CrossRef] [PubMed]
29. Tang, B.C.; Dawson, M.; Lai, S.K.; Wang, Y.-Y.; Suk, J.S.; Yang, M.; Zeitlin, P.; Boyle, M.P.; Fu, J.; Hanes, J. Biodegradable polymer nanoparticles that rapidly penetrate the human mucus barrier. *Proc. Natl. Acad. Sci. USA* **2009**, *106*, 19268–19273. [CrossRef] [PubMed]
30. Kumari, A.; Yadav, S.K.; Yadav, S.C. Biodegradable polymeric nanoparticles based drug delivery systems. *Colloids Surf. B Biointerfaces* **2010**, *75*, 1–18. [CrossRef] [PubMed]
31. Chung, C.; Lampe, K.J.; Heilshorn, S.C. Tetrakis(hydroxymethyl) Phosphonium Chloride as a Covalent Cross-Linking Agent for Cell Encapsulation within Protein-Based Hydrogels. *Biomacromolecules* **2012**, *13*, 3912–3916. [CrossRef] [PubMed]
32. Gulka, C.P.; Wong, A.C.; Wright, D.W. Spontaneous Self-Assembly and Disassembly of Colloidal Gold Nanoparticles Induced by Tetrakis(hydroxymethyl) Phosphonium Chloride. *Chem. Commun. (Camb. Engl.)* **2016**, *52*, 1266–1269. [CrossRef] [PubMed]
33. Hueso, J.L.; Sebastian, V.; Mayoral, A.; Uson, L.; Arruebo, M.; Santamaria, J. Beyond gold: Rediscovering tetrakis-(hydroxymethyl)-phosphonium chloride (THPC) as an effective agent for the synthesis of ultra-small noble metal nanoparticles and Pt-containing nanoalloys. *RSC Adv.* **2013**, *3*, 10427–10433. [CrossRef]
34. Homan, K.A.; Chen, J.; Schiano, A.; Mohamed, M.; Willets, K.A.; Murugesan, S.; Stevenson, K.J.; Emelianov, S. Silver-Polymer Composite Stars: Synthesis and Applications. *Adv. Funct. Mater.* **2011**, *21*, 1673–1680. [CrossRef] [PubMed]

35. Stevanovic, M.; Uskokovic, D. Poly(lactide-*co*-glycolide)-based Micro and Nanoparticles for the Controlled Drug Delivery of Vitamins. *Curr. Nanosci.* **2009**, *5*, 1–14.

36. Pasparakis, G.; Manouras, T.; Vamvakaki, M.; Argitis, P. Harnessing photochemical internalization with dual degradable nanoparticles for combinatorial photo-chemotherapy. *Nat. Commun.* **2014**, *5*. [CrossRef] [PubMed]

37. Duff, D.G.; Baiker, A.; Edwards, P.P. A new hydrosol of gold clusters. 1. Formation and particle size variation. *Langmuir* **1993**, *9*, 2301–2309. [CrossRef]

38. Cairns, R.; Papandreou, I.; Denko, N. Overcoming Physiologic Barriers to Cancer Treatment by Molecularly Targeting the Tumor Microenvironment. *Mol. Cancer Res.* **2006**, *4*, 61–70. [CrossRef] [PubMed]

39. Duan, H.; Nie, S. Cell-penetrating quantum dots based on multivalent and endosome-disrupting surface coatings. *J. Am. Chem. Soc.* **2007**, *129*. [CrossRef] [PubMed]

40. Dominska, M.; Dykxhoorn, D.M. Breaking down the barriers: siRNA delivery and endosome escape. *J. Cell Sci.* **2010**, *123*, 1183–1189. [CrossRef] [PubMed]

41. Panyam, J.; Zhou, W.-Z.; Prabha, S.; Sahoo, S.K.; Labhasetwar, V. Rapid endo-lysosomal escape of poly(DL-lactide-co-glycolide) nanoparticles: implications for drug and gene delivery. *FASEB J.* **2002**, *16*, 1217–1226. [CrossRef] [PubMed]

42. Zhou, Y.; Liu, S. (64)Cu-Labeled Phosphonium Cations as PET Radiotracers for Tumor Imaging. *Bioconjugate Chem.* **2011**, *22*, 1459–1472. [CrossRef] [PubMed]

43. Ross, M.F.; Kelso, G.F.; Blaikie, F.H.; James, A.M.; Cocheme, H.M.; Filipovska, A.; da Ros, T.; Hurd, T.R.; Smith, R.A.J.; Murphy, M.P. Lipophilic triphenylphosphonium cations as tools in mitochondrial bioenergetics and free radical biology. *Biochemistry (Moscow)* **2005**, *70*, 222–230. [CrossRef]

44. Sharom, F.J. The P-glycoprotein efflux pump: How does it transport drugs? *J. Membr. Biol.* **1997**, *160*, 161–175. [CrossRef] [PubMed]

45. Roepe, P.D. The P-glycoprotein efflux pump: How does it transport drugs? *J. Membr. Biol.* **1998**, *166*, 71–72. [CrossRef] [PubMed]

46. Kirtane, A.; Kalscheuer, S.; Panyam, J. Exploiting Nanotechnology to Overcome Tumor Drug Resistance: Challenges and Opportunities. *Adv. Drug Deliv. Rev.* **2013**, *65*. [CrossRef] [PubMed]

47. Bao, L.; Hazari, S.; Mehra, S.; Kaushal, D.; Moroz, K.; Dash, S. Increased Expression of P-Glycoprotein and Doxorubicin Chemoresistance of Metastatic Breast Cancer Is Regulated by miR-298. *Am. J. Pathol.* **2012**, *180*, 2490–2503. [CrossRef] [PubMed]

48. Xue, X.; Liang, X.-J. Overcoming drug efflux-based multidrug resistance in cancer with nanotechnology. *Chin. J. Cancer* **2012**, *31*, 100–109. [CrossRef] [PubMed]

49. Punfa, W.; Yodkeeree, S.; Pitchakarn, P.; Ampasavate, C.; Limtrakul, P. Enhancement of cellular uptake and cytotoxicity of curcumin-loaded PLGA nanoparticles by conjugation with anti-P-glycoprotein in drug resistance cancer cells. *Acta Pharmacol. Sin.* **2012**, *33*, 823–831. [CrossRef] [PubMed]

50. Omelyanenko, V.; Kopečková, P.; Gentry, C.; Kopeček, J. Targetable HPMA copolymer-adriamycin conjugates. Recognition, internalization, and subcellular fate. *J. Control. Release* **1998**, *53*, 25–37. [CrossRef]

51. Gong, J.; Chen, M.; Zheng, Y.; Wang, S.; Wang, Y. Polymeric micelles drug delivery system in oncology. *J. Control. Release* **2012**, *159*, 312–323. [CrossRef] [PubMed]

52. Chen, Y.; Zhang, W.; Gu, J.; Ren, Q.; Fan, Z.; Zhong, W.; Fang, X.; Sha, X. Enhanced antitumor efficacy by methotrexate conjugated Pluronic mixed micelles against KBv multidrug resistant cancer. *Int. J. Pharm.* **2013**, *452*, 421–433. [CrossRef] [PubMed]

53. Kang, K.W.; Chun, M.-K.; Kim, O.; Subedi, R.K.; Ahn, S.-G.; Yoon, J.-H.; Choi, H.-K. Doxorubicin-loaded solid lipid nanoparticles to overcome multidrug resistance in cancer therapy. *Nanomed. Nanotechnol. Biol. Med.* **2010**, *6*, 210–213. [CrossRef] [PubMed]

54. Zhang, J.; Wang, L.; Chan, H.F.; Xie, W.; Chen, S.; He, C.; Wang, Y.; Chen, M. Co-delivery of paclitaxel and tetrandrine via iRGD peptide conjugated lipid-polymer hybrid nanoparticles overcome multidrug resistance in cancer cells. *Sci. Rep.* **2017**, *7*, 46057. [CrossRef] [PubMed]

55. Thierry, A.R.; Vige, D.; Coughlin, S.S.; Belli, J.A.; Dritschilo, A.; Rahman, A. Modulation of doxorubicin resistance in multidrug-resistant cells by liposomes. *FASEB J.* **1993**, *7*, 572–579. [PubMed]

56. Li, B.; Xu, H.; Li, Z.; Yao, M.; Xie, M.; Shen, H.; Shen, S.; Wang, X.; Jin, Y. Bypassing multidrug resistance in human breast cancer cells with lipid/polymer particle assemblies. *Int. J. Nanomed.* **2012**, *7*, 187–197.

57. Susa, M.; Iyer, A.K.; Ryu, K.; Choy, E.; Hornicek, F.J.; Mankin, H.; Milane, L.; Amiji, M.M.; Duan, Z. Inhibition of ABCB1 (MDR1) Expression by an siRNA Nanoparticulate Delivery System to Overcome Drug Resistance in Osteosarcoma. *PLoS ONE* **2010**, *5*, e10764. [CrossRef] [PubMed]
58. Pramanik, D.; Campbell, N.R.; Das, S.; Gupta, S.; Chenna, V.; Bisht, S.; Sysa-Shah, P.; Bedja, D.; Karikari, C.; Steenbergen, C.; et al. A composite polymer nanoparticle overcomes multidrug resistance and ameliorates doxorubicin-associated cardiomyopathy. *Oncotarget* **2012**, *3*, 640–650. [CrossRef] [PubMed]

MDPI

St. Alban-Anlage 66

4052 Basel

Switzerland

Tel. +41 61 683 77 34

Fax +41 61 302 89 18

www.mdpi.com

Nanomaterials Editorial Office

E-mail: nanomaterials@mdpi.com

www.mdpi.com/journal/nanomaterials

www.ingramcontent.com/pod-product-compliance
Lightning Source LLC
Chambersburg PA
CBHW051838210326
41597CB00033B/5701